本书英文版于2023年由世界卫生组织（World Health Organization）出版，书名为：*WHO Study Group on Tobacco Product Regulation: Report on the Scientific Basis of Tobacco Product Regulation: Ninth Report of a WHO Study Group (WHO Technical Report Series, No. 1047)*

WHO技术报告系列 1047

WHO烟草制品管制研究小组

烟草制品管制科学基础报告

WHO研究组第九份报告

胡清源 侯宏卫 陈 欢 主译

科 学 出 版 社
北 京

内 容 简 介

本报告介绍了 WHO 烟草制品管制研究小组第十一次会议的结论和建议。该次会议讨论了烟草制品管制中的新兴问题，例如如何通过新的方式将非治疗性烟碱产品推广至包括儿童和青少年在内的不同年龄群体，涉及以下专题：①促进吸入的添加剂（凉味剂、烟碱盐和调味剂）；②合成烟碱：科学基础、全球法律环境及监管考量；③尼古丁袋：特性、用途、危害及监管；④电子烟碱传输系统和加热型烟草制品的暴露、效应和易感性的生物标志物及其优先次序评估；⑤烟草和非治疗性烟碱产品的互联网营销及相关监管考量。研究组关于每个主题的建议在相关章节末尾列出，最后一章为总体建议。

本书会引起吸烟与健康、烟草化学和公共卫生学诸多应用领域科学家的兴趣，为客观评价烟草制品的管制和披露措施提供必要的参考。

图书在版编目（CIP）数据

烟草制品管制科学基础报告 . WHO 研究组第九份报告 / WHO 烟草制品管制研究小组著 ; 胡清源 , 侯宏卫 , 陈欢主译 . -- 2025. 2. -- (WHO 技术报告系列). -- ISBN 978-7-03-081234-6

Ⅰ . TS45

中国国家版本馆 CIP 数据核字第 2025K5Z777 号

责任编辑：刘 冉 / 责任校对：杜子昂

责任印制：徐晓晨 / 封面设计：北京图阅盛世

科学出版社出版

北京东黄城根北街 16 号

邮政编码：100717

http://www.sciencep.com

北京华宇信诺印刷有限公司印刷

科学出版社发行 各地新华书店经销

*

2025 年 2 月第 一 版 开本：720 × 1000 1/16

2025 年 2 月第一次印刷 印张：21 1/4

字数：430 000

定价：160.00 元

（如有印装质量问题，我社负责调换）

WHO Technical Report Series 1047

WHO Study Group on Tobacco Product Regulation

Report on the Scientific Basis of Tobacco Product Regulation: Ninth Report of a WHO Study Group

译者名单

主　译： 胡清源　侯宏卫　陈　欢

副主译： 牟文君　李　晓　喻　昊　王高歌

译　者： 胡清源　侯宏卫　陈　欢　牟文君

李　晓　喻　昊　王高歌　杨发琉

李　乾　尹长锋　倪　震　吴夏青

韩书磊　付亚宁　王红娟　田雨闪

崔利利

译 者 序

2003年5月，第56届世界卫生大会*通过了《烟草控制框架公约》（FCTC），迄今已有包括我国在内的180个缔约方。根据FCTC第9条和第10条的规定，授权世界卫生组织（WHO）烟草制品管制研究小组（TobReg）对可能造成重要公众健康问题的烟草制品管制措施进行鉴别，提供科学合理的、有根据的建议，用于指导成员国进行烟草制品管制。

自2007年起，WHO陆续出版了九份烟草制品管制科学基础报告，分别是945、951、955、967、989、1001、1015、1029和1047。WHO烟草制品管制科学基础系列报告阐述了降低烟草制品的吸引力、致瘾性和毒性等烟草制品管制相关主题的科学依据，内容涉及烟草化学、代谢组学、毒理学、吸烟与健康等烟草制品管制的多学科交叉领域，是一系列以科学研究为依据、对烟草管制发展和决策有重大影响意义的技术报告。将其引进并翻译出版，可以为相关烟草科学研究的科技工作者提供科学性参考。希望引起吸烟与健康、烟草化学和公共卫生学等诸多应用领域科学家的兴趣，为客观评价烟草制品的管制和披露措施提供必要的参考。

第一份报告（945）由胡清源、侯宏卫、韩书磊、陈欢、刘彤、付亚宁翻译，全书由韩书磊负责统稿；

第二份报告（951）由胡清源、侯宏卫、刘彤、付亚宁、陈欢、韩书磊翻译，全书由刘彤负责统稿；

第三份报告（955）由胡清源、侯宏卫、付亚宁、陈欢、韩书磊、刘彤翻译，全书由付亚宁负责统稿；

第四份报告（967）由胡清源、侯宏卫、陈欢、刘彤、韩书磊、付亚宁翻译，全书由陈欢负责统稿；

第五份报告（989）由胡清源、侯宏卫、陈欢、刘彤、韩书磊、付亚宁翻译，全书由陈欢负责统稿；

第六份报告（1001）由胡清源、侯宏卫、韩书磊、陈欢、刘彤、付亚宁、王

---

* 世界卫生大会 (World Health Assembly，WHA) 是世界卫生组织的最高决策机构，每年召开一次。

红娟翻译，全书由韩书磊负责统稿；

第七份报告（1015）由胡清源、侯宏卫、陈欢、张小涛、田永峰、李国宇、刘彤、韩书磊、付亚宁、王红娟、田雨闪翻译，全书由刘彤负责统稿；

第八份报告（1029）由胡清源、侯宏卫、陈欢、李晓、崔利利、周静、韩书磊、付亚宁、王红娟、田雨闪、张远、任培培、苗瑞娟、王永秀翻译。

第九份报告（1047）由胡清源、侯宏卫、陈欢、牟文君、李晓、喻昊、王高歌、杨发琉、李乾、尹长锋、倪震、吴夏青、韩书磊、付亚宁、王红娟、田雨闪、崔利利翻译。

由于译者学识水平有限，本中文版难免有错漏和不当之处，敬请读者批评指正。

译　者

2024年12月

WHO 烟草制品管制研究小组第十一次会议

格鲁吉亚第比利斯，2022年12月13~15日

参加者

David L. Ashley教授，美国公共卫生服务部退役海军少将；美国佐治亚州立大学（亚特兰大）人口健康科学系研究教授

Lekan Ayo-Yusuf教授，南非比勒陀利亚大学卫生系统和公共卫生学院教授、主席兼院长

Alan Boobis教授（线上参会），英国伦敦帝国理工学院医学系名誉教授

Mike Daube教授，澳大利亚科廷大学（珀斯）健康科学学院名誉教授

Prakash C. Gupta博士，印度马哈佩治疗中心（新孟买）主任

S. Katharine Hammond教授，美国加利福尼亚大学伯克利分校公共卫生学院环境卫生学教授，工业卫生计划主任，学术事务副院长

Dorothy Hatsukami教授，美国明尼苏达大学（明尼阿波利斯）医学院和共济会癌症中心精神病学教授；癌症预防福斯特家族主席

Antoon Opperhuizen博士（线上参会），荷兰食品和消费品安全局（乌得勒支）风险评估和研究办公室主任

Ghazi Zaatari教授，黎巴嫩贝鲁特美国大学医学院病理学与实验医学系教授兼主任

WHO全球烟草管制机构论坛

Martijn Martena博士，荷兰卫生部（海牙）营养、福利和运动健康、保护和预防司烟草控制小组协调员

WHO烟草实验室网络

Nuan Ping Cheah博士，新加坡卫生科学局应用科学司制药部药物、化妆品和卷烟检测实验室主任

主要作者

Micah Berman教授，美国俄亥俄州立大学（哥伦布）公共卫生学院卫生服务管理和政策与法学院联合教授

Becky Freeman博士，澳大利亚悉尼大学医学与健康学院公共卫生副教授

Charlotte GGM Pauwels博士，荷兰国家公共卫生与环境研究所（比尔托芬）健康防护中心

Irina Stepanov教授，美国明尼苏达大学（明尼阿波利斯）公共卫生学院和共济会癌症中心环境卫生科学系教授兼梅奥公共卫生教授

Reinskje Talhout博士，荷兰国家公共卫生与环境研究所（比尔托芬）健康防护中心高级科学家

资源专家

Ivan Berlin博士，法国Pitié-Salpêtrière医院（巴黎）医学药理学部

Douglas Bettcher博士，世界卫生组织（瑞士日内瓦）高级顾问

Thomas Eissenberg博士，美国弗吉尼亚联邦大学（里士满）烟草制品研究中心心理学教授兼联合主任

Najat A. Saliba教授，黎巴嫩议员；黎巴嫩贝鲁特美国大学化学系分析化学教授

合作者

Stella Bialous博士，美国加利福尼亚大学旧金山分校护理学院和烟草控制研究与教育中心

Jennifer Brown博士，美国约翰·霍普金斯大学（马里兰州巴尔的摩）公共卫生学院

Rula Cavaco Dias女士，世界卫生组织（瑞士日内瓦）健康促进部健康人口司无烟草行动组

Ranti Fayokun博士，世界卫生组织（瑞士日内瓦）健康促进部健康人口司无烟草行动组

Geoffrey Ferris Wayne博士，美国俄勒冈州波特兰公共卫生政策和研究独立研究员

Steven Hecht教授，美国明尼苏达大学（明尼阿波利斯）共济会癌症研究中心检验医学和病理学系癌症预防教授

Sven-Eric Jordt博士，杜克大学达勒姆医学院麻醉学、药理学和癌症生物学系；美国耶鲁大学（康涅狄格州纽黑文）医学院精神病学系烟草管制科学中心

Adam Leventhal博士（线上参会），美国南加利福尼亚州大学（洛杉矶）健康科学

校区柯克医学院人口和公共卫生科学教授，成瘾科学研究所所长

Pamela Ling博士，美国加利福尼亚州大学旧金山分校医学系及烟草控制研究与教育中心

Patricia Zettler教授，美国俄亥俄州立大学（哥伦布）莫里茨法学院法学教授

欧盟及其他国家专家

Denis Choinière先生，加拿大卫生部（安大略渥太华）烟草控制局烟草制品管制办公室主任

Matus Ferech博士，比利时欧盟委员会（布鲁塞尔）健康与食品安全、跨境医疗保健和烟草控制总局

WHO非洲地区办事处

William Maina博士，WHO非洲地区办事处（乌干达坎帕拉）全民健康覆盖和更健康人口司烟草和其他非传染性疾病风险因素组高级项目官员

Fikru Tullu博士，WHO非洲地区办事处（刚果布拉柴维尔）全民健康覆盖和更健康人口司烟草和其他非传染性疾病风险因素组组长

WHO美洲地区办事处

Adriana Bacelar Gomez女士，WHO美洲地区办事处（美国华盛顿）烟草控制监测官员

WHO东南亚地区办事处

Jagdish Kaur博士，WHO东南亚地区办事处（印度新德里）健康风险因素和健康促进区域顾问

WHO欧洲地区办事处

Angela Ciobanu博士，WHO欧洲地区办事处（丹麦哥本哈根联合国大楼）技术官员

WHO地中海东部地区办事处

Fatimah El-Awa博士（线上参会），WHO东地中海地区办事处（埃及开罗纳赛尔城）无烟草行动组顾问

WHO西太平洋地区办事处

Xi Yin女士，WHO西太平洋地区办事处（菲律宾马尼拉）无烟草行动组协调员

WHO FCTC秘书处（瑞士日内瓦）

Adriana Blanco博士，主管

Kate Lannan女士，高级法律事务官员

WHO总部（瑞士日内瓦）

Rula Cavaco Dias女士，健康促进司顾问

Priscilla Cleland女士，健康促进司主管助理

Ranti Fayokun博士，健康促进部全民健康覆盖和更健康人口司科学家

Vinayak Mohan Prasad博士，健康促进部全民健康覆盖和更健康人口司无烟草行动组负责人

Moira Sy女士，健康促进部顾问

Participants in the eleventh meeting of the WHO Study Group on Tobacco Product Regulation

Tbilisi, Georgia
13–15 December 2022

Members

Professor David L. Ashley, Rear-Admiral (retired) US Public Health Service, Research Professor, Department of Population Health Sciences, Georgia State University, Atlanta, Georgia, United States of America (USA)

Professor Lekan Ayo-Yusuf, Professor, Chairperson and Head, School of Health Systems and Public Health, University of Pretoria, Pretoria, South Africa

Professor Alan Boobis (participated virtually), Professor Emeritus, Department of Medicine, Imperial College London, London, United Kingdom

Emeritus Professor Mike Daube, Faculty of Health Sciences, Curtin University, Perth, Australia

Dr Prakash C. Gupta, Director, Healis, Mahape, Navi Mumbai, India

Professor S. Katharine Hammond, Professor of Environmental Health Sciences, Director, Industrial Hygiene Program, Associate Dean of Academic Affairs, School of Public Health, University of California, Berkeley, California, USA

Professor Dorothy Hatsukami, Professor of Psychiatry and Forster Family Chair in Cancer Prevention, University of Minnesota Medical School and Masonic Cancer Center, Minneapolis, Minnesota, USA

Dr Antoon Opperhuizen (participated virtually), Director, Office for Risk Assessment and Research, Food and Consumer Product Safety Authority, Utrecht, Netherlands (Kingdom of the)

Professor Ghazi Zaatari, Professor and Chairman, Department of Pathology and Laboratory Medicine, Faculty of Medicine, American University of Beirut, Beirut, Lebanon

WHO Global Tobacco Regulators Forum

Dr Martijn Martena, Coordinator, Tobacco Control Team, Department for Nutrition, Welfare and Sport Health, Protection and Prevention, Ministry of Health, The Hague, Netherlands (Kingdom of the)

WHO Tobacco Laboratory Network

Dr Nuan Ping Cheah, Director, Cosmetics and Cigarette Testing Laboratory, Pharmaceutical Division, Applied Sciences Group, Health Sciences Authority, Singapore

Lead authors

Professor Micah Berman JD, Professor; Joint Appointment, College of Law, Health Services Management and Policy, The Ohio State University, College of Public Health, Columbus, Ohio, USA

Dr Becky Freeman, Associate Professor, Public Health, Faculty of Medicine and Health, University of Sydney, Sydney, Australia

Dr Charlotte GGM Pauwels, National Institute for Public Health and the Environment, Centre for Health Protection, Bilthoven, Netherlands (Kingdom of the)

Professor Irina Stepanov, Professor of Environmental Health Sciences and Mayo Professor in Public Health, University of Minnesota School of Public Health and Masonic Cancer Center, Minneapolis, Minnesota, USA

Dr Reinskje Talhout, Senior scientist, Center for Health Protection, National Institute for Public Health and the Environment, Bilthoven, Netherlands (Kingdom of the)

Other participants who attended in person

Resource experts

Dr Ivan Berlin, Department of Medical Pharmacology, Pitié-Salpêtrière Hospital, Paris, France

Dr Douglas Bettcher, Senior Adviser, World Health Organization, Geneva, Switzerland

Dr Thomas Eissenberg, Professor of Psychology and Co-Director, Center for the Study of Tobacco Products, Virginia Commonwealth University, Richmond, Virginia, USA

Professor Najat A. Saliba, Member of Parliament, Lebanon, Professor of Analytical Chemistry, Department of Chemistry, American University of Beirut, Beirut, Lebanon

Co-authors

Dr Stella Bialous, School of Nursing and Center for Tobacco Control Research and Education, University of California, San Francisco (CA), USA

Dr Jennifer Brown, Johns Hopkins Bloomberg School of Public Health, Baltimore (MD), USA

Ms Rula Cavaco Dias, Healthier Populations Division, Health Promotion Department, No Tobacco Unit, World Health Organization, Geneva, Switzerland

Dr Ranti Fayokun, Healthier Populations Division, Health Promotion Department, No Tobacco Unit, World Health Organization, Geneva, Switzerland

Dr Geoffrey Ferris Wayne, independent researcher in public health policy and research, Portland, Oregon, USA

Professor Steven Hecht, Professor of Cancer Prevention, Department of Laboratory Medicine and Pathology, Masonic Cancer Research Center, University of Minnesota, Minneapolis, Minnesota, USA

Dr Sven-Eric Jordt, Departments of Anesthesiology, Pharmacology and Cancer Biology, Duke University School of Medicine, Durham (NC) and Tobacco Center of Regulatory Science, Department of Psychiatry, Yale School of Medicine, New Haven (CT), USA

Dr Adam Leventhal (attended virtually), Professor of Population and Public Health Sciences, Director, Institute for Addiction Science, Kirk School of Medicine, Health Sciences Campus, University of Southern California, Los Angeles, California, USA

Dr Pamela Ling, MD, Department of Medicine and Center for Tobacco Control Research and Education, University of California, San Francisco (CA), USA

Professor Patricia Zettler, Professor of Law, Moritz College of Law, Ohio State University, Columbus, Ohio, USA

European Union and selected countries

Mr Denis Choinière, Director, Tobacco Products Regulatory Office, Tobacco Control Directorate, Health Canada, Ottawa, Ontario, Canada

Dr Matus Ferech, Directorate General for Health and Food Safety, Cross-border Health Care and Tobacco Control, European Commission, Brussels, Belgium

WHO Regional Office for Africa

Dr William Maina, Senior Project Officer, Tobacco and Other NCD Risk Factors Unit, Universal Health Coverage/Healthy Populations Cluster, WHO Regional Office for Africa, Kampala, Uganda

Dr Fikru Tullu, Team lead, Tobacco and Other NCD Risk Factors, Universal Health Coverage/ Healthier Populations Cluster, WHO Regional Office for Africa, Brazzaville, Congo

WHO Regional Office for the Americas

Ms Adriana Bacelar Gomez, Surveillance Officer, Tobacco Control, WHO Regional Office for the Americas, Washington DC, USA

WHO Regional Office for South-East Asia

Dr Jagdish Kaur, Regional Adviser, Health Risk Factors and Health Promotion, WHO Regional Office for South-East Asia, New Delhi, India

WHO Regional Office for Europe

Dr Angela Ciobanu, Technical Officer, WHO Regional Office for Europe, UN City, Copenhagen, Denmark

WHO Regional Office for the Eastern Mediterranean

Dr Fatimah El-Awa (attended virtually), Regional Adviser, Tobacco Free Initiative, WHO Regional Office for the Eastern Mediterranean, Nasr City, Cairo, Egypt

WHO Regional Office for the Western Pacific

Ms Xi Yin, Coordinator, Tobacco Free Initiative, WHO Regional Office for the Western Pacific, Manila, Philippines

Secretariat of the WHO Framework Convention on Tobacco Control, Geneva, Switzerland

Dr Adriana Blanco, Head

Ms Kate Lannan, Senior Legal Affairs Officer

WHO headquarters, Geneva, Switzerland

Ms Rula Cavaco Dias, Consultant, Department of Health Promotion

Mrs Priscilla Cleland, Assistant to Unit Head, Department of Health Promotion

Dr Ranti Fayokun, Scientist, Health Promotion Department, Division of Universal Health Coverage and Healthier Populations

Dr Vinayak Mohan Prasad, Unit Head, No Tobacco, Health Promotion Department, Division of Universal Health Coverage and Healthier Populations

Ms Moira Sy, Consultant, Department of Health Promotion

致　　谢

世界卫生组织（WHO）烟草制品管制研究小组（TobReg）对提供本报告基础背景文件的作者表示感谢。本报告在无烟草行动组负责人Vinayak Prasad博士和WHO健康促进部主任Ruediger Krech博士的监督和支持下，由Ranti Fayokun博士多方协调得以完成。

感谢以下世界卫生组织人员提供的行政支持：Priscilla Cleland女士、Rula Cavaco Dias女士和Moira Sy女士。

感谢所有背景文件的作者，他们以其专业知识对本报告的编写作出了贡献。

感谢比尔及梅林达·盖茨基金会和彭博慈善基金会资助TobReg第十一次会议背景文件的编写。

感谢世界卫生组织《烟草控制框架公约》秘书处协助起草缔约方会议的相关要求，作为背景文件的基础，并为研究组的审议工作作出了贡献。

还要感谢WHO成员国，因为本报告中的大多数文件是应成员国提出的技术指导请求而撰写的。

Acknowledgements

The WHO Study Group on Tobacco Product Regulation (TobReg) expresses its gratitude to the authors of the background papers used as the basis for this report. Production of the report was coordinated by Dr Ranti Fayokun, with the supervision and support of Dr Vinayak Prasad, Unit Head, No Tobacco Unit, Department of Health Promotion, and Dr Ruediger Krech, Director of the WHO Department of Health Promotion.

Administrative support was provided by the following WHO personnel: Mrs Priscilla Cleland, Mrs Rula Cavaco Dias and Ms Moira Sy.

TobReg acknowledges all authors of the background papers, as listed, for their expertise and contribution to development of the report.

TobReg expresses its gratitude to the Bill & Melinda Gates Foundation and Bloomberg Philanthropies for providing funds to support the preparation of the background papers for the Eleventh meeting of TobReg.

The Study Group also thanks the Secretariat of the WHO Framework Convention on Tobacco Control (WHO FCTC) for facilitating drafting of the relevant requests of the Conference of the Parties to the WHO FCTC, which served as the basis for some of the background papers and contributed to the deliberations of the Study Group.

WHO also wishes to acknowledge its Member States, as most of the papers in this report were written in response to their requests to WHO for technical guidance.

缩 略 语

AHRR	芳香烃受体阻遏物
BaP	苯并[a]芘
BAT	英美烟草公司
CC	传统卷烟
CC16	棒状细胞16 kDa蛋白
CEMA	氰乙基巯基尿酸
CI	置信区间
CO	一氧化碳
COHb	碳氧血红蛋白
COP	缔约方会议
COPD	慢性阻塞性肺病
CRP	C反应蛋白
CVD	心血管疾病
DLCO	肺一氧化碳弥散量
ENDS	电子烟碱传输系统
ENNDS	电子非烟碱传输系统
EU	欧盟
FEV_1	第1秒用力呼气量
FVC	用力肺活量
GC	气相色谱
GMR	几何均值比
HDL-C	高密度脂蛋白胆固醇
1-HOP	1-羟基芘
HTP	加热型烟草制品
IARC	国际癌症研究机构
IF	吸入促进
ISO	国际标准化组织
LC	液相色谱
LD_{50}	半数致死量
MS	质谱
nAChR	烟碱型乙酰胆碱受体
NGL	下一代实验室公司
NHANES	国家健康与营养调查

NMR　烟碱代谢率

NNAL　4-(甲基亚硝胺)-1-(3-吡啶基)-1-丁醇

NNK　4-(甲基亚硝胺)-1-(3-吡啶基)-1-丁酮

NNN　*N*-亚硝基降烟碱

NO　一氧化氮

OR　比值比

γ-OH-Acr-dGuo　(8*R*/*S*)-3-(2′-脱氧核糖-1′-基)-5,6,7,8-四氢-8-羟基嘧啶[1,2-*a*]嘌呤-10(3*H*)-酮，即丙烯醛-DNA加合物

PAH　多环芳烃

PATH　烟草与健康人口评估

PGEM　前列腺素E2代谢物

$PGF_{2\alpha}$　(*Z*)-7-[(1*R*,2*R*,3*R*,5*S*)-3,5-二羟基-2-[(*E*,3*S*)-3-甲基辛-1-烯基]环戊基]庚-5-烯酸，即前列腺素$F_{2\alpha}$

PheT　菲四醇

PMI　菲利普·莫里斯国际公司

RIVM　荷兰国家公共卫生与环境研究所

TAPS　烟草广告、促销和赞助

TNE　总烟碱当量

TNP　烟草和烟碱产品

TobReg　烟草制品管制研究小组

TPD　《欧盟烟草制品指令》

TRP　瞬时受体电位

USA　美国

USFDA　美国食品药品监督管理局

WHO FCTC　世界卫生组织《烟草控制框架公约》

WS　“威尔金森剑”

Abbreviations and acronyms

AHRR	aryl hydrocarbon receptor repressor
BaP	benzo[*a*]pyrene
BAT	British American Tobacco
CC	conventional cigarette
CC16	club cell 16-kDa protein
CEMA	cyanoethyl mercapturic acid
CI	confidence interval
CO	carbon monoxide
COHb	carboxyhaemoglobin
COP	Conference of the Parties
COPD	chronic obstructive pulmonary disease
CRP	C-reactive protein
CVD	cardiovascular disease
DLCO	diffusing capacity of the lung for carbon monoxide
ENDS	electronic nicotine delivery system
ENNDS	electronic non-nicotine delivery system
EU	European Union
FEV_1	forced expiratory volume in 1 s
FVC	forced vital capacity
GC	gas chromatography
GMR	geometric mean ratio
HDL-C	high-density lipoprotein cholesterol
1-HOP	1-hydroxypyrene
HTP	heated tobacco product
IARC	International Agency for Research on Cancer
IF	inhalation facilitation
ISO	International Organization for Standardization
LC	liquid chromatography
LD_{50}	lethal dose for 50% of animals
MS	mass spectrometry
nAChR	nicotinic acetylcholine receptor
NGL	Next Generation Labs
NHANES	National Health and Nutrition Examination Survey
NMR	nuclear magnetic resonance
NNAL	4-(methylnitrosamino)-1-(3-pyridyl)-1-butanol
NNK	4-(methylnitrosamino)-1-(3-pyridyl)-1-butanone
NNN	*N'*-nitrosonornicotine

NO	nitric oxide
OR	odds ratio
γ-OH-Acr-dGuo	(8*R/S*)-3-(2′-deoxyribos-1′-yl)-5,6,7,8-tetrahydro-8-hydroxypyrimido[1,2-*a*]purine-10(3*H*)-one
PAH	polycyclic aromatic hydrocarbons
PATH	Population Assessment of Tobacco and Health
PGEM	prostaglandin E2 metabolite
$PGF_{2\alpha}$	(Z)-7-[(1*R*,2*R*,3*R*,5*S*)-3,5-dihydroxy-2-[(E,3S)-3-hydroxyoct-1-enyl]cyclopentyl]hept-5-enoic acid
PheT	phenanthrene tetraol
PMI	Philip Morris International
RIVM	National Institute for Public Health and the Environment (Netherlands [Kingdom of the])
TAPS	tobacco advertising, promotion and sponsorship
TNE	total nicotine equivalents
TNP	tobacco and nicotine products
TobReg	WHO Study Group on Tobacco Product Regulation
TPD	[European] Tobacco Products Directive
TRP	transient receptor potential
US	United States [of America]
USA	United States of America
USFDA	United States Food and Drug Administration
WHO FCTC	WHO Framework Convention on Tobacco Control
WS	Wilkinson Sword

目　　录

Contents

1. 引　　言

烟草是全球公共卫生的一大威胁，每年全球有800多万人死于烟草[1]，其中约120万非吸烟者死于二手烟暴露[2]。因此，全面控烟对于解决全球烟草流行问题和防止不必要的死亡至关重要。产品监管有助于减少烟草需求，有效的烟草制品监管是全面控烟计划的重要组成部分[3]，包括通过强制性测试、披露测试结果、酌情设定限值、披露产品信息以及制定产品包装和标签标准来监管烟草制品的成分和释放物。世界卫生组织《烟草控制框架公约》（WHO FCTC）第9、10和11条[4]以及WHO FCTC第9、10条的部分实施指南[5]涵盖了烟草制品管制的内容。为支持成员国在烟草制品管制方面的工作，世界卫生组织提供了多种资源，包括《烟草制品管制：基本手册》[3]、《烟草制品管制：实验室检测能力建设》[6] 和基于这些手册的在线模块课程，这些资源均可通过WHO官网查阅[7]。此外，世界卫生组织（WHO）烟草制品管制研究小组（TobReg）发布了一系列报告和咨询说明，就烟草制品和非治疗性烟碱产品监管等多方面提供了指导。

世界卫生组织总干事于2003年正式组建了烟草制品管制研究小组（TobReg），以填补烟草制品管制空白。其任务是向WHO总干事提供有关烟草制品监管的政策建议。TobReg由来自全球各地涵盖世界卫生组织六大地区的各国科学专家组成，他们精通产品监管、烟草依赖治疗、烟草制品成分和释放物毒理学及实验室分析等领域[8]。作为世界卫生组织的正式机构，TobReg通过总干事向执行委员会提交技术报告，为烟草制品管制提供科学依据，并引起各成员国对世界卫生组织在该领域工作的关注。WHO技术报告系列不仅包括之前未发表的背景文件，还综合了已发表并经由TobReg讨论、评估和审查的科学文献。根据WHO FCTC第9、10条及WHO FCTC缔约方会议（COP）的相关决议以及提交给COP的WHO报告，TobReg报告确定了监管所有形式烟草制品和非治疗性烟碱产品的循证方法，包括新型产品，如电子烟碱传输系统（ENDS）、电子非烟碱传输系统（ENNDS）、加热型烟草制品（HTP）和尼古丁袋。这些报告是对世界卫生大会WHA 53.8（2000年）、WHA 53.17（2000年）和WHA 54.18（2001年）决议的响应，也已被视为世界卫生组织全球公共卫生产品。全球公共卫生产品是世界卫生组织制定及实施的举措，对全球许多国家或地区都有益处[9]。这对TobReg来说是一个独特的机会，能够直接与成员国接触，并影响国家、地区和全球政策。

TobReg第十一次会议于2022年12月13~15日在格鲁吉亚第比利斯举行，由格鲁吉亚国家疾病控制和公共卫生中心主办，世界卫生组织健康促进司无烟草行动组组织。约40名与会人员，包括TobReg成员、世界卫生组织工作人员、世界卫生组织《烟草控制框架公约》秘书处以及受邀专家。根据成员国向世界卫生组织提出的请求和缔约方会议的要求，讨论了与产品监管相关主题（包括新兴问题）的科学文献。会议还探讨了全球烟草管制机构论坛先前讨论的议题，如合成烟碱和尼古丁袋，最近，这些产品因其营销方式和利用监管漏洞而面临监管挑战。为响应成员国多次要求秘书处就烟草制品监管方面的新问题提供技术援助和权威性指导，本报告重点介绍了向包括儿童和青少年在内的不同年龄人群提供和宣传非治疗性烟碱产品的新方法。会议讨论了以下六份背景文件：

- 促进吸入的添加剂（凉味剂、烟碱盐和调味剂）；
- 合成烟碱：科学基础、全球法律环境及监管考量；
- 尼古丁袋：特性、用途、危害及监管；
- 电子烟碱传输系统和加热型烟草制品的暴露、效应和易感性的生物标志物及其优先次序评估；
- 烟草和非治疗性烟碱产品的互联网营销及相关监管考量；
- WHO烟草制品管制研究小组：二十年的建议——将证据转化为政策行动。

纳入第6份背景文件是为了指导研究组未来的工作，即将科学转化为政策，并将由研究组单独审议。因此，本报告包括前五份背景文件。世界卫生组织所有地区成员国的要求、秘书处和研究组在这些领域的知识以及相关文献构成了五份背景文件内容的基础。这些背景文件中的信息更新了当前的知识，将进一步推动烟碱和烟草制品的监管，为国家和全球政策提供信息。

背景文件由专家们根据世界卫生组织秘书处为每份文件拟定的职权范围或大纲编写，并由TobReg成员和世界卫生组织指定的专家评审员进行审查和修订。每篇论文都注明了文献检索的时间段；对大多数文件而言，这个时间范围是2022年第二季度或2023年第一季度。在汇编成技术报告之前，这些文件会在会议前后经过独立技术专家、世界卫生组织秘书处、世界卫生组织其他相关部门人员、地区办事处的同事以及研究组成员的多轮审查。

秘书处经与研究组协商，邀请了一些专家参加讨论，并就所审议的专题提供了最新的科学实证和规章。TobReg关于烟草制品管制科学基础的第九份报告旨在指导成员国实现最有效的循证手段，以弥合烟草控制的监管差距，并制定协调一致的烟草制品管制框架。此外，还确定了今后的工作领域，重点关注各国的监管需求，从而继续为成员国提供持续的技术支持战略。所有专家和其他参会者，包括研究组成员，都必须填写利益声明，并由世界卫生组织进行评估。

本报告包括引言以及关于烟草控制条例和民间社会组织相关专题的五份文

件，并在每章结尾给出建议和总结。这些建议综合了复杂的研究和证据，促进了监管的国际协调和产品监管最佳实践的采用，以及世界卫生组织所有区域的产品监管能力建设，是成员国在可靠科学基础上实施世界卫生组织《烟草控制框架公约》的现成资源。鉴于烟草和烟碱产品（TNP）在全球范围内的大肆营销，研究组敦促成员国继续关注世界卫生组织《烟草控制框架公约》所概述的减少烟草使用的循证措施，并避免被烟草和相关行业分散注意力。

研究组的第九份报告涉及促进吸入的添加剂，合成烟碱，尼古丁袋，用于评估ENDS、ENNDS和HTP的生物标志物以及烟草和非治疗性烟碱产品的互联网营销。然而，报告并未涵盖烟草和烟碱产品监管中的所有新问题，包括调味和设计特点，如过滤嘴和调味配件。研究组将在后续报告中继续涵盖产品监管的其他方面，包括其他相关产品（如水烟、卷烟和无烟烟草）以及直接影响烟草控制的其他新兴问题，并以各国的监管要求和烟草制品监管的相关问题为指导。因此，研究组将确保向所有国家提供持续、及时的技术支持，并广泛解决非治疗性烟草和烟碱产品以及对产品监管有影响的因素，特别是那些影响产品的吸引力、成瘾性和毒性的因素。

总而言之，TobReg的审议结果和建议将提高成员国对报告中所考虑主题（包括合成烟碱、烟草制品在线营销和尼古丁袋）证据的理解，有助于丰富产品监管知识体系，为世界卫生工作提供信息，特别是在向成员国提供技术支持方面的同时，通过各种平台使成员国、监管机构、民间社会组织、研究机构以及其他有关各方了解产品监管的最新动态。世界卫生组织《烟草控制框架公约》缔约方将通过公约秘书处向缔约方会议第十次会议提交一份关于实施世界卫生组织《烟草控制框架公约》第9、10条相关技术事项的综合报告，其中将包括本报告中的信息和建议。因此，研究组的活动将有助于实现可持续发展3.a目标：加强世界卫生组织《烟草控制框架公约》的实施[9, 10]。

参 考 文 献

[1] WHO report on the global tobacco epidemic 2021: addressing new and emerging products. Geneva: World Health Organization; 2021 (https://www.who.int/publications/i/item/9789240032095).

[2] Tobacco control to improve child health and development: thematic brief. Geneva: World Health Organization; 2021 (who.int/publications/i/item/9789240022218).

[3] Tobacco product regulation: basic handbook. Geneva: World Health Organization; 2018 (https://www.who.int/tobacco/publications/prod_regulation/basic-handbook/en/, accessed 10 January 2021).

[4] WHO Framework Convention on Tobacco Control. Geneva: World Health Organization; 2003

(http://www.who.int/fctc/en/, accessed 10 January 2021).

[5] Partial guidelines on implementation of Articles 9 and 10. Geneva: World Health Organizati-on; 2012 (https://www.who.int/fctc/guidelines/Guideliness_Articles_9_10_rev_240613.pdf, accessed January 2019).

[6] Tobacco product regulation: building laboratory testing capacity. Geneva: World Health Or-ganization; 2018 (https://www.who.int/tobacco/publications/prod_regulation/building-laboratory-testing-capacity/en/, accessed 14 January 2019).

[7] Tobacco product regulation courses. Geneva: World Health Organization (https://openwho.org/courses/TPRS-building-laboratory-testing-capacity/items/3S11LKUFGyoTksZ5RSZblD; https://openwho.org/courses/TPRS-tobacco-product-regulation-handbook/items/7zq-7S1jxAtpbUWiZdfH98l).

[8] TobReg members. In: World Health Organization [website]. Geneva: World Health Organization (https://www.who.int/groups/who-study-group-on-tobacco-product-regulation/about, accessed 10 January 2021).

[9] Feacham RGA, Sachs JD. Global public goods for health. The report of working group 2 of the Commission on Macroeconomics and Health. Geneva: World Health Organization; 2002 (https://apps.who.int/iris/bitstream/handle/10665/42518/9241590106.pdf?sequence=1, accessed 10 January 2021).

[10] Sustainable development goals. Geneva: World Health Organization (https://www.who.int/health-topics/sustainable-development-goals#tab=tab_2, accessed 10 January 2021).

2. 促进吸入的添加剂（凉味剂、烟碱盐和调味剂）

Reinskje Talhout, Centre for Health Protection, National Institute for Public Health and the Environment, Bilthoven, Netherlands (Kingdom of the)

Adam M. Leventhal, Institute for Addiction Science, University of Southern California, Los Angeles (CA), USA

摘　要

目的：一些添加剂可以中和烟草和烟碱产品（TNP）中气溶胶的刺激性和苦味，使吸入它们变得更容易。这对公众健康构成问题，因为它可能令人尤其是令年轻人接受和持续使用烟草制品。本章提供了一个评估促进吸入过程、机制和方法的概念框架，总结了烟草和烟碱产品中可能促进吸入的特定添加剂，并讨论了它们的潜在健康影响。

方法：在PubMed和其他资料来源有针对性（非系统性）地检索2022年9月之前的文献，不限制时间段，检索关键词包括与促进吸入过程相关的术语（如“刺激性”“抽吸持续时间”）、候选添加剂（如“薄荷醇”）或候选机制（如“TRPM8，瞬时受体电位阳离子通道-8”）。纳入综述的研究是由两位作者一致同意的。

结果：我们将吸入促进定义为对烟草和烟碱产品的修改，这种修改改善了使用者吸入产品气溶胶的感官体验（减少苦味和刺激性），并可能改变吸入行为，特别是更强烈的吸入（例如更深的吸入，更快的吸入，更大的吞吐量），以及恢复被吸入刺激物扰乱的呼吸模式。研究表明：①薄荷醇和合成凉味剂通过激活TRPM8和其他受体，降低烟草和烟碱产品气溶胶成分引起的刺激，可能促进无经验使用者的依赖；②酸性添加剂和糖类，在燃烧时生成酸，降低了烟草和烟碱产品气溶胶的“pH值”，导致质子化烟碱水平升高，这被认为比游离态烟碱刺激性更小，并可能增加血液中的烟碱水平；③电子烟中的甜味添加剂降低了人们对苦味的感知，可能增加使用频率，尽管对感知刺激性的影响尚无定论；④烟草中的糖带来甜味感受，但因独立于行业的数据有限，无法对其带来的吸入促进得出强有力的结论；⑤添加剂对促进吸入的某些影响在非吸烟者和年轻人群中被放大；⑥应该对吸入行为进行研究。

结论：一些添加剂可能通过改善感官体验来促进烟草烟气和/或电子烟气溶胶的吸入。促进吸入的添加剂可能会增加烟碱的血液水平和依赖性，在某些情况下还会增加吸入行为，尤其是在年轻人和非吸烟者中。进一步研究烟草和烟碱产品添加剂对感官属性和吸入行为的影响可以为监管政策提供有用的证据。

关键词：烟草和烟碱产品，产品吸引力，吸入促进剂，添加剂，清凉效果，pH降低，掩盖苦味

2.1 引　　言

对烟草行业内部文件的研究显示，卷烟制造商通过操控产品设计，包括外观、风味和烟气特性，来提升其吸引力和消费者接受度[1,2]。这些操控的机制包括增强烟碱的有效释放和吸收及促进烟气的吸入[3,4]。研究发现，超过100种卷烟添加剂可以掩盖卷烟散发的环境烟雾气味，增强或维持烟碱的有效释放和吸收，提高卷烟的成瘾性，并掩盖与吸烟相关的不良感官效应，如刺激[2]。此外，这些产品中还包括了可能从烟草和烟碱产品（如电子烟、雪茄和水烟）中促进吸入的添加剂[5,6]。

促进吸入的添加剂可能会通过多种方式影响个人和人群的健康。对于刚开始使用烟草和烟碱产品的人来说，烟碱和烟草是令人厌恶的，不愉快的感官感受可以阻止他们经常使用烟草和烟碱产品[7,8]。因此，对于尝试使用烟草和烟碱产品的年轻人来说，添加剂可能会增加产品的吸引力，从而增加他们使用产品的可能性。此外，添加剂导致的吸入促进可能会导致烟碱成瘾和长期重度使用的风险[7]。综合来看，它们构成了较高的滥用倾向。然而，电子烟可吸入性的提高也可能使它们成为一些成年吸烟者更满意的烟碱替代品，并鼓励他们戒烟，转而使用电子烟。决策者已将促进烟草和烟碱产品吸入的添加剂确定为吸烟的关键决定因素，因此是潜在的监管目标[8,9]。然而，目前缺乏一个以科学为基础的框架来指导关于烟草和烟碱产品中吸入促进的研究和监管政策。

本章阐述了吸入促进效果、背后的机制以及相关研究，评审并权衡了一组可能增强促进吸入效果的添加剂的证据，并讨论了这些发现对健康的潜在影响。首先，我们描述了烟碱和其他感官刺激物如何降低烟气的可吸入性，特别是对于新吸烟者。接下来，我们提出了促进吸入的定义和概念模型，包括添加剂以外的因素。我们还描述了可以评估添加剂吸入促进的研究设计。随后概述了促进吸入的添加剂类别及其在烟草和烟碱产品中的效果，重点讨论了烟草和烟碱产品及电子烟（可吸入的烟碱产品）。检索范围未包括其他可吸入产品（如加热型烟草制品），也未包括可能影响吸入促进的其他因素，例如物理设计（如过滤嘴通风）。最后，

我们讨论了研究结果，指出了证据中的空白，并描述了禁止或设定此类添加剂限值和现有的促进吸入添加剂立法的潜在影响。

2.2 方　　法

在PubMed等文献数据库和其他来源（例如会议记录、通用网络检索）中进行了检索，时间范围不限，截止到2022年9月（有一个例外，2023年1月的一篇论文提供了关于电子烟烟液中有机酸的额外证据）。使用了针对性（非系统性）检索策略，其中包括与吸入促进相关的检索词（如“刺激性”“抽吸持续时间”），候选添加剂（如“薄荷醇”）和候选机制（如“TRPM8受体”）。通过探索性的“滚雪球抽样”获得了论文，其中对文章的参考文献部分进行了检查，获取并审查了可能相关的文章。由于研究的纳入是基于合著者之间的共识，该综述可能并不全面。选择时优先考虑设计更严谨、相关性更高的研究。

2.3 烟碱和其他影响可吸入性的感官刺激物

通常，第一次接触烟草和烟碱产品会感到不愉快，因为卷烟烟气中含有许多刺激化学感觉神经的刺激物，导致不适的灼烧感和刺痛感，以及诸如咳嗽、打喷嚏和回避等反射反应[7,10]。由于烟碱是烟草烟气中的主要刺激物，这些效果也会在使用口服（如烟碱口香糖）和可吸入（如电子烟）烟碱产品时发生[11]。烟碱在人和动物中也具有奖赏效应，即使在低浓度下也是如此[7,10]。烟碱通过与位于中脑边缘多巴胺能通路和抗痛觉系统（减少疼痛）的烟碱型乙酰胆碱受体（nAChR）结合，激活大脑中控制奖励的系统。烟碱还会引起口腔和喉咙的不良感觉，包括刺激、疼痛、苦味、恶心和头晕[11,12]。作为反应，吸烟者会调整他们的烟碱摄入量，以便在避免不良感觉的同时体验奖励效果[13]，通过将烟气与空气混合，使吸入不会造成太多刺激[11]。对刚开始吸烟的人，烟碱和烟草最初的刺激性是令人不适的，因此可能会阻止他们经常使用烟草和烟碱产品[7,14]，但随着反复使用，与烟碱的中枢神经效应（非条件刺激）反复配对的感官刺激会获得激励性意义，并因与即将到来的烟碱奖励的关联而促进与吸烟相关的行为[1,15]。通过感官获得的信息来自各种神经反应，包括嗅觉（通过嗅觉神经）、刺激（三叉神经）和味觉（面部、舌咽部和迷走神经），这些线索可能通过学习与中枢介导的药物奖励的关联而产生激励价值[1]。下文将讨论卷烟烟气中可以加强这一效应的其他成分。

烟碱的刺激性和令人反感的苦味主要是通过激活位于痛觉神经末梢的烟碱型乙酰胆碱受体（nAChR）介导，如在口腔或鼻腔黏膜和肺部[1,11,13,16]。这些痛觉感受器激活位于三叉神经次级核尾侧和其他脑干区域的神经元[11,12]。在随后的接触中，这些神经元的活动降低，伴随外周感觉神经元脱敏和口腔刺激感逐渐减少[12]。烟碱也通过刺激味觉引起nAChR介导的苦味。在对啮齿类动物的研究中，即使添加了甜味剂，动物也会避开烟碱溶液[12]。

瞬时受体电位（TRP）阳离子通道参与烟碱引起的局部刺激和疼痛，特别是在人类口咽和喉部广泛表达的TRPV1、TRPA1和TRPM8亚家族[1,11]。一些针对这些TRP通道的化合物，如薄荷醇，可以改变烟碱引起的口腔刺激和疼痛[7,11]。TRPM5是化学感觉细胞中的一种信号介质，也是味觉传导的关键组成部分，它与烟碱的苦味有关[17]。

卷烟烟气中的其他化合物也参与烟气引起的疼痛和刺激，例如活性醛（如甲醛和丙烯醛）、酸（如乙酸）和挥发性有机碳氢化合物（如环己酮）[7]。例如，丙烯醛通过TRPA1刺激受体激活化学感知神经，乙酸和环己酮可能通过酸敏感离子通道、TRPV1受体和其他类型的感觉受体起作用[7,18]。

2.4 促进吸入的添加剂

2.4.1 定义和概念框架

我们将吸入促进（IF）定义为对烟草和烟碱产品的修改，提高了使用者吸入产品气溶胶的感官体验（减少苦味和刺激性），并可能改变他们的吸入行为（特别是更强烈的吸入，如更深的吸入、更快的吸气、更大的吞吐量，同时也恢复被吸入刺激物干扰的呼吸模式）。图2.1显示了吸入促进的概念模型，在下文中进行描述，并在第2.4.2小节回顾了支持性证据。需要指出的是，并不是所有添加剂的

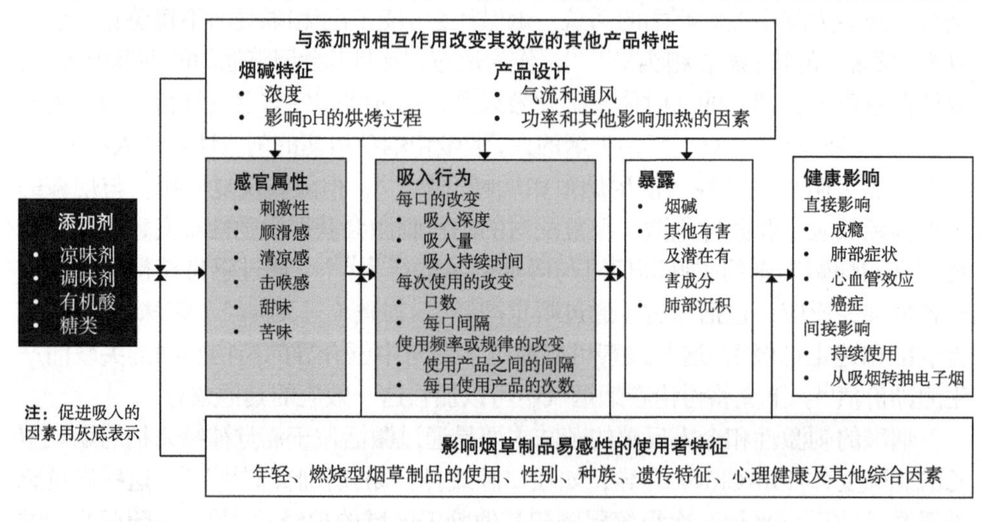

图2.1 添加剂在吸入促进方面的影响及其对健康相应影响的概念模型

概念模型中的所有因素都有可用的证据。这幅插图描绘了作者对促进吸入涉及的概念的建议。

本章重点讨论烟草和烟碱产品中的添加剂（如调味剂、凉味剂、有机酸、糖类）。其他因素也会影响吸入促进，包括从烟草中提取的化合物、烟碱浓度和设计操作（例如过滤嘴通风、气流、加热元件、烘烤过程）；然而，这些因素并没有直接涉及。

促进吸入过程

添加剂可以通过影响呼吸道感觉（增加顺滑或清凉感，减少刺激性或疼痛，以及带来令人舒适的“击喉感”）来改善吸入烟草和烟碱产品气溶胶的感官体验，从而可能使气溶胶更容易吸入。添加剂引起的嗅觉和口腔感官特征（增加甜味和减少苦味）可能通过增加产品的吸引力从而间接促进吸入，使得更多气溶胶被吸入。与吸入促进相关的吸入增加包括瞬时效果，如每次吸烟的深度、体积、速度和持续时间增加。对于那些不习惯吸入刺激性和苦味烟草及烟碱制品气溶胶的新手来说，这种影响可能更为明显。对于经验不足和年轻的使用者，具有理想感官特征的烟草和烟碱产品也可以缩短吸烟间隔和增加每次吸烟的次数（例如点燃然后熄灭卷烟），因为他们可能需要更短的时间来“恢复”在每次吸气之间的感官刺激。这可能会加剧使用和依赖性。对于重度烟碱依赖的烟草和烟碱产品每日使用者，他们习惯于维持血液中的烟碱水平并避免戒断症状，促进吸入的添加剂可能导致每次吸烟吸入更多的烟碱，从而可以促进每次更快的满足感，并减少达到烟碱满足所需的吸气次数。

促进吸入的后果

改变的吸入行为可能通过两种方式增加烟碱的摄入量。首先，与吸入促进相关的吸入行为变化可以直接增加每次吸入量、每次使用次数和每天消耗的气溶胶总量以及更深的吸入量。其次，与吸入促进相关的吸入增加可能会改变肺部沉积，允许更多的烟碱被吸收，并使肺部的某些部分更容易受到伤害。两项关于过滤嘴通风性与肺腺癌之间因果关系的证据综述表明，更深地吸入卷烟烟气可能会增加腺癌的发病率[19]。与吸入促进相关导致的烟草制品的接触增加可能会对健康产生许多直接影响，包括对心血管和肺部系统的影响，以及各种癌症风险的增加。

此外，更大的吸气量、更短的吸气间隔和更深的吸入会影响烟碱输送到血液中的速率和量，这直接对应于产品的加强作用[20,21]。烟草和烟碱产品令人愉悦的感官属性也有助于其加强效应，并与烟碱共同作用，进一步强化这种增强作用[22]。产品的加强效应直接关系到其成瘾潜力和持续使用的可能性。对于未使用烟草和烟碱产品的青少年和成年人，吸入促进在任何可吸入产品中都是有害的，因为吸入促进可能刺激产品的吸收和持续使用。然而，吸入促进对于希望转换到电子烟的成年吸烟者可能是有用的。电子烟中吸入促进的添加剂可能会增加烟碱的释放量和增强效果，从而增加人们对电子烟产品的采用和转向并促进戒烟。然而，对于完全从卷烟转向电子烟的成年人中，电子烟中吸入促进的添加剂可能会促进电子烟的持续使用，并可能增加接触有害成分的风险。在人群层面上的净效应需要更多的数据来确定。

与其他产品和使用者特征的交互

促进吸入的产品感官属性的质量可能取决于使用者的特征。一方面，年轻人和未曾吸烟者可能因为刺激和苦味而被阻碍，而吸入促进的添加剂会抑制这种阻碍。另一方面，希望转向电子烟的长期成年吸烟者可能会寻求替代卷烟感官属性的产品，并提供适宜的喉部冲击感和浓郁的味道。因此，抑制电子烟苦味和刺激性的添加剂的促进吸入可能对那些已经习惯吸入苦味、刺激性烟草烟气的吸烟者效果较小。其他使用者特征（例如遗传、心理健康、其他共病条件、种族或民族、性别）也可能影响他们接触烟碱或有害/潜在有害成分时感官的敏感性。其他产品特性可以与促进吸入的添加剂相互作用，从而增加它们对吸入行为及其影响和结果的作用。例如，在烟碱浓度很高的电子烟中，抑制烟碱刺激和苦味的添加剂可能会产生特别强烈的效果。

2.4.2 证据审查与整合

本小节总结关于几类添加剂及其在模型中描述的吸入促进过程中的作用的文献。我们将直接显示添加剂对感官体验和/或吸入行为影响的证据视为吸入促进的主要证据。对吸入促进相关作用机制及其后果（生物标志物和健康结果）的研究进行了回顾，为吸入促进科学框架的生物学合理性和健康意义提供了支持证据。

表2.1总结了添加剂及其效果和推测的机制，该表基于以下类型的证据：①添加剂对吸入促进的感觉和疼痛通路的影响的基本机制研究；②动物模型用于研究烟草和烟碱产品气溶胶对吸入行为的影响，包括感觉过程、接触（或暴露）以及吸入行为；③人体临床试验，研究服用含各种添加剂的烟草制品气溶胶对吸入促进的影响，包括感觉过程、产品吸引力、接触（或暴露）以及吸入行为；④关于使用含有添加剂的产品是否与吸入行为改变有关的观察性研究。

表2.1　吸入中涉及的添加剂类别：机制和作用

机制	添加剂	报告的感官效应	备注
TRPM8受体激活	薄荷醇	增加清凉感，降低烟碱的刺激性，薄荷味	烟草制品和电子烟都有证据证明啮齿动物的吸入行为增加，但对人类的吸入行为影响不确定
	合成凉味剂	增加清凉，减少刺激，减少苦味	证据主要来自电子烟，通常与其他口味的“冰”调味剂混合
降低pH	电子烟中的有机酸和烟碱盐	增加温和性，减少刺激	血液中烟碱含量升高
	烟草中的糖	燃烧成酸，增加温和性	主要来自行业数据
	烟草中的有机酸	增加温和性，减少刺激性	主要来自行业数据
嗅觉和口腔感觉机制	带有甜味的调味料	增加甜味，减少苦味，部分证据表明提高了顺滑感，减轻了刺激性	主要是电子烟、水烟和雪茄
	糖类	赋予甜味	主要是卷烟；主要来自行业数据

2.5　具有清凉作用的添加剂

2.5.1　薄荷醇

薄荷醇是薄荷植物（薄荷属）中天然存在的化合物。它被用作各种食品、药品和化妆品以及烟草和烟碱产品中的添加剂。薄荷醇不仅在“薄荷味”的烟草和烟碱产品中检测到，而且在没有明确标明“含薄荷醇”的产品中也能检测到[23]。薄荷醇通过激活大脑中的烟碱型乙酰胆碱受体（nAChR）来影响中枢神经系统；然而，其在吸入促进中的作用是通过其抗刺激、冷却和镇痛特性介导的[24]。薄荷醇的感官效应主要是通过其与阳离子通道内冷敏感感觉神经元中的瞬时受体电位阳离子通道-8（TRPM8）和瞬时受体电位锚定蛋白-1（TRPA1）的相互作用介导的[12]。薄荷醇可能还具有镇痛和交叉脱敏的特性，其中薄荷醇的预处理可能减少烟碱的刺激效果，即使在其急性清凉效果消失后也是如此[11]。来自啮齿动物模型的研究证据表明，烟草和烟碱产品气溶胶中呼吸道刺激物的效果可被薄荷醇抑制，导致呼吸更频繁、呼吸间隔更短和呼吸流速更快[7,25]。啮齿动物的吸入行为类似于与吸入促进相关增加的吸气次数，更短的吸气间隔和更快的吞吐速度。

美国食品药品监督管理局（USFDA）对薄荷醇在烟草卷烟中的影响进行了文献的全面审查，包括人体临床试验和观察性研究[24]。结论是，薄荷醇通过掩盖烟草烟气的刺激味道和减少与初次吸烟经历相关的厌恶感觉反应（如刺激、咳嗽）来增加卷烟的适口性，从而促进继续吸烟。这一结论在描述薄荷醇在青少年中的接受和依赖、戒烟困难和对非洲裔吸烟者不成比例的影响方面尤为明确[24]。例如，在一项观察性横断面研究中，相对年轻的成年薄荷醇和非薄荷醇卷烟的吸烟者，特别是非洲裔，报告了他们对吸烟的积极和消极主观反应；更积极的主观反应与更频繁的吸烟有关[26]。USFDA发现了关于成年人中薄荷醇与依赖性以及吸烟行为特征之间联系的混合证据[24]。由于具有稳定吸烟习惯的成年吸烟者对某些品牌的卷烟有强烈的偏好，并且存在对薄荷味的卷烟自然选择偏见，因此很难从这一人群中对薄荷味卷烟的人体临床和观察性研究中得出强有力的结论。

几项有关电子烟的临床试验表明，薄荷醇增加了对清凉和愉悦味道的感知[27-31]。四项研究表明，薄荷醇与烟碱相互作用会改变电子烟气溶胶的某些感官属性或吸引力[28-31]。例如，一项针对年轻成年电子烟使用者的研究表明，薄荷醇可以与浓度为6 mg/mL的烟碱相互作用，以抵消烟碱的不良感官特征[31]。该研究还表明，薄荷醇与烟碱的直接相互作用，增加了电子烟的吸引力，在从未使用过电子烟的人身上，比在双重使用者或既往使用者身上更为明显[29]。然而，在一项针对青少年的研究中，没有证据表明薄荷味电子烟改变了烟碱水平对感官属性或吸引力的影响；薄荷醇在两种烟碱浓度下都增加了感知的清凉感[27]。这些研究均未表明薄荷醇会影响吸烟行为或短期内接触烟碱[27,30]。

2.5.2 合成凉味剂

合成凉味剂，包括WS-3、WS-5、WS-14和WS23等化合物，已在多种烟草和烟碱产品中被检测到。烟草行业文件显示，在20世纪70~80年代，包括雷诺（RJ Reynolds）和菲利普·莫里斯（Phillip Morris）在内的主要烟草制造商测试了含有WS合成凉味剂的卷烟，但最初并未广泛推向市场[32-34]。然而，最近在德国的卷烟产品中发现了合成凉味剂[35,36]。合成凉味剂可能存在于具有清凉效果的卷烟产品中，这些产品在美国某些禁止销售薄荷醇卷烟的市场（例如加利福尼亚州）以"非薄荷醇"的形式销售[37]。在过去几年中，电子烟产品中也检测到了合成凉味剂，包括市场上以"冰"混合口味销售的产品，这种口味将含有水果、薄荷或其他特征风味的成分与合成凉味剂（如"覆盆子冰"）结合在一起[38,39]。最近的研究表明，可能含有合成凉味剂的冰混合口味电子烟在美国的青少年和年轻人中普遍使用[40-42]，其销量最近有所增加[43]。

一些合成凉味剂是基于薄荷醇的对薄荷烷结构。与薄荷醇相似，WS-3和WS-23在呼吸道和口腔内壁的TRPM8冷受体具有药理活性[7,16,25,44]。一些证据表明WS-3在TRPM8受体上比薄荷醇更活跃，能产生更强的清凉感[45,46]，并可能激活感觉刺激受体TRPA1[47-49]。

鉴于这些合成凉味剂的药理学特性，在含有这些化合物的烟草和烟碱产品中，它们可能产生清凉的感觉，掩盖烟碱的刺激性，而不提供强烈的薄荷味，这与含有薄荷醇的产品不同[41]。使用者在社交媒体和在线讨论上的报告表明，WS-23或WS-3在电子烟中具有显著的清凉感效果，而没有薄荷醇的强烈薄荷味[41]。在一项人体临床试验研究中，给烟草和烟碱产品的成年使用者提供加入烟碱盐和WS-23调味的电子烟，与无凉味剂相比，增加了电子烟的吸引力、顺滑度和清凉感，并减少了它们的苦味和刺激性[50]。此外，用WS-23调味的电子烟被认为比用薄荷醇调味的更顺滑、更凉爽、刺激性更小。凉味添加剂的效果在对比水果、烟草（2%与4%烟碱浓度）、薄荷醇或者吸烟状态之间无显著差异。凉味剂的吸入促进效果可能解释了为什么在一项观察性研究中，使用冰混合口味电子烟的相对年轻的年成人使用者报告的烟碱依赖症状比其他口味更多[40]。

2.6 降低 pH 值的添加剂

烟草行业已经开展了关于调节pH值对吸烟耐受性影响的研究[3]。在烟草和烟碱产品中，烟碱通过膜的吸收程度和烟碱引起的刺激性取决于烟碱质子化的程度[51]。质子化与非质子化（游离态）烟碱的比例取决于产品的pH值，因此可以通过添加酸性或碱性添加剂来影响。在游离状态下，烟碱渗透膜，然后转化为质子化状态，即nAChR的配体[52-54]。在高于生理水平的pH 7~12的条件下，烟碱以游离态形式存在，更容易跨膜吸收，也会带来更强烈的击喉感，感觉更刺激。尤其是在高浓度的情况下游离态烟碱可能更令人反感，因为它更容易被吸收在上呼吸道中，引起刺激，而质子化烟碱的刺激性较小，因此可以吸入更深，导致在呼吸道中沉积更深[55]。这会导致更多烟碱被吸收进入系统循环。卷烟烟气通常是微酸性的，pH值约为6，这使得烟气不那么刺鼻，比雪茄等pH值较高的烟草和烟碱产品更容易吸入[3,51]。一旦卷烟烟气到达肺泡，烟碱便从烟气中释放出来，在肺部的生理pH值下，烟碱很容易通过肺毛细血管被吸收进入系统循环[51]，这是因为肺部在pH值为7.4时吸收表面积更大，并且肺的局部缓冲能力更强[56,57]。这种效应只适用于可吸入产品，而不适用于烟碱通过口腔吸收的产品，如尼古丁袋。

因此，质子化烟碱在吸入时比游离态烟碱刺激性和苦味要小，使得高浓度的烟碱更加可口[5]。因此，在较低的pH值下，烟碱的整体释放和吸收可能更高[3]。由于游离态烟碱的刺激在很大程度上被减弱，质子化烟碱在高浓度时不那么令人反感，从而增加了产品的吸引力。此外，更多的烟碱吸收，伴随着更快、更高的血液烟碱峰值水平，可能预示着更大的滥用倾向[51]。

这一机制及其对感官吸引力和吸烟行为的影响将在下文讨论，特别是那些涉及导致电子烟烟液中烟碱盐（带有质子化烟碱）的酸性添加剂。同时，也会提及其他例子，例如，作为烟草添加剂的乙酰丙酸，以及烟草中的糖类在燃烧时导致烟气中的酸含量增加。

2.6.1 电子烟烟液中的有机酸

在美国，Juul和类似电子烟的营销导致不吸烟的年轻人使用电子烟的人数迅速增加[58,59]。这些产品含有高水平的气溶胶烟碱，并且添加的有机酸导致电子烟烟液中含有质子化烟碱而不是游离态烟碱[60]。多种有机酸被用于盐基电子烟中，包括乳酸、水杨酸、苯甲酸、乙酰丙酸、酒石酸和马来酸[61]。一些研究小组研究了质子化烟碱对血液烟碱水平的影响[62-65]。一些研究表明，与填充游离态烟碱烟液的电子烟不同，含有烟碱盐溶液的电子烟导致的烟碱血液特征与抽吸卷烟的人相似[62-65]。电子烟随机临床试验的二次数据分析也表明，转换到使用烟碱盐pod式系统电子烟（类似于Juul）的吸烟者能够维持他们血液中的烟碱水平，并转移了他们对传统卷烟的依赖性，这表明这类电子烟在增强使用者吸烟行为的潜力方面与传统卷烟相似，且能有效促进吸烟者从传统卷烟向电子烟的转换[66]。观察数据显示，使用Juul和其他含有烟碱盐的pod式电子烟的青少年对烟碱的依赖程度与使用传统卷烟的青少年相似[67]，但比使用不含烟碱盐的其他电子烟产品的年轻使用者表现出更高的依赖性[68]，这强调了有必要对这一年龄组的电子烟产品获取和市场营销进行管控。

因此，烟碱盐的使用增加了烟草和烟碱产品的成瘾性[64]，并且这种效应随着烟碱剂量的增加而增加。由Juul实验室资助的一项关于含有质子化烟碱的烟液的研究表明，与低水平相比，较高水平的质子化烟碱会显著提高血浆烟碱水平并缓解渴望[69]。在欧盟生产的一个Juul版本中含有18 mg/mL的烟碱，它释放的烟碱量较少，与传统卷烟相比，它减少了让使用者吸烟或使用电子烟的冲动[70]。美国的一个含有59 mg/mL烟碱的Juul产品给出了类似的结果[71]。

有三项研究比较了烟碱盐和游离态烟碱的感官效应。美国的一项随机临床试验表明，含有24 mg/mL烟碱盐的配方在吸引力、甜味和顺滑度方面的评分显著高于游离态烟碱，并且在苦味和刺激性方面评分较低。烟碱盐在提高顺滑度和减少刺激性方面的作用在从未吸过烟的人群中比在曾经吸烟的人群更强[72]。烟碱盐改善了感官体验，从而提高了使用电子烟的吸引力，特别是在从未吸过烟且不习惯吸入游离态烟碱的人群中。这些发现与来自英国的观察数据一致，这些数据表明，含有烟碱盐的Juul在从未吸烟者中的使用更为普遍，而通常含有游离态烟碱的罐式装置则更常被现有或吸过烟的吸烟者使用，尽管其他混杂因素（如年龄）也可以解释这种关联[73]。美国的一项临床试验研究发现，与游离态烟碱相比，含乳酸盐和苯甲酸盐（质子化）的电子烟烟液更有吸引力、更顺滑、更甜，并且比游离态烟碱的刺激性和苦味更少。有证据表明，高度质子化的电子烟烟液比中度质子化的电子烟烟液效果更强。烟碱配方的效果没有因吸烟状态或口味而异[74]。在荷兰，一项关于家庭使用的研究表明，在浓度为12 mg/mL时，烟碱盐与游离态烟碱在吸引力、刺激性和吸烟行为上没有显著差异[75]。这是唯一关于烟碱质子化状态对吸烟行为影响的研究。除了烟碱含量较低外，这项在荷兰的研究中的使用者可以自由地使用电子烟，研究人员会检测吸烟吞吐参数；而在美国的研究则采用了一套固定的吞吐方式。

2.6.2 卷烟中的乙酰丙酸和其他有机酸

在传统卷烟的生产中使用了许多不同的酸性添加剂，以增加顺滑度，减少击喉感[56,76]。例如，乳酸被用来减少刺激性和苦味，产生更甜的味道。柠檬酸添加剂不仅用于减少刺激性和改变口味，而且还用于改变烟气的pH值以及中和击喉感。此外，酒石酸和乳酸也被用来改变烟气的pH值。

对烟草行业内部文件的回顾表明，乙酰丙酸被用于增加烟碱产量，同时增强顺滑度和温和的感觉[3]。乙酰丙酸可以降低卷烟烟气的pH值，使上呼吸道脱敏，促进卷烟烟气更深地吸入肺部。行业研究还发现，在添加了乙酰丙酸的超轻型卷烟中，吸烟者的血浆烟碱峰值水平显著提高。

2.6.3 糖类

烟气的pH值也会受到烟草中的糖类的影响。一份行业文件提到，可以通过添加合适的有机酸或增加烟草中的糖含量来降低刺激性[76]。在卷烟中，烟草中0.5%的糖被转移到主流烟气中，其中大部分被燃烧或热解[77-79]。据报道，在卷烟中添加糖类可增加烟气的酸度[77,78,80]；然而，吸烟过程中糖类的燃烧产生的酸会降低pH值[81]，从而减少烟气的刺激性[82,83]，增加产品的适口性并促进吸入。烟草行业将糖类称为“改良剂”,用来“中和烟草的刺激性和苦味和/或消除刺鼻的气味”[84]。

2.7 可掩盖苦味的调味剂

2.7.1 甜味添加剂

在各种类型的可吸入烟草和烟碱产品中已经鉴定出数百种调味成分[85,86]，其中许多被归类为甜味（如水果味、薄荷味、甜品味）添加剂[85,86]。考虑到这些成分种类繁多，很难确定甜味影响的一种或几种生物学途径。从心理感觉的角度来看，有一些证据表明，具有甜味特征的烟草和烟碱产品可能通过嗅觉发挥作用，而不仅仅是通过它们的感官影响[87,88]。

研究对于甜味添加剂对吸入促进过程的影响提供了一些可能效果的证据，但具体结果并不一致。烟草制品中鉴定出的各种各样的甜味添加剂(如角豆提取物、甘草）可能通过糖类的热解间接促进吸入，这将在2.7.2小节有更详细的内容[9]。

一项对定性研究的系统综述表明，雪茄、水烟、电子烟和卷烟中的甜味降低了对刺激性的感觉，使产品更容易被接受[89]。临床试验显示，电子烟中的甜味添加剂（如水果味）可以降低人们对苦味的感知[17]。甜味添加剂在减少电子烟的刺激性和增加电子烟的顺滑度方面的效果，结果是不一致的[31,90-92]。也有证据表明，甜味添加剂减少了电子烟中烟碱的苦味增强作用[31,90]。在一项针对青少年电子烟使用者的临床研究中，青苹果味电子烟比薄荷或无味电子烟更能增加急性抽吸次数和抽吸持续时间[27]。一项关于青少年电子烟使用者的纵向观察队列研究表明，与烟草或薄荷味相比，甜味或水果味与6个月后自我报告中的每次使用电子烟的抽吸口数更多有关，但与每天使用的次数无关[93]；在该研究收集的数据中，它们的关联未见报道。在一项针对美国居民某一段时间的人群观察性研究中，甜味电子烟的使用在青少年和年轻成年人中比在老年人中更常见[94]。

在一项针对14名成年电子烟使用者进行的临床研究中，评估了使用草莓味而非烟草味电子烟（烟碱19~20 mg/mL）的影响[95,96]。在执行规定的标准15次抽吸操作中，虽然吸入和系统内保留的烟碱量相似，血液中的烟碱水平显著更高。在自由吸烟的方案中，草莓电子烟烟液的平均抽吸持续时间显著长于烟草电子烟烟液。两种口味在滥用倾向的主观衡量标准上没有差异。尽管结论受到样本量小的限制，但这项研究提供了一些证据，表明甜味电子烟烟液可能与增加的烟碱接触和吸入行为有关。

在一项基于人群的观察研究中，2015~2016年在美国对211名报告在过去24小时内使用过其产品的电子烟使用者的生物标志物暴露进行了研究，仅水果味电子烟使用者体内的丙烯腈生物标志物水平高于其他非烟草口味（薄荷、丁香、巧克力或其他）的使用者；然而，丙烯腈的浓度没有差异。暴露于烟碱（可替宁）、苯和丙烯醛的生物标志物浓度在不同口味组之间没有显著差异[97]。由于这是一项观察性研究，没有考虑到使用者行为（如使用电子烟的频率）、设备或电子烟烟液（如烟碱浓度）的差异，因此很难确定不同口味使用者的暴露生物标志物是否受到这些外部因素的影响。

2.7.2 糖类和甜味剂

除了对pH值的影响（见2.6.3小节），糖类还有助于烟草和烟碱产品[78, 83, 98-100]和电子烟[101,102]的味道。在烟草制品中，糖类燃烧产生的焦糖味能让使用者和非使用者接收到的烟草烟气的味道和气味得到改善[78, 82, 103-105]。此外，在烟草的烘烤和抽吸过程中，糖类可以参与美拉德反应，产生具有特色的木质、焦糖和烘焙风味的调味剂[15,106]。糖类通过美拉德反应产生的一类化合物是吡嗪类化合物，它也被用作烟草添加剂，特别是在可可、坚果或爆米花口味的低焦油卷烟中[15, 106]。有假设认为它们可能会减少有害的感觉，如上呼吸道的刺激，或者具有增强吸烟习惯行为的化学感觉效应[15]。甜味可能还降低吸烟者的咳嗽阈值。用蔗糖溶液漱口可以调节对咳嗽反射的敏感性，一些人认为这是内源性阿片类物质在对甜味的反应中释放所致[107]。

行业文件表明，吸烟者对烟草烟气的接受程度与烟草中的糖分水平含量成正比，这可能是由于它们的风味和它们对pH值的影响（见第2.6.3小节）[78, 99]。当糖和烟草生物碱（如烟碱）的比例增加时，影响会降低，而“喜好度”会增加到一定的最佳值[76]。在卷烟中添加糖以增强卷烟烟气的感官属性并鼓励开始并持续吸烟已被业界讨论作为其营销策略的一部分[108, 109]。在电子烟中，添加人造甜味剂蔗糖素增加了整体风味和甜味，但对刺激性或不适感没有显著影响[87]。高强度甜味剂如糖精和甘草酸也被添加到诸如小雪茄的烟草制品的烟嘴和包装中[110]。

2.8 讨　　论

2.8.1 主要发现

综上所述，本章所回顾的文献部分验证了所提出的吸入促进框架。我们发现有证据表明，多种添加剂通过增强感官体验促进烟草烟气和/或电子烟气溶胶的吸入。同时，一些提高烟草和烟碱产品感官属性的添加剂可能导致更高的烟碱血液水平或维持烟碱依赖。关于添加剂对吸入行为的客观测量影响的研究较为稀少。我们还发现了一些证据，表明多种添加剂影响了感觉和呼吸相关的途径，以及该框架的生物合理性。这些发现表明，研究烟草和烟碱产品中添加剂对感官属性和吸入行为的影响可能为监管政策提供有用的证据。特别是关于人体对刺激性和顺滑性感受的研究，以及动物模型中气道刺激的生物途径的研究。

薄荷醇和合成凉味剂被发现可以减少对卷烟和电子烟的不良感官反应。它们对吸入促进的生物合理性基于研究表明，凉味剂的感官效果主要是通过它们与呼吸道内冷敏感感觉神经元中的TRPM8和TRPA1的相互作用来介导的。在一项关于啮齿类动物吸入研究中，薄荷醇导致更深层的卷烟烟气吸入和更高的血液中可替宁含量。然而，人类吸气行为变化的证据则表现出混杂性。在一项关于电子烟的研究中观察到合成凉味剂效果的直接实验证据。合成凉味剂和薄荷醇效果的相似性在生物学上是合理的，因为它们有着共同的底层机制，尽管合成凉味剂和薄荷醇的效力不同，合成凉味剂对凉爽感和愉快的呼吸感觉的潜在影响更大。一些证据表明，薄荷醇对吸入促进的影响在不经常吸烟的年轻人群中更为显著，这些人群在其他人口特征（如性别、种族）上也有所不同。

关于降低pH值的添加剂，许多研究表明，电子烟中的酸性添加剂有助于电子烟气溶胶的吸入。尽管还需要进行更多的研究，但已有研究表明，在烟碱浓度较高（> 20 mg/mL）的情况下，电子烟烟液中的质子化烟碱（使用有机酸添加剂）比相同浓度的游离态烟碱增强了几个吸入促进过程，包括更理想的感官属性，这反过来又导致血液中烟碱水平升高以及维持烟碱依赖。需要更多的信息来确定酸性添加剂是否会导致更强烈或以其他方式改变吸入行为。据报道，酸性添加剂和糖类在燃烧时产生酸性物质，降低卷烟烟气的pH值，其他研究表明这类化合物减少了烟草烟气的刺激性，并增加了烟草烟气的顺滑度。关于对人类感知影响的大部分证据都是在较早的行业内部文件中找到的。因此，尽管2.6.1小节中电子烟的类似效果可能也会在卷烟中发现，因为它们具有相同的降低pH值的机制，但还是有必要进行额外的独立研究。一些证据表明，pH值对非吸烟者和年轻人群的吸入促进影响可能更大。

在电子烟中添加甜味（如水果味）添加剂能一直降低人们对苦味的感觉，但关于它们是否能降低刺激性的感知以及增加顺滑度的证据尚未得出明确结论，关于嗅觉和味觉效果对此的相对贡献的证据同样不明确。两项设计不同的研究发现，使用水果味电子烟的青少年的吸入行为水平增加，包括吸烟持续时间和次数。在电子烟中，添加三氯蔗糖（一种人造甜味剂）增加了整体味道的强度和甜度，但对刺激性没有影响。据报道，糖会给卷烟烟气带来甜味或焦糖味，但大多数关于糖对人类感知影响的证据都是在较早的行业内部文件中。因此，尽管可能会发现类似于甜味添加剂的效果，但独立研究还是必要的。虽然所有具有调味特性的添加剂都发现了对理想感官属性有一定影响，但缺乏对烟碱血液水平、维持烟碱依赖性和吸烟强度的影响的数据。

2.8.2 欧盟和北美对促进吸入的添加剂的监管机制

《欧盟烟草制品指令》（TPD）第7.6.d条规定欧盟成员国应禁止将含有促进吸入或烟碱吸收添加剂的烟草制品和电子烟投放市场[111]。然而，TPD没有提供吸入促进或烟碱摄取促进的定义。比利时[112]和德国[113]已经禁止使用任何含量的薄荷醇，因为它具有吸入促进特性，这一禁令得到了欧盟联合行动烟草控制的支持[114]，他们得出结论，所有薄荷醇类似物，包括香叶醇，都具有依赖TRPM8的清凉效果，并且可能会有累积作用。由于这种效应是这些化合物的固有特性，任何含有薄荷醇及其类似物的产品都不符合TPD第7.6.d条的规定，即使它们在烟草中的使用水平未引起可测量的影响。比利时已经禁用了所有激活TRPM8温度感受器的激活剂，德国也禁用了其他特定的TRPM8激活剂。

加拿大不允许使用任何具有调味特性的添加剂、甜味剂、着色剂或其他增加烟草制品吸引力的化合物，尽管有少数例外（瓜尔胶、酒精味）[115]。美国已经制定了联邦产品标准，禁止在卷烟和雪茄中使用任何调味剂，包括薄荷醇[8, 116]。这些规定是否会延伸到非薄荷醇合成凉味剂还不清楚。美国没有其他特定的产品标准来禁止可能促进吸入的其他类型的添加剂。美国食品药品监督管理局（USFDA）根据每个品牌和产品线的情况逐案决定电子烟和其他新型产品的合法市场营销。美国《烟草控制法案》规定，任何监管决定都要考虑到对整个人群的影响。因此，管制限制的设计应尽量减少年轻人使用烟草和烟碱产品，并应该尽可能不阻止成年吸烟者放弃使用传统烟草产品。迄今为止，美国已经拒绝批准许多具有特色口味和甜味特征为市场定位的电子烟产品的申请，理由是有证据表明它们吸引了年轻人[117]。关于销售薄荷味电子烟的决定尚未有定论。USFDA已经批准了几种含有有机酸和质子化烟碱的电子烟产品的市场销售[118, 119]。

2.9 推荐研究

为了为监管决策提供可操作的证据，有必要扩大关于大多数添加剂的证据基础。鉴于证据的不足，我们建议进行以下工作：

- 对添加剂对吸入行为的影响进行临床研究，例如通过在烟草和烟碱产品上安装吸入行为监测设备进行测量；
- 对使用者进行前瞻性纵向研究，以确定使用含有添加剂的烟草和烟碱产品是否与更愉快的感官知觉和/或吸入行为指标的增加有关；
- 用动物模型进行临床前研究，以解决无法在人类中进行研究的特定问题，例如将未使用烟草和烟碱产品的研究对象引入到含有或不含添加剂的烟草和烟碱产品中的影响；
- 比较电子烟中添加剂对成年吸烟者、成年非吸烟者和年轻人的吸入促进影响；
- 研究卷烟和电子烟以外的更广泛产品的促进吸入影响，包括水烟、雪茄和加热型烟草制品。

还应对可能含有吸入促进添加剂的产品进行研究，包括特定品牌和口味的产品，这些产品可以与这些品牌的添加剂的测量值进行三角验证。调查工具可以用来询问参与者使用哪种口味，他们喜欢的产品的感官属性（例如使用时的刺激程度），以及电子烟的设备类型或烟碱配方（烟碱盐或游离态烟碱）。

为了确定添加剂对吸入促进过程的影响，对潜在危害较小的烟草和烟碱产品（特别是电子烟）的研究应包括对年轻非吸烟人群和年长成年吸烟者的影响的比较。例如，电子烟中的添加剂会促进年轻非吸烟者的吸入和不良接触，但不鼓励成年吸烟者转向电子烟，这应该成为监管限制的优先事项。

关于构成吸入促进的几个基本方面还需进一步研究。目前还不清楚增加甜味或减少苦味的添加剂是否直接增加了吸入行为，还是仅仅使产品更具吸引力。在实验研究中，探讨添加剂的增甜或减苦特性（如阻断嗅觉，苦味受体敲除的啮齿动物模型）对吸入行为的影响可能是有益的。目前尚不清楚哪种研究设计最适合评估添加剂是否增加吸入行为。对现有使用者的吸入行为的研究存在选择偏差的风险，因为参与者有预先存在的偏好。在其中对烟草和烟碱产品暴露受控的动物模型的吸入行为可能特别有用，尽管这与减少动物研究的目标相悖。应进行研究以确定增加吸入是否必要或足以增加烟碱和其他有害成分的接触，并增加包括成瘾在内的不良健康后果的风险。比较改变吸烟持续时间、次数、速度、体积、抽吸间隔时间和使用期间间隔时间对暴露和结果的影响将是有用的。这样的研究将表明哪些感官和吸入行为结果对于评估新添加剂的影响至关重要。

测试可以通过要求研究参与者按预期使用产品并报告他们在使用过程中的感官体验来进行。采用非盲测和盲测的研究可能是有价值的，在这些研究中，参与者不知道产品的名称，也没有看到营销材料，从而在服用产品时引出主观的刺激性、甜味、凉爽度或其他感官属性。与吸入相关的其他结果（如抽烟持续时间、速度、体积、抽吸间隔）也是有用的。这些数据（连同科学文献）可以与成分清单和市场营销材料进行三角验证，以确定产品是否违反了限制吸入促进的添加剂的禁令或产品标准。

2.10 政策建议

我们就所有可吸入性烟草和烟碱产品向政策制定者提出以下建议：

- 禁止使用有助于促进吸入的成分，因为这些成分促进了吸入型烟草制品（卷烟、雪茄、水烟、加热型烟草制品或其他含有烟草的吸入产品）的使用。没有理由允许使用使烟草产品更具吸引力的成分，如调味剂。

世界卫生组织《烟草控制框架公约》第9、10条的部分实施指南[120]指出，从公共卫生的角度来看，没有理由允许使用诸如调味剂之类的成分，这些成分使烟草和烟碱产品更具吸引力。因此，部分实施指南建议“缔约方应规范管制所有增加烟草和烟碱产品吸引力的设计特征，以降低烟草和烟碱产品的吸引力”。因此，鉴于世界卫生组织对吸引力的定义（如味道、气味和其他感官属性、使用便捷性、剂量系统的灵活性、成本、声誉或形象、假定的风险和益处以及其他旨在刺激使用的产品的其他特征等因素），决策者应禁止使用有助于促进吸入的成分，因为这些成分会促进产品的使用。此类禁令包含在《欧盟烟草制品指令》适用于烟草制品和电子烟的第7.6.d条中。

这里回顾的证据支持了我们提出的吸入促进定义，建议政策制定者使用此定义以规范烟草和烟碱产品中允许使用的添加剂：吸入促进定义为对烟草和烟碱产品的修改，改善使用者吸入产品气溶胶的感官体验（减少苦味和刺激性），并可能改变吸入行为（特别是更强烈的吸入，如更深的吸入、更快的吸入、更大的吸入量），还可以恢复通常被吸入性刺激物扰乱的呼吸模式。

除了禁止使用有利于促进吸入的成分外，建议政策制定者包含一份不限定的，尤其是针对可吸入烟草制品的此类化合物的清单。建议将该清单列入立法，以便在新的科学见解需要时，可以轻松地对列表进行调整。具体化合物的清单将有助于监督和执法。例如，比利时[112]和德国[113]已经禁止使用任何水平的薄荷醇，因为它具有吸入促进的特性。比利时随后禁用了所有激活TRPM8温度感受器的激活剂[112]。另一种直接的方法是提供一份清单，列出烟草和烟碱产品中允许使用的添加剂，但不包括任何具有吸入促进效应的化合物，这与加拿大的政策类似[115]。

建议禁止在传统卷烟中使用有利于吸入促进的成分，因为使用传统卷烟对任何类型的使用者都没有好处，无论是吸烟者还是非吸烟者。电子烟和其他可能比传统卷烟危害小的可吸入产品的立法可能取决于每个国家的情况和政策目标。决策者可以考虑人口水平的影响，并权衡证据，一方面，判断吸入促进的添加剂是否可以使这些产品成为一些成年吸烟者更满意的烟碱替代品，另一方面，判断它们是否会增加对年轻人和非吸烟者的吸引力、依赖风险和其他不良后果。以下建议的目的是防止年轻人和从不吸烟的人开始使用任何类型的可吸入烟草和烟碱产品，包括电子烟。因此，我们建议禁止在所有可吸入烟草和烟碱产品中添加任何有利于吸入促进的添加剂。

- 禁止在烟草和烟碱产品中添加任何水平的薄荷醇，也禁止在烟草和烟碱产品中添加具有类似化学结构或生理和感官效应的合成（如WS）和天然凉味剂（如香叶醇）化学品，以避免替代使用。

之前的一项详尽的综述[24]提供了充分的证据表明薄荷醇添加剂促进了卷烟中的吸入，而这一结论在本章的证据中得到了加强。我们建议任何与《欧盟烟草制品指令》（TPD）及类似立法框架保持一致的监管机构禁止在所有可吸入烟草和烟碱产品中添加任何水平的薄荷醇。具有类似化学结构或类似生理和感官效应的化学品应列入禁令，以避免替代。这些将包括合成类似物，如具有与薄荷醇类似的清凉效果的WS化合物，以及具有类似性能并可能吸入促进的天然化合物，如香叶醇。鉴于有证据表明，电子烟中的凉味剂在年轻人群和非烟草产品使用者中更强烈地促进吸入，因此，对可吸入的烟草和烟碱产品凉味剂添加剂的监管限制值得考虑。

- 禁止电子烟烟液中烟碱盐含量超过20 mg/mL，以保护儿童、青少年和非吸烟者。在电子烟液和烟草产品中设定最低pH值将降低烟碱的生物利用度，并降低产品的成瘾性。

通过对添加剂降低电子烟pH值的证据进行三角验证，并放大其对年轻人群和非吸烟者在吸入促进的影响，监管机构应考虑禁止在烟碱水平> 20 mg/mL的电子烟烟液中使用酸性添加剂和烟碱盐。虽然对于烟碱含量< 20 mg/mL的充分证据

尚不可用，但作为预防措施，监管机构可能会考虑在任何烟碱含量禁止此类添加剂。此外，在电子烟烟液中设置最低pH值可能是这些产品监管的实际应用。还应考虑禁止在卷烟中添加酸性或产生酸性的添加剂的措施。有些烟草产品可能含有其他降低pH值的添加剂或改动，如某些烟草叶的烘干过程或糖类添加剂。根据pH值而不是特定添加剂的存在对可吸入烟草产品的监管限制也值得考虑。

- 禁止在所有烟草和烟碱产品中使用包括糖类在内的所有甜味添加剂。

我们的研究发现，添加剂的调味特性可以掩盖苦味，这表明所有赋予甜味的调味剂都应该被禁止，包括糖。在电子烟烟液中，这指的是在任何添加水平下吸入促进的调味剂，包括任何具有非烟草特征风味的电子烟中使用的所有成分。在烟草制品中，监管机构可以考虑禁止在赋予烟草以外的特征风味的水平上使用此类调味剂，尽管调味剂在低于明显可感知的烟草以外的风味的浓度下也可能产生影响。对风味特征的评估需要感官小组进行监督和执行；一种更省时的方法是禁止在任何程度上添加这种调味料。虽然目前的论文关注的是吸入促进的添加剂，但天然烟草叶也可能含有糖类和调味剂。例如，糖类天然存在于许多烟草中。监管机构还可以考虑禁止烟草中天然存在的糖类和调味剂，因为它们也会赋予风味。对于消费者来说，糖类或调味剂是添加还是天然存在并不重要。

参考文献

[1] Megerdichian CL, Rees VW, Wayne GF, Connolly GN. Internal tobacco industry research on olfactory and trigeminal nerve response to nicotine and other smoke components. Nicotine Tob Res. 2007;9(11):1119-29. doi:10.1080/14622200701648458.

[2] Rabinof M, Caskey N, Rissling A, Park C. Pharmacological and chemical effects of cigarette additives. Am J Public Health. 2007;97(11):1981-91. PMID: 17666709.

[3] Keithly L, Ferris Wayne G, Cullen DM, Connolly GN. Industry research on the use and effects of levulinic acid: a case study in cigarette additives. Nicotine Tob Res. 2005;7(5):761-71. doi:10.1080/14622200500259820.

[4] Ferris Wayne G, Connolly GN. Application, function, and effects of menthol in cigarettes: a survey of tobacco industry documents. Nicotine Tob Res. 2004;6 Suppl 1:S43-54. doi:10.1080/14622203310001649513.

[5] Duell AK, Pankow JF, Peyton DH. Nicotine in tobacco product aerosols: It's deja vu all over again". Tob Control. 2020;29(6):656-62. doi:1136/tobaccocontrol-2019-055275.

[6] Stanton CA, Villanti AC, Watson C, Delnevo CD. Flavoured tobacco products in the USA: synthesis of recent multidiscipline studies with implications for advancing tobacco regulatory science. Tob Control. 2016;25(Suppl 2):ii1-3. doi:10.1136/tobaccocontrol-2016-053486.

[7] Willis DN, Liu B, Ha MA, Jordt SE, Morris JB. Menthol attenuates respiratory irritation responses to multiple cigarette smoke irritants. FASEB J. 2011;25(12):4434-44. Doi:10.1096/f.11-188383.

[8] Proposed rule on tobacco product standard for menthol in cigarettes. Silver Spring (MD): Food and Drug Administration; 2022.

[9] Nair JN. Additives in tobacco products: contribution of carob bean extract, cellulose fibre, guar gum, liquorice, menthol, prune juice concentrate and vanillin to attractiveness, addictiveness and toxicity of tobacco smoking. Heidelberg: German Cancer Research Center; 2012 (https://www.researchgate.net/publication/272180111_Additives_in_Tobacco_Products_Contribution_of_Carob_Bean_Extract_Cellulose_Fibre_Guar_Gum_Liquorice_Menthol_Prune_Juice_Concentrate_and_Vanillin_to_Attractiveness_Addictiveness_and_Toxicity_of_Tobacco).

[10] Alarie Y. Sensory irritation by airborne chemicals. CRC Crit Rev Toxicol. 1973;2(3):299-363. doi:10.3109/10408447309082020.

[11] Arendt-Nielsen L, Carstens E, Proctor G, Boucher Y, Clavé P, Nielsen KA et al. The role of TRP channels in nicotinic provoked pain and irritation from the oral cavity and throat: translating animal data to humans. Nicotine Tob Res. 2022;24(12):1849-60. doi:10.1093/ntr/ntac054.

[12] Carstens E, Carstens MI. Sensory effects of nicotine and tobacco. Nicotine Tob Res. 2022;24(3):306-15. PMID: 33955474.

[13] De Biasi M, Dani JA. Reward, addiction, withdrawal to nicotine. Annu Rev Neurosci. 2011;34:105-30. doi:10.1146/annurev-neuro-061010-113734.

[14] Mead EL, Dufy V, Oncken C, Litt MD. E-cigarette palatability in smokers as a function of flavorings, nicotine content and propylthiouracil (PROP) taster phenotype. Addict Behav. 2019;91:37-44. doi:10.1016/j.addbeh.2018.11.014.

[15] Alpert HR, Agaku IT, Connolly GN. A study of pyrazines in cigarettes and how additives might be used to enhance tobacco addiction. Tob Control. 2016;25(4):444-50. doi:10.1136/tobaccocontrol-2014-051943.

[16] Fan L, Balakrishna S, Jabba SV, Bonner PE, Taylor SR, Picciotto MR et al. Menthol decreases oral nicotine aversion in C57BL/6 mice through a TRPM8-dependent mechanism. Tob Control. 2016;25(Suppl 2):ii50-4. doi:10.1136/tobaccocontrol-2016-053209.

[17] Johnson NL, Patten T, Ma M, De Biasi M, Wesson DW. Chemosensory contributions of e-cigarette additives on nicotine use. Front Neurosci. 2022;16:893587. doi:10.3389/fnins.2022.893587.

[18] Kichko TI, Kobal G, Reeh PW. Cigarette smoke has sensory effects through nicotinic and TRPA1 but not TRPV1 receptors on the isolated mouse trachea and larynx. Am J Physiol Lung Cell Mol Physiol. 2015;309(8):L812-20. doi:10.1152/ajplung.00164.2015.

[19] Song MA, Benowitz NL, Berman M, Brasky TM, Cummings KM, Hatsukami DK et al. Cigarette filter ventilation and its relationship to increasing rates of lung adenocarcinoma. J Natl Cancer Inst. 2017;109(12). doi:10.1093/jnci/djx075.

[20] Benowitz NL. Pharmacology of nicotine: addiction, smoking-induced disease, and therapeutics. Annu Rev Pharmacol Toxicol. 2009;49:57-71. doi:10.1146/annurev.pharmtox.48.113006.094742.

[21] Benowitz NL. Pharmacokinetic considerations in understanding nicotine dependence. Ciba Found Symp. 1990;152:186-200;200-9. doi:10.1002/9780470513965.ch11.

[22] Chaudhri N, Caggiula AR, Donny EC, Palmatier MI, Liu X, Sved AF. Complex interactions between nicotine and nonpharmacological stimuli reveal multiple roles for nicotine in reinforcement. Psychopharmacology (Berl). 2006;184(3-4):353-66. doi:10.1007/s00213-005-0178-1.

[23] Schneller LM, Bansal-Travers M, Mahoney MC, McCann SE, O' Connor RJ. Menthol, nico-

tine, and favoring content of capsule cigarettes in the US. Tob Regul Sci. 2020;6(3):196-204. doi:10.18001/trs.6.3.4.

[24] Scientific review of the effects of menthol in cigarettes on tobacco addiction: 1980-2021. Rockville (MD): Department of Health and Health Services; 2021 (https://www.fda.gov/media/157642/download).

[25] Ha MA, Smith GJ, Cichocki JA, Fan L, Liu YS, Caceres AI et al. Menthol attenuates respiratory irritation and elevates blood cotinine in cigarette smoke exposed mice. PLoS One. 2015;10(2):e0117128. doi:10.1371/journal.pone.0117128.

[26] Cohn AM, Alexander AC, Ehlke SJ. Affirming the abuse liability and addiction potential of menthol: differences in subjective appeal to smoking menthol versus non-menthol cigarettes across African American and white young adult smokers. Nicotine Tob Res. 2022;24(1):20-7. doi:10.1093/ntr/ntab137.

[27] Jackson A, Green B, Erythropel HC, Kong G, Cavallo DA, Eid T et al. Influence of menthol and green apple e-liquids containing different nicotine concentrations among youth e-cigarette users. Exp Clin Psychopharmacol. 2021;29(4):355-65. doi:10.1037/pha0000368.

[28] Rosbrook K, Green BG. Sensory effects of menthol and nicotine in an e-cigarette. Nicotine Tob Res. 2016;18(7):1588-95. doi:10.1093/ntr/ntw019.

[29] Leventhal AM, Goldenson NI, Barrington-Trimis JL, Pang RD, Kirkpatrick MG. Effects of non-to-bacco favors and nicotine on e-cigarette product appeal among young adult never, former, and current smokers. Drug Alcohol Depend. 2019;203:99-106.

[30] Krishnan-Sarin S, Green BG, Kong G, Cavallo DA, Jatlow P, Gueorguieva R, et al. Studying the interactive effects of menthol and nicotine among youth: An examination using e-cigarettes. Drug Alcohol Depend. 2017;180:193-9.

[31] Leventhal AM, Goldenson NI, Barrington-Trimis JL, Pang RD, Kirkpatrick MG. Effects of non-tobacco favors and nicotine on e-cigarette product appeal among young adult never, former, and current smokers. Drug Alcohol Depend. 2019;203:99-106. doi:10.1016/j.drugalcdep.2019.05.020.

[32] Leffingwell JC. Wilkinson Sword “CWM”, 1975. Los Angeles (CA): University of California at Los Angeles, Industry Documents Library (https://www.industrydocuments.ucsf.edu/docs/tkjn0089).

[33] Newman FS. Memorandum from Philip Morris in-house counsel (Newman) to Philip Morris in-house counsel (Holtzman) providing legal advice and analysis regarding content of draft press release for new product introduction. Northwind comments on 3rd tier inquires, 1981. Los Angeles (CA): University of California at Los Angeles, Industry Documents Library; 2021 (https://www.industrydocuments.ucsf.edu/docs/kgxy0101).

[34] Daylor FL. Accomplishments of 81000, 1982. Los Angeles (CA): University of California at Los Angeles, Industry Documents Library; 2022 (https://www.industrydocuments.ucsf.edu/docs/xyjx0112).

[35] Reger L, Moß J, Hahn H, Hahn J. Analysis of menthol, menthol-like, and other tobacco favoring compounds in cigarettes and in electrically heated tobacco products. Beitr Tabakforsch Int. 2018;28(2):93-102. doi:10.2478/cttr-2018-0010.

[36] Jabba SV, Jordt SE. Turbocharged Juul device challenges European tobacco regulators. Eur

Respir J. 2020;56(2).

[37] Jewett C, Baumgaertner E. R.J. Reynolds pivots to new cigarette pitches as favor ban takes effect. New York Times, 11 January 2023 (https://www.nytimes.com/2023/01/11/health/cigarettes-favor-ban-california.html).

[38] Erythropel HC, Anastas PT, Krishnan-Sarin S, O'Malley SS, Jordt SE, Zimmerman JB. Differences in flavourant levels and synthetic coolant use between USA, EU and Canadian Juul products. Tob Control. 2020;30(4):453-5. doi:10.1136/tobaccocontrol-2019-055500.

[39] Jabba SV, Erythropel HC, Torres DG, Delgado LA, Anastas PT, Zimmerman JB et al. Synthetic cooling agents in US-marketed e-cigarette refill liquids and disposable e-cigarettes: chemical analysis and risk assessment. Nicotine Tob Res. 2022;24(7):1037-46. doi:10.1093/ntr/ntac046.

[40] Leventhal A, Dai H, Barrington-Trimis J, Sussman S. "Ice" flavoured e-cigarette use among young adults. Tob Control. 2023;32(1):114-7. doi:10.1136/tobaccocontrol-2020-056416.

[41] Leventhal AM, Tackett AP, Whitted L, Jordt SE, Jabba SV. Ice flavours and non-menthol synthetic cooling agents in e-cigarette products: a review. Tob Control. 2022. doi:10.1136/tobaccocontrol-2021-057073.

[42] Chafee BW, Halpern-Felsher B, Croker JA, Werts M, Couch ET, Cheng J. Preferences, use, and perceived access to favored e-cigarettes among United States adolescents and young adults. Drug Alcohol Depend Rep. 2022;3:100068. doi:10.1016/j.dadr.2022.100068.

[43] Ali FRM, Seaman EL, Diaz MC, Ajose J, King BA. Trends in unit sales of cooling flavoured e-cigarettes, USA, 2017–2021. Tob Control. 2022. doi:10.1136/tc-2022-057395.

[44] Lemon CH, Norris JE, Heldmann BA. The TRPA1 ion channel contributes to sensory-guided avoidance of menthol in mice. eNeuro. 2019;6(6). doi:10.1523/ENEURO.0304-19.2019.

[45] Leffingwell J, Rowsell D. Wilkinson Sword cooling compounds: from the beginnning to now. Perfumer Flavorist. 2014;39:34-43 (https://img.perfumerfavorist.com/fles/base/allured/all/document/2014/02/pf.PF_39_03_034_10.pdf).

[46] Symcool® cooling agents. Infinite sensations. Brunswick (GA): Symrise, Inc.' undated (https://www.symrise.com/fleadmin/symrise/Marketing/Scent_and_care/Aroma_molecules/symrise-symcool-A4-pages-eng.pdf).

[47] Johnson S, Tian M, Sheldon G, Dowd E. Trigeminal receptor study of high-intensity cooling agents. J Agric Food Chem. 2018;66(10):2319-23. doi:10.1021/acs.jafc.6b04838.

[48] Bandell M, Story GM, Hwang SW, Viswanath V, Eid SR, Petrus MJ et al. Noxious cold ion channel TRPA1 is activated by pungent compounds and bradykinin. Neuron. 2004;41(6):849-57. doi:10.1016/s0896-6273(04)00150-3.

[49] Klein AH, Iodi Carstens M, McCluskey TS, Blancher G, Simons CT, Slack JP et al. Novel menthol-derived cooling compounds activate primary and second-order trigeminal sensory neurons and modulate lingual thermosensitivity. Chem Senses. 2011;36(7):649-58. doi:10.1093/chemse/bjr029.

[50] Tackett AP, Han DH, Peraza N, Whaley R, Leventhal AM. Effects of "Ice" flavored e-cigarettes with synthetic cooling agent WS-23 or menthol on user-reported appeal and sensory attributes. In: Annual meeting of the Society for Research on Nicotine and Tobacco, San Antonio (TX); 2023.

[51] Benowitz NL. The central role of pH in the clinical pharmacology of nicotine: implications

for abuse liability, cigarette harm reduction and FDA regulation. Clin Pharmacol Ther. 2022;111(5):1004-6. doi:10.1002/cpt.2555.

[52] Wittenberg RE, Wolfman SL, De Biasi M, Dani JA. Nicotinic acetylcholine receptors and nicotine addiction: A brief introduction. Neuropharmacology. 2020;177:108256. doi:10.1016/j.neuropharm.2020.108256.

[53] Gholap VV, Kosmider L, Golshahi L, Halquist MS. Nicotine forms: Why and how do they matter in nicotine delivery from electronic cigarettes? Expert Opin Drug Deliv. 2020;17(12):1727-36. doi:10.1080/17425247.2020.1814736.

[54] Xiu X, Puskar NL, Shanata JA, Lester HA, Dougherty DA. Nicotine binding to brain receptors requires a strong cation-pi interaction. Nature. 2009;458(7237):534-7. doi:10.1038/nature07768.

[55] Pankow JF. A consideration of the role of gas/particle partitioning in the deposition of nicotine and other tobacco smoke compounds in the respiratory tract. Chem Res Toxicol. 2001;14(11):1465-81. doi:10.1021/tx0100901.

[56] Addictiveness and attractiveness of tobacco additives. Brussels: Scientific Committee on Emerging and Newly Identified Health Risks; 2010 (https://ec.europa.eu/health/scientifc_committees/emerging/docs/scenihr_o_029.pdf).

[57] Willems EW, Rambali B, Vleeming W, Opperhuizen A, van Amsterdam JG. Significance of ammonium compounds on nicotine exposure to cigarette smokers. Food Chem Toxicol. 2006;44(5):678-88. doi:10.1016/j.fct.2005.09.007.

[58] Ramamurthi D, Chau C, Jackler RK. JUUL and other stealth vaporisers: hiding the habit from parents and teachers. Tob Control. 2018. doi:10.1136/tobaccocontrol-2018-054455.

[59] Kavuluru R, Han S, Hahn EJ. On the popularity of the USB flash drive-shaped electronic cigarette Juul. Tob Control. 2019;28(1):110-2. doi:10.1136/tobaccocontrol-2018-054259.

[60] Jackler RK, Ramamurthi D. Nicotine arms race: JUUL and the high-nicotine product market. Tob Control. 2019;28(6):623-8. doi:10.1136/tobaccocontrol-2018-054796.

[61] Pennings JLA, Havermans A, Pauwels C, Krüsemann EJZ, Visser WF, Talhout R. Comprehensive Dutch market data analysis shows that e-liquids with nicotine salts have both higher nicotine and flavour concentrations than those with free-base nicotine. Tob Control. 2022. doi:10.1136/tobaccocontrol-2021-056952.

[62] Hajek P, Pittaccio K, Pesola F, Myers Smith K, Phillips-Waller A, Przulj D. Nicotine delivery and users' reactions to Juul compared with cigarettes and other e-cigarette products. Addiction. 2020;115(6):1141-8. doi:10.1111/add.14936.

[63] O' Connell G, Pritchard JD, Prue C, Thompson J, Verron T, Graf D et al. A randomised, open-label, cross-over clinical study to evaluate the pharmacokinetic profiles of cigarettes and e-cigarettes with nicotine salt formulations in US adult smokers. Intern Emerg Med. 2019;14(6):853-61. doi:10.1007/s11739-019-02025-3.

[64] Prochaska JJ, Vogel EA, Benowitz N. Nicotine delivery and cigarette equivalents from vaping a JUULpod. Tob Control. 2021. doi:10.1136/tobaccocontrol-2020-056367.

[65] Reilly SM, Bitzer ZT, Goel R, Trushin N, Richie JP. Free radical, carbonyl, and nicotine levels produced by Juul electronic cigarettes. Nicotine Tob Res. 2019;21(9):1274-8. doi:10.1093/ntr/nty221.

[66] Leavens ELS, Nollen NL, Ahluwalia JS, Mayo MS, Rice M, Brett EI, et al. Changes in depen-

dence, withdrawal, and craving among adult smokers who switch to nicotine salt pod-based e-cigarettes. Addiction. 2022;117(1):207-15. doi:10.1111/add.15597.

[67] Kechter A, Cho J, Miech RA, Barrington-Trimis JL, Leventhal AM. Nicotine dependence symptoms in US youth who use JUUL e-cigarettes. Drug Alcohol Depend. 2021;227:108941. doi:10.1016/j.drugalcdep.2021.108941.

[68] Tackett AP, Hebert ET, Smith CE, Wallace SW, Barrington-Trimis JL, Norris JE et al. Youth use of e-cigarettes: Does dependence vary by device type? Addict Behav. 2021;119:106918. doi:10.1016/j.addbeh.2021.106918.

[69] Goldenson NI, Fearon IM, Buchhalter AR, Henningfeld JE. An open-label, randomized, controlled, crossover study to assess nicotine pharmacokinetics and subjective effects of the JUUL system with three nicotine concentrations relative to combustible cigarettes in adult smokers. Nicotine Tob Res. 2021;23(6):947-55. doi:10.1093/ntr/ntab001.

[70] Mallock N, Rabenstein A, Gernun S, Laux P, Hutzler C, Karch S et al. Nicotine delivery and relief of craving after consumption of European JUUL e-cigarettes prior and after pod modification. Sci Rep. 2021;11(1):12078. doi:10.1038/s41598-021-91593-6.

[71] Phillips-Waller A, Przulj D, Smith KM, Pesola F, Hajek P. Nicotine delivery and user reactions to Juul EU (20 mg/ml) compared with Juul US (59 mg/ml), cigarettes and other e-cigarette products. Psychopharmacology (Berl). 2021;238(3):825-31. doi:10.1007/s00213-020-05734-2.

[72] Leventhal AM, Madden DR, Peraza N, Schiff SJ, Lebovitz L, Whitted L et al. Effect of exposure to e-cigarettes with salt vs free-base nicotine on the appeal and sensory experience of vaping: a randomized clinical trial. JAMA Netw Open. 2021;4(1):e2032757. doi:10.1001/jama-networkopen.2020.32757.

[73] Tattan-Birch H, Brown J, Shahab L, Jackson SE. Trends in use of e-cigarette device types and heated tobacco products from 2016 to 2020 in England. Sci Rep. 2021;11(1):13203. doi:10.1038/s41598-021-92617-x.

[74] Han DH, Wong M, Peraza N, Vogel EA, Cahn R, Mason TB et al. Dose-response effects of two nicotine salt formulations on electronic cigarette appeal and sensory attributes. Tob Control. 2023. doi:10.1136/tc-2022-057553.

[75] Pauwels CGGM, Pennings JLA, Boer K, Baloe EP, Hartendorp APT, van Tiel L et al. Sensory appeal and puffing intensity of e-cigarette use: influence of nicotine salts versus free-base nicotine in e-liquids. submitted. 2022.

[76] Hale R, Christatis K, Lin SS, Wynn R. Basic favor investigation: low tar/high favor literature review, 1990. Los Angeles (CA): University of California at Los Angeles, Industry Documents Library; 2009 (https://www.industrydocuments.ucsf.edu/docs/myjj0045).

[77] Klus H, Scherer G, Muller L. Infuence of additives on cigarette related health risks. Beitr. Ta-bakforsch Int. 2012;25(3):412-93. doi:10.2478/cttr-2013-0921.

[78] Talhout R, Opperhuizen A, van Amsterdam JG. Sugars as tobacco ingredient: Effects on mainstream smoke composition. Food Chem Toxicol. 2006;44(11):1789-98. doi:10.1016/j.fct.2006.06.016.

[79] Gager FL JR, Nedlock JW, Martin WJ. Tobacco additives and cigarette smoke. Part I. Transfer of d-glucose, sucrose, and their degradation products to the smoke. Carbohydrate Res. 1971;17(2):327-33. doi:10.1016/s0008-6215(00)82540-9.

[80] Seeman JI, Dixon M, Haussmann HJ. Acetaldehyde in mainstream tobacco smoke: formation and occurrence in smoke and bioavailability in the smoker. Chem Res Toxicol. 2002;15(11):1331-50. doi:10.1021/tx020069f.

[81] Elson LA, Betts TE. Sugar content of the tobacco and pH of the smoke in relation to lung cancer risks of cigarette smoking. J Natl Cancer Inst. 1972;48(6):1885-90. PMID: 5056275.

[82] Rodgman A. Some studies of the effects of additives on cigarette mainstream smoke properties. II. Casing materials and humectants. Beitr Tabakforsch Int. 2002;20(4):279-99. doi.org/10.2478/cttr-2013-0742.

[83] Leffingwell JC. Leaf chemistry: basic chemical constituents of tobacco leaf and differences among tobacco types. In: Davis DL, Nielsen MT, editors. Tobacco: production, chemistry and technology. Oxford: Blackwell Science; 1999:265-84 (http://www.lefngwell.com/download/Leffingwell%20-%20Tobacco%20production%20chemistry%20and%20technology.pdf).

[84] Jenkins CR, Boham N, Burden AJ, Dixon M, Dowle M, Fiebelkorn RT et al. British American Tobacco Company Limited product seminar. Tobacco documents 760076123-760076408. 1997:760076123-408.

[85] Krusemann EJZ, Pennings JLA, Cremers J, Bakker F, Boesveldt S, Talhout R. GC-MS analysis of e-cigarette refill solutions: A comparison of flavoring composition between flavor categories. J Pharm Biomed Anal. 2020;188:113364. doi:10.1016/j.jpba.2020.113364.

[86] Bakker't Hart IME, Bakker F, Pennings JLA, Weibolt N, Eising S, Talhout R. Flavours and flavourings in waterpipe products: a comparison between tobacco, herbal molasses and steam stones. Tob Control. 2022. doi:10.1136/tobaccocontrol-2021-056955.

[87] Rosbrook K, Erythropel HC, DeWinter TM, Falinski M, O' Malley S, Krishnan-Sarin S et al. The effect of sucralose on favor sweetness in electronic cigarettes varies between delivery devices. PLoS One. 2017;12(10):e0185334. doi:10.1371/journal.pone.0185334.

[88] Krusemann EJZ, Wenng FM, Pennings JLA, de Graaf K, Talhout R, Boesveldt S. Sensory evaluation of e-liquid flavors by smelling and Vvaping yields similar results. Nicotine Tob Res. 2020;22(5):798-805. doi:10.1093/ntr/ntz155.

[89] Kowitt SD, Meernik C, Baker HM, Osman A, Huang LL, Goldstein AO. Perceptions and experiences with flavored non-menthol tobacco products: a systematic review of qualitative studies. Int J Environ Res Public Health. 2017;14(4). doi:10.3390/ijerph14040338.

[90] Pullicin AJ, Kim H, Brinkman MC, Buehler SS, Clark PI, Lim J. Impacts of nicotine and flavoring on the sensory perception of e-cigarette aerosol. Nicotine Tob Res. 2020;22(5):806-13. doi:10.1093/ntr/ntz058.

[91] Hayes JE, Baker AN. Flavor science in the context of research on electronic cigarettes. Front Neurosci. 2022;16:918082. doi:10.3389/fnins.2022.918082.

[92] Baker AN, Bakke AJ, Branstetter SA, Hayes JE. Harsh and sweet sensations predict acute liking of electronic cigarettes, but favor does not affect acute nicotine intake: a pilot laboratory study in men. Nicotine Tob Res. 2021;23(4):687-93. doi:10.1093/ntr/ntaa209.

[93] Leventhal AM, Goldenson NI, Cho J, Kirkpatrick MG, McConnell RS, Stone MD et al. Flavored e-cigarette use and progression of vaping in adolescents. Pediatrics. 2019;144(5). doi:10.1542/peds.2019-0789.

[94] Soneji SS, Knutzen KE, Villanti AC. Use of favored e-cigarettes among adolescents, young

adults, and older adults: findings from the Population Assessment for Tobacco and Health Study. Public Health Rep. 2019;134(3):282-92. doi:10.1177/0033354919830967.

[95] St Helen G, Dempsey DA, Havel CM, Jacob P 3rd, Benowitz NL. Impact of e-liquid favors on nicotine intake and pharmacology of e-cigarettes. Drug Alcohol Depend. 2017;178:391-8. doi:10.1016/j.drugalcdep.2017.05.042.

[96] St Helen G, Shahid M, Chu S, Benowitz NL. Impact of e-liquid flavors on e-cigarette vaping behavior. Drug Alcohol Depend. 2018;189:42-8. doi:10.1016/j.drugalcdep.2018.04.032.

[97] Smith DM, Schneller LM, O' Connor RJ, Goniewicz ML. Are e-cigarette favors associated with exposure to nicotine and toxicants? Findings from wave 2 of the Population Assessment of Tobacco and Health (PATH) study. Int J Environ Res Public Health. 2019;16(24). doi:10.3390/ijerph16245055.

[98] Seeman JI, Lafoon SW, Kassman AJ. Evaluation of relationships between mainstream smoke acetaldehyde and "tar" and carbon monoxide yields in tobacco smoke and reducing sugars in tobacco blends of US commercial cigarettes. Inhal Toxicol. 2003;15(4):373-95. doi:10.1080/08958370304461.

[99] Bernasek PF, Furin OP, Shelar GR. Sugar/nicotine study. Industry documents library. 1992;ATP 92-210:22. (sljb0079–Truth Tobacco Industry Documents) (ucsf.edu).

[100] Weeks WW. Relationship between leaf chemistry and organoleptic properties of tobacco smoke. In: Davis DL, Nielson MT, editors. Tobacco: production, chemistry and technology. Oxord: Blackwell Science; 1999:304-12 (https://www.wiley.com/en-us/Tobacco%3A+Produ-ction%2C+Chemistry+and+Technology-p-9780632047918).

[101] Fagan P, Pokhrel P, Herzog TA, Moolchan ET, Cassel KD, Franke AA et al. Sugar and aldehyde content in favored electronic cigarette liquids. Nicotine Tob Res. 2018;20(8):985-92. doi:10.1093/ntr/ntx234.

[102] Patten T, De Biasi M. History repeats itself: Role of characterizing flavors on nicotine use and abuse. Neuropharmacology. 2020;177:108162. doi:10.1016/j.neuropharm.2020.108162.

[103] Final opinion on additives used in tobacco products. Brussels: Scientific Committee on Emerging and Newly Identified Health Risks; 2016 (https://health.ec.europa.eu/latest-updates/scenihr-fnal-opinion-additives-used-tobacco-products-2016-01-29-1_en).

[104] Bates C, Jarvis M, Connolly G. Tobacco additives. Cigarette engineering and nicotine addiction. London: Action on Smoking and Health, Imperial Cancer Research Fund; 1999 (https://www.researchgate.net/publication/242598454_Tobacco_Additives_Cigarette_Engine-ering_and_Nicotine_Addiction).

[105] Fowles J. Chemical factors influencing the addictiveness and attractiveness of cigarettes in ew Zealand. Auckland: Ministry of Health Libaray; 2001 (http://www.moh.govt.nz/note-book/nbbooks.nsf/0/D0B68B2D9CB811ABCC257B81000D9959?OpenDocument).

[106] Banožić M, Jokić S, Ačkar Đ, Blažić M, Šubarić D. Carbohydrates-key players in tobacco aroma formation and quality determination. Molecules. 2020;25(7):1734. doi:10.3390/molecules25071734.

[107] Wise PM, Breslin PA, Dalton P. Sweet taste and menthol increase cough reflex thresholds. Pulm Pharmacol Ther. 2012;25(3):236–41. doi:10.1016/j.pupt.2012.03.005.

[108] Truth tobacco industry documents. Los Angeles (CA): University of California at Los Angeles,

Industry Documents Library (https://www.industrydocumentslibrary.ucsf.edu/tobacco/).
[109] Ferreira CG, Silveira D, Hatsukami DK, Paumgartten FJ, Fong GT, Gloria MB et al. The effect of tobacco additives on smoking initiation and maintenance. Cad Saude Publica. 2015;31(2):223-5. doi:10.1590/0102-311XPE010215.
[110] Erythropel HC, Kong G, deWinter TM, O' Malley SS, Jordt SE, Anastas PT et al. Presence of high-in-tensity sweeteners in popular cigarillos of varying flavor profiles. JAMA. 2018;320(13):1380-3. doi:10.1001/jama.2018.11187.
[111] Directive 2014//40/Eu of the European Parliament and of the Council of 3 April 2014 on the approximation of the laws, regulations and administrative provisions of the Member States concerning the manufacture, presentation and sale of tobacco and related products and repealing Directive 2001/37/EC. Off J Eur Unuon. 2014;L127/1 (https://eurlex.europa.eu/legal-content/EN/TXT/PDF/?uri=CELEX:32014L0040).
[112] No more menthol in cigarettes and smoking tobacco. Berlin: Federal Institute for Risk Assessment; 2020 (https://www.bfr.bund.de/en/press_information/2020/19/no_more_menthol_in_cigarettes_and_smoking_tobacco-246948.html)
[113] Verduidelijking van artikel 5 van het Koninkilijk besluit van 5/02/2016 [Clarifcation of Article 5 of the Royal Decree of 5/02/2016]. The Hague: Federale Overheidsdienst Volksgezondheid; 2022 (https://www.health.belgium.be/sites/default/fles/uploads/felds/fpshealth_theme_fle/nl_art5_inhalation_facilitation.pdf).
[114] Agreement No. 761297-JATC-JA-03-2016. WP9: D9.3 Report on the peer review of the enhanced reporting information on priority additives. Brussels: European Union, Joint Action on Tobacco Control; 2020) (https://jaotc.eu/wp-content/uploads/2021/04/D9.3-Report-on-the-peer-review-of-the-enhanced-reporting-information-on-priority-additives.pdf).
[115] Regulating tobacco and vaping products: Tobacco regulations. Ottawa: Government of Canada; 2020 (https://www.canada.ca/en/health-canada/services/smoking-tobacco/regulating-tobacco-vaping/tobacco.html).
[116] Tobacco product standard for characterizing favors in cigars (FDA-2021-N-1309). Silver Spring (MD): Food and Drug Administration; 2020 (https://fda.report/media/158013/Tobacco+Product+Standard+for+Characterizing+Flavors+i-n+Cigars.pdf).
[117] Decision summary for marketing denial of flavored e-cigarettes. Silver Spring (MD): Food and Drug Administration; 2021.
[118] Marketing order and post-authorization marketing restrictions and requirements LogicTobacco Products. Silver Spring (MD): Food and Drug Administration; 2022.
[119] Marketing order and post-authorization marketing restrictions and requirements Vuse Tobacco Products. Silver Spring (MD): Food and Drug Administration; 2021.
[120] WHO Framework Convention on Tobacco Control. Partial guidelines for implementation of Articles 9 and 10 (FCTC/16.3.2017). Geneva: World Health Organization; 2017 (https://fctc.who.int/publications/m/item/regulation-of-the-contents-of-tobacco-products-and-regulation-of-tobacco-product-disclosures#:~:text=Whereas%20Article%209%20deals%20with,governm-ental%20authorities%20and%20the%20public).

3. 合成烟碱：科学基础、全球法律环境及监管考量

Micah L. Berman, College of Public Health, Moritz College of Law and Cancer Control Program, James Comprehensive Cancer Center, The Ohio State University, Columbus (OH), USA

Patricia J. Zettler, Moritz College of Law and Cancer Control Program, James Comprehensive Cancer Center, The Ohio State University, Columbus (OH), USA

Sven-Eric Jordt, Departments of Anesthesiology, Pharmacology and Cancer Biology, Duke University School of Medicine, Durham (NC) and Tobacco Center of Regulatory Science, Department of Psychiatry, Yale School of Medicine, New Haven (CT), USA

主 要 发 现

- 合成烟碱产品，包括尼古丁袋、电子烟烟液、一次性电子烟、烟碱口香糖、烟碱牙签和注入式燃烧产品，正在全球范围内推广和销售。
- 合成烟碱产品在销售时附带营销宣传（例如“无烟草”），这可能暗示它们比含天然烟草来源烟碱的产品更安全，而且一些产品还随附可能吸引年轻人口味的概念（如“巧克力梦”“粉红柠檬水”）。
- 合成烟碱以*S*-烟碱和*R*-烟碱两种形式添加到市场产品中。*S*-烟碱是烟草植物中的主要烟碱形式。然而，*R*-烟碱以及*R*-烟碱和*S*-烟碱混合物的药理、代谢和毒理效应尚不清楚。
- 目前尚无用于合成烟碱的化学分析的标准方法，且产品被掺杂天然烟草源烟碱是一个值得关注的问题。
- 合成烟碱产品是否受到现行的烟草管制法规的监管，取决于法律如何定义法规所涵盖的产品。仅适用于“烟草制品”或“烟草衍生”产品的法律可能不足以涵盖合成烟碱产品，因为合成烟碱并非来源于烟草植物。
- 烟草公司意识到某些烟草管制法律不涵盖合成烟碱产品，因此试图利用

这些监管漏洞。

- 一些国家已修改其烟草控制法律，使其适用于含有非烟草制造或衍生的烟碱（如合成烟碱）的产品。然而，许多国家的烟草控制法律并未明确适用于此类产品，或者未覆盖当前市场上的全部产品范畴。

3.1 引　　言

烟草公司正在日益推广各种合成烟碱产品，这些产品包含被宣传为含有通过化学合成而非烟草植物提取的烟碱。尽管它们的市场营销有时声称或暗示它们存在更少的风险，但这些产品尚未被证明比含烟草来源烟碱的产品风险更低。在许多国家，合成烟碱产品并未明确受到现行烟草控制法规的管辖，尽管在某些国家它们可能受到消费者保护等其他法律的管辖。然而，在其他一些国家，烟草控制法已经得到更新，以不同方式覆盖这些产品。各国应考虑法律调整，以填补合成烟碱产品的监管漏洞，覆盖目前市场上的产品以及未来可能出现的产品。

3.1.1 背景

新型烟草制品（如电子烟碱和非烟碱传输系统、仿烟草产品和尼古丁袋的崛起引领了包括年轻人在内的新形式的烟碱使用。烟草控制措施的成功以及与抽吸传统烟草制品（包括卷烟、雪茄、水烟和无烟烟草制品）所带来的社会污名，促使行业开发电子烟和其他与传统产品不同的新型产品。近来，公司已经开始销售这些新型或非传统产品的版本，并声称它们含有合成烟碱而不是烟草来源的烟碱[1]。这些产品的销售有时带有吸引年轻人的口味。此外，尽管目前没有证据表明含有合成烟碱的产品在健康效应或成瘾性方面与含有烟草来源烟碱的产品不同，但合成烟碱产品仍以可能暗示它们比烟草来源烟碱产品更安全的营销宣传进行销售[2]。

最近，在全球互联网销售市场上，出现了一些宣传含有合成烟碱或“无烟草”烟碱的产品，这引发了许多世界卫生组织（WHO）成员国考虑分享关于这一主题的监管信息。许多成员国已经向WHO寻求技术支持，以解决这个问题，并提供有关处理声称含有合成烟碱的产品的现有可用证据和权威建议的综合信息。这份由WHO委托的报告旨在阐明这些问题。

该报告涵盖了针对娱乐用途而非医疗用途（如戒烟）销售的合成烟碱产品。报告提供了正在销售的各种合成烟碱产品的概述，营销宣传内容以及合成烟碱制造、毒理学、药理学和检测的科学知识。此外，报告还提供了关于全球合成烟碱产品的法律背景的信息，重点关注烟草控制法律。具体来说，我们在烟草控制法律网站对210个国家和欧盟（EU）的烟草制品指令进行了审查和编码。在总共

211个司法管辖区中，21个没有相关法律或无英文翻译。在其余的190个司法管辖区中，有52个法律定义范围宽泛，至少涵盖某些合成烟碱产品（例如电子烟，但不包括其他合成烟碱产品），29个提供了更广泛范围的合成烟碱产品定义，92个的定义不适用于任何类型的合成烟碱产品，而在17个司法管辖区中，法律是否涵盖合成烟碱产品尚不清楚。

3.1.2 合成烟碱产品的种类

新闻报道[3]表明，美国（USA）目前是合成烟碱产品的最大市场，尽管这可能会因为2022年3月对美国法律的修订而发生变化，该修订将合成烟碱产品列入美国食品药品监督管理局（USFDA）的烟草制品管理范围。第二大市场是韩国[3]。目前，大多数市场上标榜含有合成烟碱的产品要么是电子烟、电子烟烟液，要么是尼古丁袋。然而，这并不是唯一正在销售的合成烟碱制品类型[2]。例如，一些公司销售口香糖产品，声称含有合成或“无烟草”烟碱[4,5]；至少有两家公司正在市场推广合成烟碱牙签[6,7]；一家加拿大公司PODA宣布计划在2021年推出一种含有“以合成烟碱浸泡的颗粒茶叶”的“非燃烧产品”[8]。该公司后来被菲利普·莫里斯购买，尚不清楚其产品是否会上市。至少有一家公司Outlaw Dip提供“100%无烟草”的湿嚼烟，“不含任何烟草成分”[9]，还有至少一家其他公司，Ronin，销售含有“非烟草烟碱”的燃烧型大麻二酚卷烟[10]。因此，有各种各样的产品被销售为含有合成烟碱而非烟草来源的烟碱，新型产品可能会继续涌现。

此外，许多这些合成或“无烟草”烟碱产品含有可能吸引年轻人的口味。例如，有些牙签带有口味，如“奶油糖果蛋糕”和“草莓芝士蛋糕”[7]，某些一次性电子烟带有口味概念，如“香蕉冰”和“冰蓝柑”[11]。

3.1.3 合成烟碱产品的营销和推广

许多销售合成烟碱产品的公司提出可能隐含或明示暗示其产品比含烟草来源烟碱的产品更“安全”的声明。这些声明包括合成烟碱含有的杂质相较于烟草来源烟碱少，以及合成烟碱等同于药用级烟碱。公司还声称，合成烟碱产品相对于含有烟草来源烟碱的产品具有其他优势，如它们提供更多的满足感和更好的口味体验，以及它们更环保。一些合成烟碱产品被推销为戒烟的有效辅助手段，或等同于获批准的烟碱替代疗法，有时附带免责声明，表示该产品不是戒烟产品。表3.1提供了一些示例。

表3.1　关于合成烟碱产品的促销声明示例

产品	制造商	宣传
Juice Head pouches	Juice Head（美国）	"……可能提供更高的烟碱满足感，潜在风险可能比烟草来源的烟碱小。此外，虽然烟草烟碱通常具有浓烈的气味和味道，但合成烟碱几乎是无色无味的。" "……值得注意的是，烟草种植（通常享受很大的补贴）可能对环境造成严重损害，通常是一个高度劳动密集、烦琐和浪费资源的过程。" "需要指出的是，烟草来源的烟碱可能伴随着更多的副作用风险，而不含烟草来源烟碱则风险更低。"[12]
Pacha Mama vape pen	Charlie's Holdings, Inc.（美国）	"相较传统采收的烟碱，纯度和一致性更高"[13]
Outlaw Dip	Outlaw Dip Company（美国）	"药品级"[9]
Bidi Pouch	Kaival Brands Innovations Group, Inc.（美国）	"旨在帮助成年吸烟者迈出无烟第一步"[14]
ZIA gum	Next Generation Labs LLC（美国）	"唯一使用合成烟碱开发的烟碱口香糖"[4]"提供与含烟碱的任何烟草来源产品相同的烟碱满足感"[4]"ZIA™口香糖不旨在帮助戒烟"[15]
VaporX e-juice and disposable e-cigarettes	Vaporex Co., Ltd（韩国）	"我们致力于通过为吸烟者提供有价值和适当的电子烟体验来保护他们的健康"[16]

3.2　合成烟碱的科学基础

在美国和其他国家，合成烟碱产品（如电子烟、口服袋和其他产品类别）的快速、监管不严的引入引发了关于它们的安全性以及合成烟碱的成瘾和强化特性的潜在差异的问题。本节我们将回顾化学合成的策略、产品中不同形式的合成烟碱、制造商和专利情况，以及合成烟碱的毒理学、药理学和代谢特性。

3.2.1　方法

在PubMed和Web of Science等研究数据库中，使用诸如"合成烟碱""*R*-烟碱""(+)-烟碱""L-烟碱""D-烟碱""外消旋混合烟碱""烟碱合成"等术语，检索了关于合成烟碱和烟碱对映体（下文定义）效应的期刊文章。专利可通过patents.google.com与"烟碱"、"合成"和/或"立体选择性"、"对映选择性"等术语的组合来查找。在www.industrydocuments.ucsf.edu/tobacco/的烟草遗留数据库中，使用诸如"合成烟碱""烟碱合成""烟碱的合成""*R*-烟碱"等术语进行了检索。

3.2.2 结果

合成烟碱：它是什么，与烟草来源烟碱有何不同

烟碱存在于两种化学形式中，这两种形式是结构上的镜像。这两种形式被称为对映体，即*S*-烟碱和*R*-烟碱（图3.1A）。烟草植物中的烟碱主要由*S*-烟碱组成（> 99%），只含有极小量的*R*-烟碱[17]。化学家首次在1904年合成了烟碱[18]，结果产生了一种包含*S*-烟碱和*R*-烟碱的50 ：50比例[18,19]的混合物，称为外消旋混合烟碱。这种混合物与烟草来源的烟碱不同，因为它含有更多的*R*-烟碱和较少的*S*-烟碱。

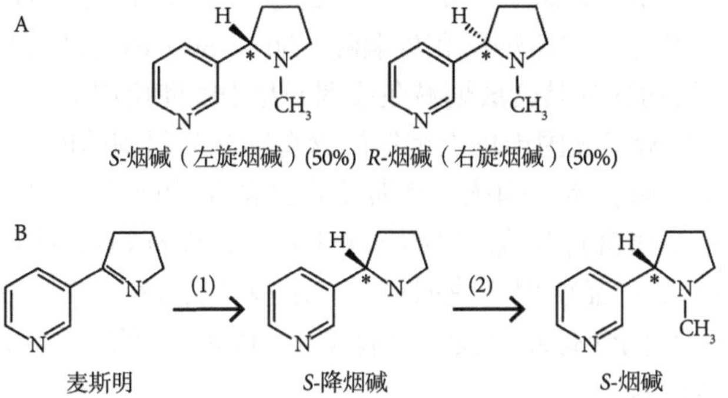

A

S-烟碱（左旋烟碱）(50%)　*R*-烟碱（右旋烟碱）(50%)

B

麦斯明　(1)　*S*-降烟碱　(2)　*S*-烟碱

图3.1　合成烟碱的结构和化学性质

A. *S*-烟碱和*R*-烟碱的结构。这些化合物在用星号标记的碳原子处具有构型差异，该碳原子是一个手性中心。在烟草叶中，大于99%的烟碱以*S*-烟碱的形式存在。下一代实验室（NGL）公司推广的合成“无烟草烟碱”是外消旋混合烟碱，含有50%的*S*-烟碱和50%的*R*-烟碱。纯*S*-烟碱在化学上与从烟草中提取的*S*-烟碱无法区分。B. Zanoprima公司关于*S*-烟碱合成的专利中，涉及生物技术步骤。起始材料是麦斯明（myosmine），首先通过重组酶[1]，一种NADH/NADPH依赖的亚胺还原酶，进行立体选择性反应，转化为*S*-降烟碱。然后，*S*-降烟碱通过甲基化[2]转化为*S*-烟碱

在“烟草行业内部公司文件真相数据库”（该数据库是在美国的诉讼中编制的烟草行业内部公司文件的数据库）中，使用“合成烟碱”一词进行检索，发现烟草行业在20世纪60年代已经考虑过使用合成烟碱。英美烟草的员工提出了将合成烟碱添加到燃烧型卷烟中，以增加烟碱相对于焦油的比例[20]；然而，因为担心合成烟碱仅以外消旋混合烟碱的形式可获得，其健康影响未知，该提议未进一步推进。此外，合成烟碱的价格要高得多，远高于烟草来源的烟碱[20]。RJ Reynolds和Liggett & Myers的员工也考虑使用合成烟碱来调整卷烟中的烟碱含量；但出于同样的原因，这个想法被放弃了[21,22]。“烟草行业内部公司文件真相数据库”在1978年之后没有进一步的证据表明美国主要烟草公司考虑使用合成烟碱。随后，

化学家们开发了新的合成烟碱策略，包括生产纯*S*-烟碱的方法，这是烟草叶中普遍存在的烟碱形式[19,23]。

合成烟碱市场：制造商、专利和定价

2015年，下一代实验室（NGL）公司开始在美国推广合成烟碱，使用商业产品的商标TFN®（无烟草烟碱）和制药产品的商标PHARMANIC®。同年，NGL公司提交了一项名为"(*RS*)-烟碱制备方法"的美国和全球专利申请[24]。美国于2017年授予了NGL公司该专利，描述了以烟酸乙酯作为起始物质的合成途径。烟酸乙酯源自烟酸，这是一种合成化学品，由石化来源生产。它与*N*-乙烯基-2-吡咯烷酮反应形成一种烟草生物碱——麦斯明。麦斯明然后转化为降烟碱。降烟碱随后的甲基化反应产生了一个*S*-烟碱和*R*-烟碱（50 ∶ 50）的外消旋烟碱混合物（图3.1A）[24]。NGL公司还为其合成烟碱在戒烟产品中的使用申请了专利[25]。2016年，Hellinghausen等分析了美国市场上销售的含有NGL公司制造的TFN品牌合成烟碱的电子烟烟液的烟碱含量，确认了其为外消旋混合烟碱[26]。虽然自2015年以来在美国市场上已经销售含有合成烟碱的电子烟产品，但直到2021年，当知名电子烟公司Puff Bar宣布其产品使用合成烟碱时，这些产品才引起公众关注[27]。对这些产品的分析显示它们含有外消旋混合烟碱[28]。Puff Bar产品中合成外消旋混合烟碱的来源尚未被披露。

同时，化学领域的进步导致了生产纯*S*-烟碱策略的优化。几家公司已经申请了合成*S*-烟碱的专利。德国Contraf-Nicotex-Tobacco是世界上最大的药品级烟碱供应商，开发了一种合成烟酸乙酯和*N*-乙烯基吡咯烷酮烟酸产生外消旋混合烟碱的方法；随后的选择性纯化丰富了该化合物，以生产纯*S*-烟碱[29,30]。含有Contraf-Nicotex-Tobacco合成*S*-烟碱的电子烟产品自2020年以来在美国市场上销售[31,32]。Zanoprima生命科学有限公司（英国伦敦）也合成制造*S*-烟碱[33]，并于2021年获得了一项涉及生物技术步骤的合成*S*-烟碱的美国专利[34]。起始材料是麦斯明，首先通过商业可获得的重组酶，一种NADH/NADPH依赖的亚胺还原酶，将其转化为*S*-降烟碱。然后，*S*-降烟碱通过甲基化转化为*S*-烟碱（图3.1B）。这种产品目前在SyNic品牌名下销售[33]。Hangsen国际公司是一个主要的电子烟设备和电子烟烟液制造商，申请了一项针对类似工艺的中国和全球专利，并在"Motivo"品牌名下销售合成*S*-烟碱[35,36]。NJOY（一家重要的电子烟制造商，即将被卷烟制造商Altria [37]收购），也获得了烟碱合成和纯化的专利[38]。一些专利中将得到的烟碱纯度描述为"> 99.9%"，手性纯度为99.6%以上的*S*-烟碱或更高。批发产品的纯度列为99.9%的*S*-烟碱[39]。

2019年，NGL公司的一名代表表示，该公司的外消旋合成混合烟碱产品，即*R*-烟碱和*S*-烟碱的混合物，"其成本只有烟草衍生烟碱的价格的3~4倍"[40]。截至

2023年3月21日，每升NGL TFN混合型合成烟碱的批发价格为1800美元，而同一批发商提供的每升烟草衍生烟碱的价格为229.99~429.99美元，具体取决于品牌。因此，合成版本的价格是烟草衍生烟碱的4~8倍[41-43]。Zanoprima的SyNic合成*S*-烟碱的销售价格为每升999.99美元，而同一卖家报价烟草衍生烟碱的价格为229.99美元，相差约4倍[39,44]。因此，尽管合成烟碱的价格仍然远高于烟草衍生烟碱，但合成烟碱产品仍在市场上销售，包括电子烟产品和口服尼古丁袋，也被称为"white snus"。这些产品通常在进行广告推广时宣传它们比含有烟草衍生烟碱的产品更纯净和更健康。

合成烟碱的制造商（表3.2）已经开始执行他们的知识产权，导致法律冲突和市场整合。中国政府最近确认了NGL公司的知识产权，使该公司能够在中国实施其专利，中国生产了绝大多数电子烟产品[45]。Zanoprima在美国地方法院起诉了一家大型电子烟及烟液制造商Hangsen，指控侵犯其专利[46]。2021年，Hangsen制造的烟碱被添加到在美国销售的"Geekbar"产品中；然而，在诉讼提交后，Hangsen停止在美国销售其"Motivo"品牌的合成*S*-烟碱，但在美国以外地区继续销售[35,47]。

表3.2　合成烟碱的主要制造商及其合成路线

制造商	反应物	产物	立体选择性步骤
Next Generation Labs LLC (NGL)	烟酸乙酯	外消旋*S*-烟碱：*R*-烟碱 (50 ： 50)	不适用
Contraf-Nicotex-Tobacco	烟酸乙酯	*S*-烟碱	立体选择性再结晶
Zanoprima Lifesciences Ltd	麦斯明	*S*-烟碱	酶的立体选择性步骤
Hangsen International Group	麦斯明	*S*-烟碱	酶的立体选择性步骤
NJOY LLC	外消旋*S*-烟碱：*R*-烟碱 (50 ： 50)	*S*-烟碱	立体选择性再结晶

合成烟碱制造商的健康声明

与电子烟制造商一样，生产合成烟碱的公司也以与健康相关的声明来推广其产品。NGL公司声称"TFN不含烟草源烟碱含有的许多残留杂质……TFN几乎是无味的……无需掩盖烟草源烟碱的异常味道和气味"[48]。NGL公司还声称"'*R*'和'*S*'异构体的特定比例可能提供令人满意但不上瘾或较少上瘾的烟碱使用"。Contraf-Nicotex-Tobacco反对这一观点，声称其合成*S*-烟碱优于外消旋版本，并表示"如果您查看欧洲和美国药典，烟碱中*S*-异构体的百分比必须高于99%"[40,49]。Zanoprima声称其合成*S*-烟碱"不含相关的烟草生物碱、烟草特有亚硝胺（TSNA）、异味和刺激的味道"[33]。这些声明可能代表它们的合成烟碱具有更优越的、类似药物的属性。公司还声称他们采用可持续的“绿色化学”方法进行生产，从环境上来看，这比农业烟草生产更环保，后者需要农药、化肥、大量的土地使用和危险的生产方法。

合成烟碱的毒理学、药理学和代谢特性

如上所述，目前市场上销售的合成烟碱产品存在两种形式，*S*-烟碱和外消旋混合烟碱，后者由50%的*S*-烟碱和50%的*R*-烟碱组成。由于合成*S*-烟碱在化学上与烟草源*S*-烟碱完全相同，其毒理学、代谢学和药理学性质应该是相同的，尤其是如果它们以主要合成产品制造商所宣称的纯度（> 99.9%）添加的话。然而，即使在这种高纯度下，仍可能存在来自化学过程中的其他化学物质的微量残留，这需要进一步关注。

如果消费者使用含有合成外消旋烟碱的产品，那么他们烟碱的摄入量中有50%是*R*-烟碱。关于*R*-烟碱的毒理学、代谢学和药理学效应了解得较少，而关于*S*-烟碱的则了解较多。一项关于小鼠的研究确定了静脉注射后60分钟内对50%的动物产生致死效应所需的剂量（LD_{50}），*S*-烟碱为0.33 mg/kg，*R*-烟碱为6.15 mg/kg，后者高出18倍，表明在这些条件下*R*-烟碱的急性毒性较低[50]。这项研究还确定了诱发惊厥所需的*R*-烟碱剂量较*S*-烟碱更高。

药理学研究表明，*R*-烟碱作为烟碱受体激动剂的效力约为*S*-烟碱的十分之一[51]。一项对大鼠大脑中烟碱结合情况的研究表明，*S*-烟碱的效力比*R*-烟碱高出10倍或更多[52]。长期给予*S*-烟碱或*R*-烟碱，都会增加大鼠大脑中烟碱受体的数量[53]。

在一项关于大鼠的操作性行为研究中，用于检测大鼠是否能够区分注射的*R*-烟碱或*S*-烟碱与生理盐水的能力，结果显示*S*-烟碱的效力比*R*-烟碱高出9倍[54]。一项用于表征大鼠烟碱引发的运动刺激作用的研究显示，*S*-烟碱至少比*R*-烟碱高出10倍的效力，能够更强烈地促进运动[55]。在大鼠的有条件的味觉厌恶实验中，*S*-烟碱比*R*-烟碱强效4~5倍[56]。与*S*-烟碱相反，*R*-烟碱不会引起大鼠体重减轻，也不会引发肾上腺素释放[51,53]。在标准烟碱药理学实验范式中，对对映体的药理学研究显示，在烟草中主要存在的烟碱对映体（> 99%）*S*-烟碱，要比*R*-烟碱强大4~28倍，而合成外消旋烟碱中*R*-烟碱的含量高达50% [51, 52, 54, 55, 57-60]。

S-烟碱和*R*-烟碱在代谢方面也存在差异。在豚鼠中的研究表明，*S*-烟碱只形成氧化代谢物，而*R*-烟碱既形成氧化代谢物，也形成*N*-甲基化代谢物[61]。所形成的*S*-可替宁和*R*-可替宁的降解动力学也不同。对不同实验动物种类的代谢研究显示，*S*-烟碱和*R*-烟碱在降解和排泄方面存在显著差异，*R*-烟碱代谢中还存在性别差异[60-62]。人类、大鼠和豚鼠肝细胞质提取物中的*S*-烟碱和*R*-烟碱*N*-甲基化也存在种属差异[63]。虽然人类提取物催化了两种形式的烟碱*N*-甲基化，但大鼠提取物不会形成任何*N*-甲基化产物，而豚鼠提取物只会转化*R*-烟碱而不转化*S*-烟碱[63]。

目前尚不清楚这些*N*-甲基化产物是否具有生物活性，以及*S*-烟碱和*R*-烟碱的甲基化产物是否有不同的作用方式。这些发现表明需要进一步研究*R*-烟碱的人体代谢和其行为效应，并且不应仅基于动物模型来预测*R*-烟碱消耗的毒理学结果。缺少此类重要数据和观察到的种属差异使得无法评估*R*-烟碱对人体的毒理风险。

除烟碱受体介导的药理学效应外，*R*-烟碱和*S*-烟碱对其他药理靶点也具有不同的作用。例如，一项由烟草行业赞助的关于乙酰胆碱酯酶的研究显示（乙酰胆碱酯酶是分解突触间隙中的神经递质乙酰胆碱以终止神经传导的酶），*R*-烟碱是一种比*S*-烟碱更强大的酶抑制剂，它结合到酶蛋白上的不同位点[64]。实验是以电鳗分离出的乙酰胆碱酯酶为研究对象进行的，且在浓度远高于吸烟者接受的烟碱浓度下进行的。需要进一步研究这些效应是否会在人类身上发生以及它们如何影响乙酰胆碱水平和神经传导。两种形式的烟碱都会干扰某些涉及炎症调控的脂质介质的产生，且具有相似的效力，这表明某些生物过程同样受到两种形式的烟碱的影响[65]。

神经物理学研究

通过神经物理学研究以确定*R*-烟碱和*S*-烟碱是否引发不同的气味或刺激感觉。当人们通过鼻子暴露于烟碱蒸气时会感觉到厌恶。在较高浓度下，烟碱蒸气会引起鼻部刺激，包括刺痛和灼烧感，这是由三叉神经介导的，三叉神经将疼痛信号传输到大脑。测试对象报告检测*S*-烟碱的阈值低于*R*-烟碱，并且刺痛和灼烧的强度更大，而嗅觉感知在相似的水平上被引发。在黏膜电位的电子记录中，*S*-烟碱引发的反应比*R*-烟碱强烈。对比非吸烟者，吸烟者可能会将*S*-烟碱感知为更愉悦，这可能是由于以前吸烟的经验[66]。迄今为止，这似乎是唯一系统性研究人类对*S*-烟碱和*R*-烟碱反应的研究。实验非常短暂，因为单个蒸气刺激仅持续了250毫秒。

合成烟碱的分析检测

Hellinghausen等[26]开发了一种方法来验证标有合成外消旋烟碱的电子烟产品。他们使用手性固定相通过高压液相色谱分离*R*-烟碱和*S*-烟碱，然后进行圆二色性检测和电喷雾电离质谱检测。他们报告表明,其中一个产品的总烟碱含量（*R*-烟碱和*S*-烟碱之和）比产品标签上所述的含量要高两倍，产品标签上事实上只列出了*S*-烟碱的强度，而其他标签上列出的烟碱含量等于测量的含量，其中一半是*S*-烟碱。这些观察结果表明，监管机构应强制实施统一的产品标签规定，以防止不知情的使用者无意中接触到比他们习惯更高水平剂量的*R*-烟碱或更低水平的*S*-烟碱的剂量。不适当的烟碱含量标签可能会促使消费者购买总烟碱含量更高的产品，可能导致显著增加的*S*-烟碱摄入量。作者还检测到需要进一步鉴定的杂质[26]。对

Puff Bar电子烟产品中S-烟碱和R-烟碱的存在进行^{1}H核磁共振光谱、旋光度和气相色谱-质谱（GC/MS）分析，证实了两种烟碱形式的存在，但S-烟碱的含量略高于R-烟碱。作者推测制造商可能添加了来自烟草的烟碱，尽管需要进一步的分析才能确定[28]。

有几种方法已被提出来区分烟草提取烟碱和合成烟碱。由于现在可以获得高纯度的合成S-烟碱，且这些化合物在化学上是相同的，因此无法通过标准的分析技术进行区分。碳同位素分析被提出作为一种解决方案。碳有三种同位素：^{12}C、^{13}C和^{14}C。^{14}C的半衰期为5700年，这一特性被用于生物材料的放射性碳定年。^{14}C受到太阳辐射不断补充，然后被集成到植物物质中，包括烟草植物及其天然产品，如烟碱。合成烟碱是从数百万年前形成的石油化学前体中制备的，其^{14}C含量要低得多。例如，已经开发了一种^{14}C分析方法来区分天然和化石化学衍生的香草醛，一种常见的风味化学品[67]。根据涉及的代谢途径不同，天然产品也可能含有更高的^{13}C：^{12}C。高温液相色谱联用同位素比质谱（HT RPLC/IRMS）已成为鉴定掺有合成添加剂的食品以及区分天然和合成咖啡因、乙醇、糖和其他化学品的标准方法[67,68]。Cheetham等[69]使用了一种^{14}C方法来比较烟草提取和合成烟碱的样本，并发现烟草提取的样本含有100%的“现代”生物碳，其碳同位素分布与地球大气中当前分布相同。合成烟碱样品只含有约35%的生物碳，表明它们的合成过程中可能使用了一些天然前体。经商业纯化的合成烟碱制剂被发现具有高纯度（> 99.9%烟碱含量），仅含有微量的烟碱衍生物和降解物，符合美国药典对药用级烟碱的要求[70]。商业纯化的烟草衍生烟碱样品也具有类似的高纯度，也符合美国药典对药用级烟碱的要求。作者还设计了一种用于识别含有合成和烟草提取烟碱混合物的产品以及从电子烟烟液中提取烟碱的方法，这是分析市售产品以检测碳基溶剂（丙二醇、甘油）、风味化学品和其他添加剂是否存在的重要的第一步。

氢同位素（氕和氘）和氮同位素（^{15}N）在烟碱中的定性和定量分析方法也揭示了来自不同地点的烟草提取烟碱与合成烟碱之间存在显著差异[71,72]。因此，尽管在区分合成烟碱和烟草提取烟碱的分析方法方面取得了重要进展，但目前尚无标准方法。应用这些方法所需的仪器和技能成本高昂，只有少数国家具备这种能力。

3.2.3 总结与讨论

制造商已开发出几种更高效、更经济的合成烟碱生产方法。目前，市场上添加的合成烟碱有两种形式——外消旋烟碱（包含50%的S-烟碱和50%的R-烟碱）以及纯S-烟碱。合成烟碱的价格仍然明显高于烟草提取烟碱。使用含有外消旋烟碱产品的消费者吸入的R-烟碱量远远高于使用烟草提取烟碱或纯S-烟碱的使用者，这引发了对这类产品的长期安全性的疑虑。尽管在标准药理实验和行为测试中，R-

烟碱的效力明显低于*S*-烟碱，但唯一关于*R*-烟碱效应的毒理学研究是关于急性效应的研究。有证据表明*R*-烟碱对其他药理学和毒理学靶点有差异性影响，存在可能引发意外毒理效应的担忧。在已发表的药理学研究中，没有一项动物暴露时间超过1~2周，也没有一项研究检查了随后的病理效应。没有一项已发表的报道研究了外消旋烟碱的效应，其中包含*R*-烟碱和*S*-烟碱，也没有比较吸入和摄入后的效果，吸入和摄入是消费者使用烟碱的两种途径。大多数关于*R*-烟碱和*S*-烟碱效应的研究是在20世纪70~90年代发表的。毒理学方法自那时以来已经取得了显著进展，应用于研究*R*-烟碱摄入的长期效应。化学分析方法可以区分合成烟碱和烟草衍生烟碱；然而，这些方法尚未标准化，需要投资大量资金用于先进设备和培训。分析研究引发了对市售产品标签不准确和未披露添加烟草衍生烟碱的担忧，后者可能是为了增加成瘾性和提高利润而添加的，同时继续维持合成烟碱的健康声明。经过测试的商业制剂，包括纯化的烟草衍生烟碱和合成烟碱，都符合美国药典对药用级烟碱纯度的要求；但并非所有目前市售的制剂都进行了比较。鉴于合成烟碱和烟草衍生制剂的纯度非常相似，应基于强有力的科学证据来提出合成烟碱和纯化的烟草衍生烟碱对健康的健康声明。

如果监管机构限制在市场产品中使用合成烟碱，制造商可以迅速修改其已开发的化学合成方法来生成烟碱类似物[19]。烟草行业长期以来一直在研究与烟碱相关的烟草生物碱的成瘾和强化效应，包括新烟碱、降烟碱、新烟草碱、可替宁和麦斯明[19, 59, 73-76]。监管机构应注意，这些类似物可能用于替代市售产品中的烟碱。

3.3 法 律 环 境

未受监管的合成烟碱产品市场可能会破坏公共卫生在减轻烟草使用危害方面取得的进展[2,77,78]。例如，美国的立法者在2021年11月致函给美国食品药品监督管理局（USFDA），表达了对未受监管的合成烟碱产品销售“破坏减少年轻人吸电子烟的努力”的担忧[79]。此外，对于某些合成产品的市场营销宣称可能会误导使用这些产品的人，例如，暗示这些产品比烟草来源的烟碱产品更安全，尽管没有证据支持这一主张。

如果合成烟碱产品继续不受监管，公司很可能会选择出售含有合成烟碱而不是烟草来源的烟碱的产品（或者至少声称如此），从而破坏了对新型烟草和烟碱产品的全面监管[77]。公司意识到，一些国家的烟草控制法律不约束于合成烟碱产品。两家全球主要的合成烟碱供应商Hangsen和NGL都宣扬合成烟碱的一个主要优势是“新市场推出的限制较少”[48]。在美国法律最近的变化之前，美国的一名投资分析师将合成烟碱称为潜在的“金票”，因为使用合成烟碱而不是烟草来源的烟碱可能意味着“无FDA监管、无烟草税、无口味限制和无限制的面向消费者的电子商务[1]”。Puff Bar是生产一次性电子烟的公司，其产品在年轻人中很受欢迎，该公司曾利用美国以前的监管漏洞，在2021年初美国食品药品监督管理局（USFDA）采取执法行动后，重新推出其产品，声称其使用合成烟碱使其不受烟草制品监管的规定[80]。

因此，对于政策制定者来说，一个关键问题是，包含合成烟碱（或其他非烟草来源的烟碱替代品）的产品是否受到现有烟草制品的监管框架的约束。这取决于相关法律中使用的术语的定义，以及这些定义是否特定于（并仅限于）烟草提取产品。世界卫生组织《烟草控制框架公约》（WHO FCTC）将“烟草制品”定义为“完全或部分由烟草烟叶作为原材料制成，用于抽吸、吸吮、咀嚼或鼻吸的产品”，这似乎排除了非烟草合成烟碱产品[81]。然而，WHO FCTC的措辞不妨碍成员国在其国家法律中将合成烟碱产品纳入“烟草制品”的定义中，或以其他方式将包含合成烟碱的产品纳入国家的烟草控制法律中。值得注意的是，世界卫生组织《烟草控制框架公约》的缔约方会议已要求该公约的秘书处“根据需要就新型烟草制品的适当分类提供建议和咨询，以支持监管工作和定义新的产品类别的必要性”[81]。

为了更好地了解全球合成烟碱产品的法律环境以及成员国如何修改现有的监管定义，包括遵守国际义务[81]，我们对210个国家和欧盟的烟草控制法律进行了调查，以确定这些法律是否以及以何种方式适用于合成烟碱产品。

3.3.1 方法

我们的审查涵盖了市场准入要求的烟草控制法规（如营销前的注册）、销售限制（如销售的年龄限制或零售商放置烟草制品的地点的限制）、包装和标识要求（如特定警示声明或图像的要求）以及广告法规（如电视广告的限制）。我们排除了其他类型的与烟草相关的法律，如税收法、无烟法和对调味剂的监管。

大多数法律都可以在烟草控制法律网站[82]上找到，该网站包含了210个国家和欧盟的法律。美国在2022年3月对“烟草制品”的定义进行的修正尚未在该网站上提供，我们从其他地方获取了相关信息[83-85]。因此，分析中包括了总共211个司法管辖区的法律。

在这211个司法管辖区中，有21个司法管辖区没有可用的法律，或者没有英文版本。三名受过美国法律培训的英语使用者（包括两位作者MLB和PJZ，以及Annamarie Beckmeyer）审查了其余190个司法管辖区（189个国家和欧盟）的相关法律。欧盟指令不是有约束力的法律，而是“设定了欧盟所有国家必须实现的目标的立法行为……，留给各个国家制定自己的法律以实现这些目标”。由于《欧

1 Lavery MS. Tobacco synthetic nicotine bursts on to the scene. 2021（可向作者索取）

盟烟草制品指令》对欧洲烟草政策制定至关重要，我们将欧盟视为一个独立的司法管辖区[86]。

法律研究软件MonQcle [87]，被用来为任何合成烟碱产品编码法律的适用性。如果法律适用于合成烟碱产品，那么它们将被编码，以确定它们是否适用于电子烟之外的任何合成烟碱产品，以及确定这些受管制的合成烟碱产品是否需要遵守市场准入要求、销售限制、包装和标识要求以及广告限制。

3.3.2 结果

一些国家的法律措辞足够广泛，足以覆盖某些合成烟碱产品，或者更广泛地覆盖此类产品。然而，许多国家的烟草控制法律并未明确适用于这类产品（表3.3）。

表3.3 211个司法管辖区的烟草控制法律对含有合成烟碱的产品的适用性

含有合成烟碱的产品的覆盖范围	司法管辖区数量	特征
不适用	92	“烟草制品”被定义为含有来自烟草植物的成分的产品
对某些产品的明确覆盖范围	52	电子烟（和其他特定产品类型）被定义为包括来自任何来源的烟碱，但“烟草制品”在其他情况下仅限于由烟草植物制成的产品
更广泛的覆盖范围	29	“烟草制品”被定义为明确包括合成烟碱或来自任何来源的烟碱
覆盖范围不明确	17	产品定义是指烟草植物或烟气，但没有明确要求这些产品必须由烟草生产或衍生
暂无	21	

不覆盖合成烟碱产品的法律

在对这190项法律进行编码时，有92项法律不适用于任何类型的合成烟碱产品。这个类别中的许多司法管辖区根据其烟草含量来定义所涵盖的产品。例如，在2022年3月之前，美国法律将“烟草制品”（用于联邦监管的目的）定义为“由烟草制成或来源于烟草的产品”[84, 85, 88]。

这个类别中的一些法律并没有明确包括相关术语，但这些术语本身暗示着合成烟碱产品可能不在范围之内。例如，在一些法律中，术语“烟草制品”被使用，但没有明确定义。WHO非洲地区的国家拥有不适用于任何类型的合成烟碱产品的法律可能性最大。

明确仅适用于某些合成烟碱产品的法律

在52个司法管辖区，法律包括足够广泛的定义以涵盖某些合成烟碱产品，通常是电子烟，但不包括其他目前市场上销售的合成烟碱产品，如尼古丁袋、烟碱牙签和烟碱口香糖。其中许多法律根据烟草含量来定义“烟草制品”（如上一类别中所述），然后单独定义“电子烟”或类似的术语，而不指定烟碱的来源或含量。

这个类别中的其他法律没有定义相关术语，但包括可能涵盖含有合成烟碱的电子烟的术语。例如，在一些法律中，术语如“电子烟”或“电子烟碱传输系统”被使用，而没有明确定义其范围。因此，这些术语可能足够广泛，以涵盖含有合成烟碱而非烟草来源烟碱的电子烟。

一些司法管辖区的法律在一定程度上适用于电子烟以外的产品。例如，一些法律还包括不含烟草的草本烟产品，这可能为包括含有合成烟碱的燃烧型产品留下了余地。

在那些仅覆盖某些合成烟碱产品的法律中，有少数完全禁止销售和分发电子烟。对于仅覆盖某些合成烟碱产品的法律的司法管辖区，没有明显的地理分布模式。然而，似乎存在一种时间趋势，因为这一类别中的大多数法律在2017年或之后颁布。

涵盖更广泛合成烟碱产品的法律

29个司法管辖区的法律起草得足够宽泛，可以涵盖当前市场上的所有或大多数合成烟碱产品，以及可能出现的产品（表3.4）。这些法律中只有少数几个完全禁止更广泛范围的合成烟碱产品。

表3.4　包括涵盖所有合成烟碱产品的产品定义的国家法律实例

国家	年份	评论和定义
摩尔多瓦共和国	2015	摩尔多瓦共和国对其烟草控制法进行了全面修订，以履行其根据世界卫生组织《烟草控制框架公约》的义务，并根据摩尔多瓦与欧盟的关联协议将其政策与欧盟政策协调。在对“烟草制品”进行监管的同时，经修订的法律还规定了“相关产品”的监管，其中“相关产品”定义为包括“由植物制成的草本烟产品和含烟碱的产品，包括电子烟”[重点强调][89]
新加坡	2010	自2010年以来，新加坡的法律已经包括对“烟草替代品”的监管。尽管定义随着时间的推移进行了修订，但它一直被用作一个总称术语，以监管含有烟碱的产品（不论来源如何），但这些产品不属于《烟草法》中的其他明确定义的类别[90]
美国	2022	为了应对那些声称不属于美国《烟草控制法案》管辖范围的合成烟碱产品的引入，美国对该法中的“烟草制品”的定义进行了修改以包括“任何产品……含有任何来源的烟碱，以供人使用”[91]

世界卫生组织欧洲地区各国最有可能制定涵盖更广泛的合成烟碱产品的法律。大多数法律都是在2016年之后通过的。例如，摩尔多瓦共和国的法律将“相关产品”与“烟草制品”区分开来，包括“由植物制成的草本烟产品和含烟碱的产品，包括电子烟”，从而提供了对现有和新兴合成烟碱产品的广泛覆盖[89]。新加坡的法律包括一个定义，用以广泛覆盖合成烟碱产品，将“烟草制品”定义为

包括“烟草替代品”，而“烟草替代品”则被定义为“含有烟碱的任何物品、物体或事物”，无需烟碱来自烟草。该法明确排除了从“烟草替代品”中的“①卷烟或雪茄，或者任何其他形式的烟草；②烟草衍生品；③含有任何形式的烟草或烟草衍生品的混合物；④根据《健康产品法案》注册的治疗产品”[90]。

美国现在对合成烟碱产品进行监管，但没有明确禁止[92]。2022年3月，美国《联邦食品、药物和化妆品法案》经过修订，将合成烟碱产品纳入美国食品药品监督管理局（USFDA）的烟草制品权限范围。现在，“烟草制品”的定义包括“由烟草制成或来源于烟草的任何产品，或含有来源于任何途径的烟碱，以供人使用，包括烟草制品的任何组成部分、零件或配件”[重点强调]。因此，该定义涵盖了所有或大多数目前市场上销售的合成烟碱产品，以及可能出现的产品。根据这项法律，合成烟碱产品现在需要在合法销售前获得USFDA的前期市场授权。目前尚未有合成烟碱产品获得这样的授权，但USFDA报告称已收到来自200多家公司的超过100万个市场申请。USFDA拒绝了92.5万个申请，接受了8600个进行进一步审查[92]。

合成烟碱产品覆盖范围不明确的法律

有17个司法管辖区的法律并不清楚其定义是否涵盖合成烟碱产品。例如，一些法律在定义烟碱产品时提及了烟草，但并未说明法律的适用范围是否仅限于来自烟草的烟碱。对于一些国家的法律，烟草控制法律网站上的信息不足以让作者确定法律是否适用于含有合成烟碱的产品。

3.3.3 讨论

各种法律调整可以将合成烟碱产品纳入烟草控制法规的范围。一些国家采取的方法仅覆盖电子烟或电子烟烟液中合成烟碱产品。这些方法不包括对目前市场上销售或可能出现的其他类型的合成烟碱产品的潜在监管，这将削弱新型烟草和烟碱产品的全面监管。如摩尔多瓦共和国、新加坡以及最近的美国等国家的方法所显示的，法律调整可以在烟草控制法规下涵盖目前市场上销售的所有合成烟碱产品和可能出现的产品。

尽管我们的分析仅限于合成烟碱，但它提供了烟草行业可能会寻求利用法律中的漏洞或不确定性来推广新产品或规避与烟草相关的法规的示例。关于烟碱类似物法律格局的进一步研究可能有助于帮助各国制定适当的监管方法[73]。

关于合成烟碱产品的全球法律环境的描述存在一些限制。只编码了英文版本的法律，只包括了覆盖市场准入、销售限制、包装和标识以及广告的烟草控制法。未提供英文版本和其他类型的烟草法律，如税收法，也可能包括合成烟碱产品。编码的法律通常不包括次国家层面的司法管辖区，这些地方法律可能对术语的定义有所不同。烟草控制法律网站可能不完整，因为它可能不包括最近颁布的法律或影响法律解释或可执行性的法律裁决。此外，法律的英文版本可能无法完全反映原始版本。

重要的是，我们编码的法律不包括用于规范烟草制品以外的其他产品的法律。即使合成烟碱产品不在某国烟草制品的监管体系范围内，它们可能受到药物（或药物-器械组合产品）或其他消费者保护法的监管。这些法律可以为各国提供在不改变其烟草控制法律的情况下监管合成烟碱产品的机会[76]。例如，在美国法律于2022年3月进行修订以将合成烟碱产品纳入法律之前，公共卫生团体敦促美国食品药品监督管理局将合成烟碱产品视为药物[93]。Hangsen和NGL等合成烟碱制造商声称他们的产品“提供与吸烟者从烟碱中寻求的相同满足感”，这隐含承认他们销售的产品被用作成瘾药物的效应[48]。大多数产品网站都包括健康警示或免责声明，承认其产品中的烟碱是成瘾性的，并可能具有危险性。此外，澳大利亚药物管理局要求购买含烟碱的电子烟需要处方[94]。尽管此要求似乎包括合成烟碱产品，但它不是通过澳大利亚的烟草控制法规范的，因此不在本分析的范围之内。

最后，这项分析未能评估通过烟草控制法或其他类型的法律所规定的要求是否得以执行。执法可能在国家内部和国家之间有所不同。

3.4 政策建议

- 那些在合成烟碱产品方面存在法规漏洞的国家（与来源于烟草的烟碱产品相比）应考虑修改其烟草控制法，以确保将合成烟碱产品纳入法规范围。
- 那些选择修改其烟草控制法以覆盖合成烟碱产品的国家应考虑法律调整，将法律的适用范围扩大到目前市场上销售的所有合成烟碱产品以及未来可能出现的产品。这些产品可能包括含有合成烟碱类似物、具有类似性质的其他化学物质或在现场生成烟碱或类似物的化学系统的产品。
- 各国应执行有关合成烟碱产品纯度的标准，最好采用欧洲和美国的药典标准。监管机构应考虑实施产品标准，禁止在市售产品中混合烟草源性烟碱和合成烟碱。
- 建议决策者应执行统一的含烟碱产品标签规定，无论是天然还是合成烟碱，都应分别声明*S*-烟碱的含量，以及*R*-烟碱和任何其他烟碱类似物或任何具有类似性质的化学物质的含量。
- 各国应考虑禁止含有*R*-烟碱或除*S*-烟碱以外的任何烟碱类似物的合成烟碱产品，禁止*R*-烟碱及其他烟碱类似物含量超过烟草产品中的含量，直到这些化学物质在此类产品中的使用安全得到确认。

- 监管机构应考虑限制推广合成烟碱产品的营销做法，除非提供了支持此类声明的科学证据，否则不得普遍使用其“无味”、“更纯净”或“比烟草来源烟草更健康”等宣传语。

3.5 结 论

各公司正在推广越来越多种类的合成烟碱产品，如果不受监管，可能会破坏减少烟草和烟碱成瘾的工作，以及世界卫生组织成员国全面监管烟草和烟碱产品的工作及努力。关于合成烟碱在不同类型的消费产品中对人类健康的影响的知识仍然不全面。尽管合成烟碱产品在许多国家的现行烟草控制立法下没有明确的监管，但一些国家的法律已经更新，以覆盖这些产品。上述信息显示，各国可以进行各种法律调整来填补合成烟碱产品的监管漏洞，这些调整可涵盖目前市场上的各种产品范围以及未来可能出现的产品。

参 考 文 献

[1] What you need to know about new synthetic nicotine products. Washington DC: Truth Initiative; 2021 (https://truthinitiative.org/research-resources/harmful-effects-tobacco/what-you-need-know-about-new-synthetic-nicotine-products).

[2] Ramamurthi D, Chau C, Lu Z, Rughoobur I, Sanaie K, Krishna P et al. Marketing of “tobacco-free” and “synthetic nicotine” products (white paper). Palo Alto (CA): Stanford University; 2022 (https://tobacco-img.stanford.edu/wp-content/uploads/2022/03/13161808/Synthetic-Nicotine-White-Paper-3-8-2022F.pdf).

[3] Schmid T. A real up and comer: synthetic nicotine. Tobacco Asia, 14 February 2021 (https://www.tobaccoasia.com/features/a-real-up-and-comer-synthetic-nicotine/).

[4] Ray K, Schuman V. Next Generation Labs CEO Vincent Schuman announces ZIA™ gum with TFN® synthetic nicotine. Chicago (IL): Cision® PR Web; 2017 (https://www.prweb.com/releases/2017/11/prweb14861999.htm).

[5] Lucy TM Gum. Midletown (PA): Lucy Goods; 2023 (https://lucy.co/products/chewpark).

[6] NicotinePicks. Kirksville (MO); 2023 (https://nicotinepicks.com/).

[7] Crave Nicotine ToothPicks. Stockton (CA): Crave Nicotine; 2023 (https://web.archive.org/web/20210609132039/https://podalifestyle.com/).

[8] Poda and our flagship Beyond Burn™ Poda Pods are set to revolutionize the heat-no-t-burn industry. Vancouver (BC): Poda Lifestyle and Wellness Ltd; 2021 (https://web.archive.org/web/20210609132039/https://podalifestyle.com/).

[9] Outlaw Dip. Deer Park (NY): Outlaw Dip Co.; 2023 (https://outlawdip.com/faq/).

[10] Ronin Smokes. Cambridge (Ont): Ronin Smokes; 2023 (https://www.roninsmokes.com/).

[11] The Puff Bar. New & improved. Gendale (CA): Puff Bar; 2021 (https://web.archive.org/

web/20210726184612/https://puffbar.com/collections/puff-bar).

[12] Juice Head pouches. Huntington Beach (CA): Juice Head; 2023 (https://juicehead.co/collections/juice-head-pouches).

[13] Pachamama SYN Vape Disposable-1500 puffs. Louiville (KY): VaporFi; 2023 (https://www.vaporfi.com/pachamama-synthetic-disposable-vape-pen/).

[14] BidiTM Vapor launches BidiTM Pouch, a nicotine delivery product in a tin pack. Jack-sonville Beach (FL): QRX Digital; 2020 (https://www.newswire.com/news/bidit-m-vapor-launches-biditm-pouch-a-nicotine-delivery-product-in-a-21245979).

[15] Can't light up? ZIATM. Boca Raton (FL): Next Generation Labs; 2018 (https://www.nextgenerationlabs.com/wp-content/uploads/2017/07/ZIA-gum-presentation-final.pptx).

[16] VAPORX. Seoul: VAPORX; 2022 (http://vaporx.co.kr/22).

[17] Zhang H, Pang Y, Luo Y, Li X, Chen H, Han S et al. Enantiomeric composition of nicotine in tobacco leaf, cigarette, smokeless tobacco, and e-liquid by normal phase high-performance liquid chromatography. Chirality. 2018;30(7):923-31. doi: 10.1002/chir.22866.

[18] Pictet A, Rotschy A. Synthese des Nicotines [Synthesis of nicotines]. Ber Dtsch Chem Gesells. 1904;37(2):1225-35. doi:10.1002/CBER.19040370206.

[19] Wagner FF, Comins DL. Recent advances in the synthesis of nicotine and its derivatives. Tetrahedron. 2007;63(34):8065-82. doi:10.1016/j.tet.2007.04.100.

[20] Anderson H. Manufacture of nicotine. In: British American Tobacco Records; 1964. p. 100048807-8 (https://www.industrydocuments.ucsf.edu/docs/lgcy0212).

[21] Moates RF. Feasibility of synthetic nicotine production. In: RJ Reynolds Records; Master Settlement Agreement 1967. p. 500613486-9 (https://www.industrydocuments.ucsf.edu/docs/srhn0096).

[22] Southwick E. Synthesis of nicotine. In: Myers L, editor. Liggett & Myers Records 1978. p. lg0292754-lg5 (https://www.industrydocuments.ucsf.edu/docs/gpyw0014).

[23] Ye X, Zhang Y, Song X, Liu Q. Research progress in the pharmacological effects and synthesis of nicotine. Chem Select. 2022;7(12):e202104425. doi:10.1002/slct.202104425.

[24] Arnold M, inventor; Next Generation Labs, LLC, assignee. Process for the preparation of (R,S)-nicotine. US patent 9,556,142. 2017 (https://patents.google.com/patent/US20160115150A1/en?oq=20160115150).

[25] Arnold M, inventor; Next Generation Labs LLC, Kaival Labs LLC, assignee. Nicotine replacement therapy products comprising synthetic nicotine. US patent 10,610,526. 2020.

[26] Hellinghausen G, Lee JT, Weatherly CA, Lopez DA, Armstrong DW. Evaluation of nicotine in tobacco-free-nicotine commercial products. Drug Test Anal. 2017;9(6):944-8. doi:10.1002/dta.2145.

[27] Puff Bar. Tobacco free. Los Angeles (CA): Cool Clouds Distribution Inc.; 2021 (https://puffbar.com/pages/about-puff-bar).

[28] Duell AK, Kerber PJ, Luo W, Peyton DH. Determination of (R)-(+)- and (S)-(−)-nicotine chirality in Puff Bar e-liquids by (1)H NMR spectroscopy, polarimetry, and gas chroma-tography–mass spectrometry. Chem Res Toxicol. 2021;34(7):1718-20. doi:10.1021/acs.chemrestox.1c00192.

[29] Weber B, Pan B, inventors; Siegfried AG, Contraf-Nicotex-Tobacco GmbH, assignee. En-

antiomeric separation of racemic nicotine by addition of an *O,O′*-disubstituted tartaric acid enantiomer patent US20200331883A1. 2019 06/27/2019 (https://patents.google.com/patent/WO2019121649A1/en).

[30] Weber BT, Lothschütz C, Pan B, inventors; Siegfried AG. Contraf-Nicotex-Tobacco GmbH assignee. Preparation of racemic nicotine by reaction of ethyl nicotinate with *N*-vinylpyrrolidone in the presence of an alcoholate base and subsequent process steps USA 2020 (https://patents.google.com/patent/US20200331884A1).

[31] *S*-isomer tobacco free nicotine. Irvine (CA): Five Pawns; 2020 (https://web.archive.org/web/20200811035816/https://fivepawns.com/blogs/five-pawns-news-events/s-isomer-tobacco-free-nicotine).

[32] Synthetic nicotine. Plainville (CT): Tea Time Eliquid Co.; 2021 (https://web.archive.org/web/20211028035558/https://teatimeliquid.com/pages/synthetic-nicotine).

[33] Towards a clean nicotine future. London: Zanoprima; 2023 (https://www.zanoprima.com/).

[34] McCague R, Narasimhan AS, inventors; Zanoprima Lifesciences Limited (London, GB), assignee. Process of making (S)-nicotine. USA patent 10,913,962. 2021 02/09/2021 (https://patents.google.com/patent/US10913962B2).

[35] What is MOTiVO™? Seoul: Hangsen International Group Ltd; 2023 (https://perma.cc/ BE6K-P3US).

[36] Method for preparing nicotine of high optical purity. (https://patents.google.com/patent/WO2022105482A1/en).

[37] Altria announces definitive agreement to acquire NJOY Holdings, Inc. Richmond (VA): Altria Group; 2023 (https://investor.altria.com/press-releases/news-details/2023/Altria-Announces-Definitive-Agreement-to-Acquire-NJOY-Holdings-Inc/default.aspx).

[38] Willis B, Ahmed MM, Freund W, Sawyer D, inventors; NJOY, LLC, assignee. Synthesis and resolution of nicotine. USA patent US10759776B2. 2020 09/01/2020 (https://image-ppubs.uspto.gov/dirsearch-public/print/downloadPdf/10759776).

[39] SyNic Pure Nicotine 1000mg/mL. Newbury Park (CA): Nicotine River; 2023 (https://nicotineriver.com/collections/synic%E2%84%A2-nicotine/products/synic-pure-nicotine-1000mg-ml).

[40] Rossel S. Synthetic nicotine is gaining acceptance. Tobacco Reporter, 1 December 2019 (https://tobaccoreporter.com/2019/12/01/mirror-image/).

[41] TFN® Nicotine. Not derived from tobacco leaf, stem, or waste dust. Phoenix (AZ): Liquid Nicotine Wholesalers; 2023 (https://liquidnicotinewholesalers.com/tfn-pure-liquid-nicotine.html).

[42] CNT Nicotine. Phoenix (AZ): Liquid Nicotine Wholesalers; 2023 (https://liquidnicotinewholesalers.com/cnt-nicotine.html).

[43] Cultra™ Pure Nicotine. Phoenix (AZ): Liquid Nicotine Wholesalers; 2023 (https://liquidnicotinewholesalers.com/cultra-pure-liquid-nicotine.html).

[44] PurNic Pure Nicotine 1000mg/mL. Thousand Oaks (CA): Nicotine River; 2023 (https://nicotineriver.com/collections/purnic%E2%84%A2-nicotine/products/purnic-pure-nicotine-1000mg-ml).

[45] Next Generation Labs LLC has been granted a Notice of Allowance from China for its Process for The Preparation of (R-S) Synthetic Nicotine-Patent #201580069647.2. Chicago (IL): Cision PRWeb; 2021 https://www.prweb.com/releases/next_generation_labs_llc_has_been_granted_

a_notice_of_allowance_from_china_for_its_process_for_the_preparation_of_r_s_synthetic_nicotine_patent_201580069647_2/prweb17809747.htm).
[46] Zanoprima Lifesciences Ltd v. Hangsen International Group Ltd (6:22-cv-00268) District Court, W.D. Texas; 2022 (https://perma.cc/SV6B-VYDA).
[47] Hangsen releases synthetic nicotine. Tobacco Reporter, 11 Setember 2020 (https://tobaccoreporter.com/2020/09/11/hangsen-and-geek-vape-release-synthetic-nicotine-product/).
[48] What is TFN®. Boca Raton (FL): Next Generation Labs; 2021 (http://www.nextgeneration- labs.com/).
[49] Specific nicotine isomers ratios could potentially offer nicotine use at satisfying but non-addictive levels as revealed by Next Generation Labs CEO Vincent Schuman. Chica-go (IL): Cision PRWeb; 2017 (https://www.prweb.com/releases/2017/11/prweb14911138.htm).
[50] Shimada A, Iizuka H, Kawaguchi T, Yanagita T. [Pharmacodynamic effects of d-nicotine- Comparison with l-nicotine]. Nihon Yakurigaku Zasshi. 1984;84(1):1-10. PMID:6489864.
[51] Ikushima S, Muramatsu I, Sakakibara Y, Yokotani K, Fujiwara M. The effects of D-nicotine and L-isomer on nicotinic receptors. J Pharmacol Exp Ther. 1982;222(2):463-70. PMID:7097565.
[52] Martin BR, Aceto MD. Nicotine binding sites and their localization in the central nervous system. Neurosci Biobehav Rev. 1981;5(4):473-8. doi:10.1016/0149-7634(81)90017-8.
[53] Zhang X, Gong ZH, Nordberg A. Effects of chronic treatment with (+)- and (–)-nicotine on nicotinic acetylcholine receptors and *N*-methyl-D-aspartate receptors in rat brain. Brain Res. 1994;644(1):32-9. doi:10.1016/0006-8993(94)90343-3.
[54] Meltzer LT, Rosecrans JA, Aceto MD, Harris LS. Discriminative stimulus properties of the optical isomers of nicotine. Psychopharmacology (Berl). 1980;68(3):283-6. doi:10.1007/BF00428116.
[55] Clarke PB, Kumar R. Characterization of the locomotor stimulant action of nicotine in tolerant rats. Br J Pharmacol. 1983;80(3):587-94. doi:10.1111/j.1476-5381.1983.tb10733.x.
[56] Kumar R, Pratt JA, Stolerman IP. Characteristics of conditioned taste aversion produ-ced by nicotine in rats. Br J Pharmacol. 1983;79(1):245-53. doi:10.1111/j.1476-5381.1983.tb10518.x.
[57] Romano C, Goldstein A, Jewell NP. Characterization of the receptor mediating the nico-tine discriminative stimulus. Psychopharmacology (Berl). 1981;74(4):310-5. doi:10.1007/BF00432737.
[58] Rosecrans JA, Meltzer LT. Central sites and mechanisms of action of nicotine. Neurosci Biobehav Rev. 1981;5(4):497-501. doi:10.1016/0149-7634(81)90020-8.
[59] Goldberg SR, Risner ME, Stolerman IP, Reavill C, Garcha HS. Nicotine and some related compounds: effects on schedule-controlled behaviour and discriminative properties in rats. Psychopharmacology (Berl). 1989;97(3):295-302. doi:10.1007/BF00439441.
[60] Jacob P 3rd, Benowitz NL, Copeland JR, Risner ME, Cone EJ. Disposition kinetics of nicotine and cotinine enantiomers in rabbits and beagle dogs. J Pharm Sci. 1988;77(5):396-400. doi:10.1002/jps.2600770508.
[61] Nwosu CG, Godin CS, Houdi AA, Damani LA, Crooks PA. Enantioselective metabolism during continuous administration of S-(–)- and R-(+)-nicotine isomers to guinea-pigs. J Pharm Pharmacol. 1988;40(12):862-9. doi:10.1111/j.2042-7158.1988.tb06289.x.
[62] Nwosu CG, Crooks PA. Species variation and stereoselectivity in the metabolism of nicotine enantiomers. Xenobiotica. 1988;18(12):1361-72. doi:10.3109/00498258809042260.
[63] Crooks PA, Godin CS. N-Methylation of nicotine enantiomers by human liver cytosol. J Pharm

Pharmacol. 1988;40(2):153-4. doi:10.1111/j.2042-7158.1988.tb05207.x.

[64] Yang J, Chen YK, Liu ZH, Yang L, Tang JG, Miao MM et al. Differences between the binding modes of enantiomers S/R-nicotine to acetylcholinesterase. RSC Adv. 2019;9(3):1428-40. doi:10.1039/c8ra09963d.

[65] Saareks V, Mucha I, Sievi E, Vapaatalo H, Riutta A. Nicotine stereoisomers and cotinine stimulate prostaglandin E2 but inhibit thromboxane B2 and leukotriene E4 synthesis in whole blood. Eur J Pharmacol. 1998;353(1):87-92. doi:10.1016/s0014-2999(98)00384-7.

[66] Thuerauf N, Kaegler M, Dietz R, Barocka A, Kobal G. Dose-dependent stereoselective activation of the trigeminal sensory system by nicotine in man. Psychopharmacology (Berl). 1999;142(3):236-43. doi:10.1007/s002130050885.

[67] Mao H, Wang H, Hu X, Zhang P, Xiao Z, Liu J. One-pot efficient catalytic oxidation for bio-vanillin preparation and carbon isotope analysis. ACS Omega. 2020;5(15):8794-803. doi:10.1021/acsomega.0c00370.

[68] Zhang L, Kujawinski DM, Federherr E, Schmidt TC, Jochmann MA. Caffeine in your drink: natural or synthetic? Anal Chem. 2012;84(6):2805-10. doi:10.1021/ac203197d.

[69] Cheetham AG, Plunkett S, Campbell P, Hilldrup J, Coffa BG, Gilliland S 3rd et al. Ana-lysis and differentiation of tobacco-derived and synthetic nicotine products: addressing an urgent regulatory issue. PLoS One. 2022;17(4):e0267049. doi:10.1371/journal.pone.0267049.

[70] Nicotine (USP 29-NF 24). North Bethesda (MD): United States Pharmacopeial Convention; 2020 (https://online.uspnf.com/uspnf/document/1_GUID-3D851985-2C16-408D-99E1-F241A9767168_4_en-US).

[71] Liu B, Chen Y, Ma X, Hu K. Site-specific peak intensity ratio (SPIR) from ^{1}D ^{2}H/^{1}H NMR spectra for rapid distinction between natural and synthetic nicotine and detection of possible adulteration. Anal Bioanal Chem. 2019;411(24):6427-34. doi:10.1007/s00216-019-02023-6.

[72] Han S, Cui L, Chen H, Fu Y, Hou H, Hu Q et al. Stable isotope characterization of tobacco products: a determination of synthetic or natural nicotine authenticity. Rapid Commun Mass Spectrom. 2023;37(3):e9441. doi:10.1002/rcm.9441.

[73] Vagg R, Chapman S. Nicotine analogues: a review of tobacco industry research interests. Addiction. 2005;100(5):701-12. doi:10.1111/j.1360-0443.2005.01014.x.

[74] Clemens KJ, Caillé S, Stinus L, Cador M. The addition of five minor tobacco alkaloids increases nicotine-induced hyperactivity, sensitization and intravenous self-administration in rats. Int J Neuropsychopharmacol. 2009;12(10):1355-66. doi:10.1017/S1461145709000273.

[75] Hall BJ, Wells C, Allenby C, Lin MY, Hao I, Marshall L et al. Differential effects of non-nicotine tobacco constituent compounds on nicotine self-administration in rats. Pharmacol Biochem Behav. 2014;120:103-8. doi:10.1016/j.pbb.2014.02.011.

[76] Harris AC, Tally L, Muelken P, Banal A, Schmidt CE, Cao Q et al. Effects of nicotine and minor tobacco alkaloids on intracranial-self-stimulation in rats. Drug Alcohol Depend. 2015;153:330-4. doi:10.1016/j.drugalcdep.2015.06.00.

[77] Zettler PJ, Hemmerich N, Berman ML. Closing the regulatory gap for synthetic nicotine products. Boston Coll Law Rev. 2018;59(6):1933-82. PMID:30636822.

[78] Jordt SE. Synthetic nicotine has arrived. Tob Control. 2023;32(e1):e113-7. doi:10.1136/tobaccocontrol-2021-056626.

[79] Merkley JA, Kaine T, Warren E, Brown S, Markey EJ, Baldwin T et al. [Letter 16 November 2021]. Washington DC: United States Senate; 2021 (https://www.merkley.senate.gov/imo/media/doc/21.11.16%20Signed%20Synthetic%20Nicotine%20Letter%20to%20FDA.pdf).

[80] Maloney J. Puff Bar defies FDA crackdown on fruity e-cigarettes by ditching the tobacco. The Wall Street Journal, 2 March 2021 (https://www.wsj.com/articles/puff-bar-defies-fda-crackdown-on-fruity-e-cigarettes-by-ditching-the-tobacco-11614681003).

[81] FCTC/COP8(22) Novel and emerging tobacco products. Geneva: World Health Organization; 2018 (https://fctc.who.int/who-fctc/governance/conference-of-the-parties/eight-session-of-the-conference-of-the-parties/decisions/fctc-cop8(22)-novel-and-emerging-to-bacco-products).

[82] Tobacco control laws. Washington DC: Campaign for Tobacco-Free Kids; 2023 (https://www.tobaccocontrollaws.org/).

[83] H.R.2471-Consolidated Appropriations Act, 2022. Washington DC: Congress.gov; 2022 (https://www.congress.gov/bill/117th-congress/house-bill/2471/text).

[84] 21 USC 321: Definitions; generally. In: United States Code. Washington DC: Office of the Law Revision Counsel;2022 (https://uscode.house.gov/view.xhtml?req=granu-leid:USC-prelim-title21-section321&num=0&edition=prelim).

[85] Section 101 of the Tobacco Control Act-Amendment of Federal Food, Drug, and Cosmetic Act (FDCA). Silver Spring (MD): Food and Drug Administration; 2023 (https://www.fda.gov/tobacco-products/rules-regulations-and-guidance/section-101-tobacco-control-act-amendment-federal-food-drug-and-cosmetic-act-fdca).

[86] Types of legislation. Brussels: European Union; 2023 (https://european-union.europa.eu/institutions-law-budget/law/types-legislation_en).

[87] About LawAtlas.org. Philadelphia (PA): Temple University, Beasley School of Law, Center for Public Health Law Research; 2023 (https://lawatlas.org/page/lawatlas-about).

[88] Family Smoking Prevention and Tobacco Control Act and Federal Retirement Reform. Public law 111-31-June 22, 2009. Washington DC: Government Printing Office Act 2009 (https://www.govinfo.gov/content/pkg/PLAW-111publ31/pdf/PLAW-111publ31.pdf).

[89] Parlamentul Lege Nr. 278 din 14–12-2007 privind controlul tutunului [Parliamentary law no. 278 of 14 12 2007 on tobacco]. Chişinău; Republic of Moldova; 2023 (https://www.legis.md/cautare/getResults?doc_id=128322&lang=ro).

[90] Tobacco (Control of Advertisements and Sale) Act 1993. 2020 revised edition. Singapore: Legislation Division of the Attorney-General's Chambers of Singapore; 2023 (https://sso.agc.gov.sg/Act/TCASA1993).

[91] Lipstein A, Zeller M. E-cigarette companies found a loophole in synthetic nicotine-it won't stop the FDA. The Hill, 7 April 2022 (https://thehill.com/opinion/healthcare/3260879-e-cigarette-companies-found-a-loophole-in-synthetic-nicotine-it-wont-stop-the-fda/).

[92] Regulation and enforcement of non-tobacco nicotine (NTN) products. Silver Spring (MD): Food and Drug Administration; 2023 (https://www.fda.gov/tobacco-products/products-ingredients-components/regulation-and-enforcement-non-tobacco-nicoti-ne-ntn-products).

[93] American Academy of Pediatrics, American Cancer Society Cancer Action Network, American Heart Association, American Lung Association, Campaign for Tobacco Free Kids, Parents Against Vaping E-cigarettes et al. Letter to Dr Janet Woodcock, Acting Commissioner, US Food

and Drug Administration. Re: Synthetic nicotine and Puff Bar. Washington DC: Tobacco Free Kids; 2021 (https://www.tobaccofreekids.org/assets/content/what_we_do/federal_issues/fda/regulatory/2021_03_18_puff-bar-synthetic-nicotine.pdf).

[94] TGA confirms nicotine e-cigarette access is by prescription only. Canberra (ACT): Department of Health and Aged Care, Therapeutic Goods Administration; 2020 (https://www.tga.gov.au/news/media-releases/tga-confirms-nicotine-e-cigarette-access-prescription-only).

4. 尼古丁袋：特性、用途、危害及监管

Charlotte GGM Pauwels, National Institute for Public Health and the Environment, Centre for Health Protection, Bilthoven, Netherlands (Kingdom of the)

Reinskje Talhout, National Institute for Public Health and the Environment, Centre for Health Protection, Bilthoven, Netherlands (Kingdom of the)

Jennifer Brown, Johns Hopkins Bloomberg School of Public Health, Baltimore (MD), USA

Rula Cavaco Dias, Healthier Populations Division, Health Promotion Department, No Tobacco Unit, World Health Organization, Geneva, Switzerland

Ranti Fayokun, Healthier Populations Division, Health Promotion Department, No Tobacco Unit, World Health Organization, Geneva, Switzerland

主要发现、挑战和监管影响

- 近期尼古丁袋已在全球多个市场上市，其销售也在迅速增长。
- 尼古丁袋提供足够剂量的烟碱，能诱发并维持烟碱成瘾。
- 尼古丁袋具有诱人的味道等特性，可以隐蔽使用而不会产生吸烟的羞耻感。
- 烟碱对健康有害，包括神经系统和心脏系统。
- 由于尼古丁袋上市时间较短，相关数据仍然有限。鉴于其与传统口服烟草制品（尤其是口含烟）的相似性，有必要采取审慎的应对策略。
- 尼古丁袋在某些司法管辖区未受监管或缺乏专门监管。一些国家已制定了“面向未来”的法规，将尼古丁袋纳入现行法律框架，另一些国家则近期更新了相关法律。而有些国家仍沿用仅针对传统烟草制品的定义。

关键词：烟碱产品，尼古丁袋，特性，危害，监管，监管机制

4.1 引　　言

在过去十年中，新型烟碱和烟草产品，如电子烟碱传输系统（ENDS）、电子

非烟碱传输系统（ENNDS）和加热型烟草制品（HTP），在全球市场上迅速增长。其中一些产品，如ENDS，也已通过烟草和相关行业向儿童和青少年销售和推广[1, 2]。大约自2018年以来，随着烟草行业不断扩大新型烟碱和烟草产品的种类，另一类称为尼古丁袋的产品已被引入多个市场[3]。用于描述这些产品的别称包括"无烟草的尼古丁袋"、"烟草无叶袋"和"烟草衍生的尼古丁袋"；在本书中，我们称之为"尼古丁袋"。在某些司法管辖区（如美国）称之为"白色袋子"。

尽管有报道称，美国市场上大多数含烟碱的产品仍以烟草为主要来源，但合成烟碱正日益流行[4]。尼古丁袋是一种烟碱的预分装袋。它们在某些方面与传统的无烟烟草制品（如口含烟）相似，例如外观、烟碱含量以及使用方式（将它们放在牙龈和嘴唇之间）；然而，与含有烟草的口含烟不同，据报道，尼古丁袋不含烟草，而是由纤维素粉和其他成分构成。尽管如此，烟碱仍可能从烟草中提取，因此可能含有烟草中的其他物质，类似于ENDS。如果烟碱确实来源于烟草，那么尼古丁袋"无烟草"的宣传可能会产生误导。

尼古丁袋的口味与ENDS、ENNDS及传统无烟烟草制品的口味相似。这些口味通过持续使用增强烟碱的效果，提高产品适口性，并对成年人，特别是年轻人，包括未使用烟碱的青少年具有更强的吸引力。尼古丁袋中的一些成分，如增加pH值的碱性剂，可能会增加烟碱的递送[5]。一些尼古丁袋声称含有合成烟碱，通常是*S*-烟碱和*R*-烟碱异构体的外消旋混合物；少数产品具有立体选择性，含有更多更强的*S*-异构体，该异构体在烟草植物中占主导地位（另见背景文件2）。关于*R*-烟碱在人类中的药理和代谢作用的研究仍然非常有限[6]。直到2022年，美国食品药品监督管理局（USFDA）对烟草制品的定义包括烟草衍生的烟碱，而含有合成烟碱的产品在法律上不被视为烟草制品。该定义于2022年被修订，自此美国食品药品监督管理局获得了对任何来源的且不用于治疗目的的烟碱产品的监管权[7]。在其他一些司法管辖区，烟碱产品仅在明确列入烟草法时才被视为烟草制品（参见第3.3.2小节）。

尼古丁袋最初在欧洲推出，但现已扩展至其他国家，如印度尼西亚、肯尼亚、巴基斯坦、美国以及世界卫生组织西太平洋地区的一些国家。部分国家已向WHO寻求技术援助，以应对这些产品的问题。尼古丁袋在世界许多地方的销量正在迅速增长[5, 8]，包括丹麦、挪威、瑞典和美国。例如，在美国，尼古丁袋的销售额从2016年的6.42万美元增加到2018年的5200万美元[5]，预计奥地利、克罗地亚、德国和英国等欧洲国家的销售额也将持续增长[8]。根据欧睿国际的报告，2021年全球售出了约68亿支尼古丁袋，较2018年估计的零售量增长了2000%以上。预计到2023年底，全球销量将超过110亿支[9]。

这种与传统烟草和烟碱产品非常相似的新产品的推出，对世界卫生组织各区域的监管都构成了严峻挑战。许多制造商和零售商将它们宣传为"更健康的替代品”，其广告通常以吸引年轻人为主[10, 11]。此外，网上也有大量的尼古丁袋广告[11]。

制造商试图说服监管机构将尼古丁袋归类为非烟草产品，因为这些产品是否受烟草法规约束有时并不明确，或处于监管的“灰色地带”[6]。例如，尼古丁袋常被宣传为“非烟草产品”、“白色袋子”和“无烟产品”。在一些国家，特别是中低收入国家，制造商声称其中所含的烟碱并非来自烟草，进而主张这些产品不应受到烟草控制法的约束[6]。烟草制造商还寻求对新型烟碱和烟草产品（包括尼古丁袋）进行监管豁免，如肯尼亚的Lyft案例。Lyft产品已获得该国药品监管机构的批准，并自2019年7月起在肯尼亚销售[12]。然而，多个健康倡导组织向肯尼迪内阁卫生部长递交请愿书，要求禁止Lyft尼古丁袋，指控这些产品被非法引入肯尼亚市场。当内阁部长质疑批准该产品作为药物的依据时，这些产品的销售被暂停，并告知英美烟草公司（BAT），Lyft必须遵守肯尼亚对烟草制品的监管要求。健康倡导组织现在坚持认为这些产品不应被允许上市[12]。

世界卫生组织成员国已就尼古丁袋产品的定义以及有关这些产品的知识和证据（包括与产品相关的潜在和实际风险、其特性以及各国如何对其进行监管）向世界卫生组织寻求技术援助。本章总结了尼古丁袋的已知特性、产品使用者、使用的潜在风险以及调节尼古丁袋的机制。这些信息源自科学文献、互联网检索、制造商的网页和烟碱产品的市场数据，旨在帮助监管机构更好地理解这些产品、各国的经验和所面临的监管挑战。本章还提供了关于尼古丁袋监管的指导意见和一些供各国参考的建议。

4.2 方　　法

在文献数据库PubMed和其他来源（如网络检索、欧睿国际的专栏、电子情报和烟草情报、制造商的网页和烟碱产品的市场数据）中进行了关于烟碱产品的检索，纳入了截至2023年3月的同行评审出版物。这些内容最初根据标题和摘要进行筛选,然后进一步审查全文。检索的关键词包括“烟碱产品”“尼古丁袋”“无烟叶袋”“烟草衍生尼古丁袋”“非烟草产品”“白袋”“无烟产品”“特征”“危害性”“监管”“监管机制”。

此外，在2020年，WHO向所有六个地区的顾问分发了一份调查问卷，以收集各国尼古丁袋的使用经验、现有的监管机制和监管中发现的困难。2021年，又向WHO全球烟草监管机构论坛分发了另一份调查问卷，该论坛有3周时间完成调查问卷并返回给WHO。随后制定了调查问卷，并在2022年进行了数据收集。通过电子邮件非正式地向欧盟成员国的监管机构寻求信息，并对WHO烟草控制法数据库和无烟儿童运动中的烟草控制法进行了审查。总共审查了124项国家法律。

4.3 产品特性

不同品牌的尼古丁袋和同品牌下的不同产品在重量、烟碱浓度和生物利用度、风味和袋装尺寸方面存在差异。与传统含烟草的口含烟一样，20~25个尼古丁袋通常装在一个口袋大小的罐子中（图4.1）。在一些国家，罐子可能设有隔间，用于丢弃用过的袋子（如Zyn，Ace）[13]。罐盖通常标示品牌名称、口味和袋装尺寸（例如“slim”）。在一些国家，烟碱含量以点或数值形式的“强度”表示（例如，图4.1中5个数字中的4）。烟碱的含量因品牌而异，可以mg/袋的形式表示。由于缺乏统一的标识要求，烟碱浓度的多样化表达可能让消费者感到困惑。罐子的盖子或底部通常附有警示信息，尽管若商品不受烟草制品监管，此类警示并非强制，但可能给消费者造成遵守烟草制品法规的误导性印象。

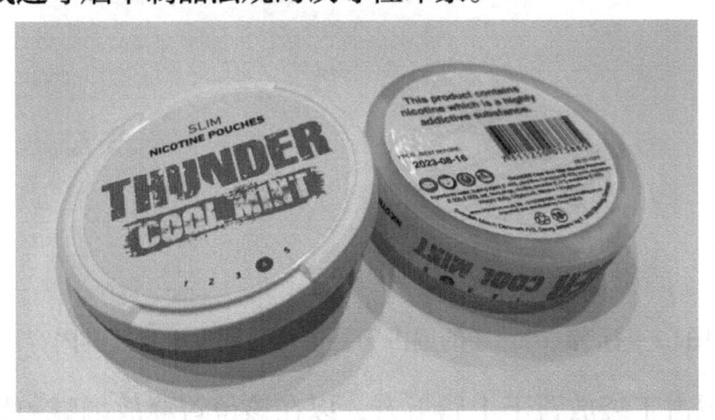

图4.1 Thunder Cool Mint尼古丁袋罐的底部和盖子，其含量或“强度”以数字尺度表示

单个尼古丁袋重149~800 mg[5, 14]，烟碱含量一般为3~50 mg/g，相当于每袋烟碱剂量为2~32.5 mg[3, 15, 16]，而传统口含烟重0.3~1.13 g，烟碱剂量为6.81~20.6 mg/g湿烟草[17]。据报道，一些尼古丁袋的烟碱含量极高，例如在爱沙尼亚市场上，某些产品的烟碱含量高达120 mg/g的尼古丁袋，其中最强的产品来自俄罗斯联邦[8]。在6家制造商的37个品牌（2~6 mg/袋）中，每袋烟碱含量为1.29~6.11 mg，含水量为1.12%~47.2%，pH值为6.9~10.1，游离态烟碱含量为79.7%~99.2%。游离态烟碱的生物利用度较高，优于质子化烟碱（见2.6节）[5]。这些产品中的烟碱来源可能是烟草提取物，也可能是合成烟碱。这些品牌提供多种口味，如水果味、甜味、咖啡味和薄荷等。有时这些口味会进行混合，例如描述为“甜、酸菠萝、奶油椰子和坚果底色的平衡结合”（Lyft），并给出了诸如“热带风”（Velo）这样的概念名称。薄荷口味在美国尤为流行，2019年占美国尼古丁袋市场总量的54.6%[18]。在2019年1月至2020年6月期间，水果味尼古丁袋的销量显著增长[19]。凉爽和水果类别占据了市场主导地位，占这些产品口味的近70%[20]。尼古丁袋有多种尺寸可

供选择，较小的产品在包装上标注为“轻薄”或“迷你”。

尼古丁袋本身是由不溶于水的材料制成的，类似于茶袋，主要由纤维素纤维制成，但可渗透唾液和烟碱[21]。袋中含有灰白色或白色粉末，其中含有由烟碱和酸组成的盐或游离态烟碱（图4.2）。除烟碱外，其他成分包括纤维素、水、盐和其他添加剂，如pH调节剂、填充剂、无热量甜味剂、稳定剂（羟丙基纤维素）和调味料[3, 22]。

图4.2　标有“slim”的尼古丁袋包装、尼古丁袋及其内容物

使用者将尼古丁袋放置于上唇下方，以在该位置释放烟碱和特定味道。品牌网站和网上商店建议将尼古丁袋保留5分钟至1小时[23]。将袋子放入口腔后不久，通常会感到由烟碱引起的刺痛感，这种感觉可持续长达15分钟[22]。根据调查和网站数据，Zyn使用者每天使用10~12份尼古丁袋[22, 24]。

4.4　市场营销

目前市场上的尼古丁袋主要由大型烟草制造商生产，如英美烟草公司（Lyft，Velo，Zonnic）[25]、美国奥驰亚集团（On!）、瑞典火柴公司（Zyn, G.4）[21]、帝国烟草公司（Skruf，Knox，ZoneX）[26,27]、菲利普·莫里斯国际公司（Shiro，Sirius），Swisher（Rogue）和日本烟草公司（Nordic Spirit）[3, 28]。在美国，联邦贸易委员会在2021年首次报告了尼古丁袋[29]，当时这些公司在美国销售了1.407亿盒此类产品，销售额为4.205亿美元。尼古丁袋的销量从2016年7月的163178套（15~20个袋）（709635美元）增加到2020年6月的45965455套（216886819美元）[19]。2020年，美国市场份额最高的是瑞典火柴公司（78.7%），其次是美国奥驰亚集团（10%）和英美烟草公司（7.6%）。一些小公司也生产尼古丁袋，如口含烟部出品的王牌超级白（丹麦）或 Microzero AB出品的N!Xs（瑞典和其他欧洲国家）[30]。

尼古丁袋在网上和烟草商那里被宣传为烟草和烟碱产品的无烟和无烟叶替代品，比传统口含烟和传统卷烟的"危害更小"[14]。这些也是使用者报告的一些原因（见下文）[22, 31]。无烟烟草使用者也认为口含烟比传统卷烟的风险更小[32, 33]。虽然尼古丁袋容器通常印有对烟碱上瘾的警示，但它们通常没有许多国家无烟烟草所要求的关于口腔癌和牙龈疾病风险的警示。因为它们的白色粉末填充物，尼古丁袋也被称为"白色烟袋"，而不是传统的棕色烟草口含烟[34]。"白色口含烟"被宣传为"更温和、更纤细、更有风味、更具视觉吸引力"[26]。有不同尺寸的尼古丁袋可供选择，较小的尼古丁袋被宣传为谨慎（"因为小袋又薄又小，没有人会看到你把它放在嘴唇下"）[35]。网络营销宣称尼古丁袋可以在"任何地方、任何时间"使用，包括在禁止吸烟的地方[23,34]。当母公司也拥有尼古丁袋品牌时，在传统卷烟的主要品牌的网站上出现了交叉广告。例如，奥驰亚与万宝路联合推广的On!尼古丁袋，并且On!和Camel的消费者收到了一封电子邮件邀请，邀请他们"探索来自 Velo 的朋友提供的尼古丁选择"[36]。电子邮件广告包括声称可以将该产品用在任何地方使用（84%的电子邮件），尼古丁袋是其他烟草制品的替代品（69%），不含烟叶（55%）并且"无唾液"（52%）或"无烟"（31%）[37]。尼古丁袋并不比卷烟贵，因为一盒尼古丁袋的价格略低于或与一包传统卷烟相当[3]。一些公司还声称，与电子烟相比，尼古丁袋不需要电池和充电设备[35,38]，而且与传统的口含烟相比，白色小袋看起来更干净，也不会弄脏牙齿[34,39]。

根据欧睿监测公司的一份报告[40]，"现代口服产品制造商已经接受了影响者营销和社交媒体平台，例如 Instagram"。新闻调查局的一篇文章总结了英美烟草新产品的营销策略，包括加热烟草和口服烟碱，在以时尚的青年为中心的广告活动中将烟碱产品描述为"酷"和"令人向往"；付钱给社交媒体影响者以推广尼古丁袋；赞助音乐和体育赛事；以及国际上提供免费的尼古丁袋样品，这似乎吸引了未成年人和非吸烟者[10]。在许多国家，尼古丁袋在音乐家、足球运动员和有影响力的人的社交媒体账户上被推广[10]。

4.5 使用者概况

一些研究报告了尼古丁袋使用的普遍性。澳大利亚（0.1%）、加拿大（0.9%）、英国（1.1%）和美国（0.7%）的10296名成年当前吸烟者或近期戒烟者中，总体使用为0.8%。在当前很少吸烟和既往吸烟者中，使用尼古丁袋的男性（1.1%）多于女性（0.5%）。在所有被研究的国家中，18~24岁人群的使用率最高（2.3%）（25~39岁，1.4%；40~54岁，0.4%；≥55岁，0.1%）[41]。

在2019年对英国3883名吸烟者、电子烟使用者、双重使用者和既往使用者的调查中，15.9%的人听说过尼古丁袋，3.1%的人见过它们出售；4.4%的人曾经使用过尼古丁袋，2.7%的人是当前使用者[42]。在一项代表荷兰人口的5805名样本的调查中，只有6.9%的人知道尼古丁袋，主要是因为他们认识使用它们的人（33%）[31]。在受访者中，0.6%曾使用过尼古丁袋，0.06%是目前的使用者。当前吸烟者的使用量高于以往平均水平（1.91%），尤其是那些喜欢吸薄荷烟的人（6.26%）。青少年（13~17岁）的认知度相对较高（9.1%），但只有0.3%的人曾经使用过尼古丁袋，也没有人报告目前使用过尼古丁袋。

在2019年对加拿大11714名16~19岁的年轻人、英国11170名和美国11838名年轻人进行的在线重复横断面调查中，1%的加拿大人、1.3%的英格兰人和1.5%的美国人在过去30天内使用过尼古丁袋[43]。来自美国2021年全国青年烟草调查的数据显示，1.9%的初中生和高中生（11~18岁）曾使用过尼古丁袋[44]。此外，0.8%的学生报告目前（过去30天）使用过尼古丁袋。在报告目前使用过尼古丁袋的学生中，大多数（63.5%）报告在过去30天内使用过1~5天，17.2%报告在过去30天内有20~30天使用过尼古丁袋。此外，61.6%的当前使用者报告说在过去30天内使用过有风味的尼古丁袋，薄荷和薄荷醇是最常见的口味。

2021年初在美国进行的一项研究对当前已确定吸烟者的美国成年人（一生中至少抽100支卷烟，现在每天或几天都抽烟）进行的一项基于网络的调查发现，29.2%的人曾经见过或听说过尼古丁袋，5.6%的人曾经使用过，16.8%的人表示有兴趣在未来6个月内使用尼古丁袋[45]。年轻的成年吸烟者比成年吸烟者更有可能看到或听说过尼古丁袋。在吸烟的成年人中，受教育程度较高的人使用尼古丁袋的概率较低，而那些在使用传统方法或曾经使用无烟烟草之前尝试戒烟的人使用尼古丁袋的概率较高。

一项基于瑞典匹配北美数据的研究调查了Zyn使用者的人口统计数据以及使用模式和原因[22]。Zyn使用者的平均年龄约为33岁，男性，白种人，高中毕业，年收入超过5万美元（即中等收入）。大多数是目前的无烟烟草使用者和以前的烟草使用者（主要是以前的卷烟-无烟烟草双重使用者）。另外两项研究报告了尼古丁袋使用者的类似情况：男性，25~34岁或25~44岁，曾吸烟和/或使用电子烟[31, 42]。Zyn的使用者发现尼古丁袋的吸引力从中等到极高。使用的原因是"比其他烟草制品对我的健康危害更小"（62%）、"易于使用"（53%）、"使用时没人能分辨"（50%）、"对我健康的危害比卷烟小"（49%）和"没有烟味并避免随地吐痰"（48%）。在荷兰受访者中，尼古丁袋主要用在"聚会上"（38%）、"与朋友一起"（38%）或"在家"（26%）[31]。使用尼古丁袋的主要原因是"出于好奇"（72%）和"它很美味和/或令人愉快"（23%），还有他们认为"它比卷烟更不健康"（23%）。只有8%的受访者表示"不同口味的可用性"是使用的一个重要原因。

4.6 产品潜在危害评估

4.6.1 吸引力

世界卫生组织《烟草控制框架公约》（WHO FCTC）[46]第9、10条的部分实施指南建议对有吸引力的产品特性进行监管，特别是减少从未吸烟的年轻人的使用。如上所述，尼古丁袋有很多吸引人的特点。例如，它们有各种水果、薄荷和其他口味（如肉桂、甘草和咖啡）可供选择[3, 5, 22]，并添加了甜味剂[22]。该产品在美国的成本略低于或与一包传统卷烟相当[3, 47]，这可能是一些但不是所有潜在使用者使用的影响因素。进一步吸引人的特点是，人们认为该产品对戒烟有效，比其他烟草制品危害更小，在禁止吸烟的地方使用起来更便捷和隐蔽[22]。

4.6.2 致瘾性

尼古丁袋中含有足够的烟碱来维持成瘾[3, 47]：Zonnic 4 mg提供2 mg烟碱[47]，这与卷烟提供的水平相似。烟碱从On!尼古丁袋释放到口腔唾液中，所有口味的烟碱释放水平相同[48]。

有两项研究探讨了尼古丁袋的药代动力学[16, 49]。在Lunell等由瑞典火柴资助和设计的一项研究中，测试了浓度为3 mg、6 mg或8 mg烟碱的尼古丁袋[16]。使用1 h后，烟碱分别释放1.6 mg（56%）、3.5 mg（60%）和3.8 mg（50%），使用过程中使用者血液中的烟碱量逐渐增加，峰值浓度分别为7.7 ng/mL、14.7 ng/mL和18.5 ng/mL。作者报告说，Zyn（6 mg和8 mg）递送烟碱的速度和程度与无烟烟草制品相似。在由奥驰亚客户服务有限责任公司资助的一项由Rensch等[49]进行的研究中，测试了各种口味和4 mg烟碱的尼古丁袋。参与者静脉血中的烟碱含量在使用30分钟内和之后的10分钟内逐渐增加，达到9.6~12.1 ng/mL的峰值浓度。使用这些烟袋可以减少吸烟的冲动或对卷烟的渴望。所有的尼古丁袋都被认为是令人愉快的，但没有自己品牌的卷烟那么多。味道似乎并不影响烟碱的药代动力学或主观反应。

相比之下，在大约5分钟内从一根卷烟中吸入1~2 mg烟碱。抽吸一支烟后静脉血血浆中的烟碱在抽吸第一口卷烟后的5~8分钟内达到10~30 ng/mL的峰值浓度[50, 51]。因此，尼古丁袋使用者暴露的烟碱量，特别是≥ 4 mg的袋，与吸烟者暴露的烟碱量相同。一个不同之处在于，吸烟期间的浓度在很短的时间内达到峰值，而使用尼古丁袋时，浓度增加得更缓慢。对口含烟也进行了类似的观察，烟碱血浆水平的上升速度比吸烟时要慢[51]。

烟碱释放较慢是一个重要的区别，因为正是烟碱快速达到峰值使吸烟如此容易上瘾。药物被吸收并到达大脑的速度越快，它引起的“兴奋”就越强，奖励效应就越强。此外，行为和“奖励”之间的短时间间隔提供了强烈的行为条件作用[52]。烟碱替代疗法产品，如在胃和肠道中吸收的尼古丁口香糖，以及通过皮肤吸收的贴片，导致烟碱的吸收非常缓慢，因此成瘾性要低得多[50]。尼古丁袋中烟碱的吸收速率似乎更接近尼古丁口香糖的速度，而不是可吸入含烟碱的产品[16, 50]。

Zyn等尼古丁袋含有pH调节剂[22]，这可能会增加产品的成瘾可能性，因为更高的pH值会产生更多所谓的“游离”态烟碱[5]，这使得产品比其他形式的烟碱更刺激，但比其他形式的烟碱更容易被口腔吸收。pH值较高的烟草口含烟制品向使用者提供更多的烟碱[53]。尼古丁袋产品在袋含量质量、水分含量（1.12%~47.2%）、碱度（pH 6.86~10.1）和游离态烟碱百分比（7.7%~99.2%）方面存在差异。总烟碱含量为1.29~6.11 mg/袋，游离态烟碱含量为0.166~6.07 mg/袋[5]。

4.6.3 毒性

尼古丁袋不含燃烧或熏烤烟叶的化学副产品，它们也不会被吸入。传统无烟品牌的烟叶中在燃烧时会产生尼古丁袋中不存在的有毒化学物质。事实上，无烟草尼古丁袋可能是所有烟草和烟碱产品中有害成分最少的[3, 54]。然而，“无烟草”一词可能具有误导性，因为许多尼古丁袋的烟碱依然是从烟草中提取的，因此并非完全没有烟草残留物。产品成分、释放物和危害的生物标志物尚未进行独立研究[3, 55]。主要风险因素是烟碱，这是一种根据欧盟REACH法规注册的已知有毒物质[56, 57]。在口服、皮肤或吸入暴露后，它被划分为急性毒性（第2类），并受到危险声明的约束H300：吞咽致命，H310：接触皮肤致命，H330：吸入致命[55]。含烟草口含烟的烟碱剂量越高，从不吸烟的使用者使用时，心率和收缩压的增加就越大[58]。与合成烟碱相关的一个问题是，*R*-烟碱的药理和代谢效应在很大程度上是未知的[6]。从烟草中提取的烟碱可能被烟草特有亚硝胺污染，这些亚硝胺是致癌的。大多数其他成分也用于食品中，因此可以假设通过口服途径相对安全[212]，尽管这还没有得到解决，这些成分应该在尼古丁袋的背景下进行研究，特别是对于局部影响。BAT对尼古丁袋中的金属、醛和烟草特定亚硝胺等有毒物质的研究显示，铬和甲醛在部分而非全部样品中水平较低[59]。

牙科医生已经警告尼古丁袋的有害影响[60, 61]，而烟草行业的一份出版物显示了少量的牙釉质染色[62]。Chaffee等综述了有关口腔和牙周影响的文献，并得出认为其缺乏证据的结论[63]。

一项由BAT进行的体外筛选试验显示，大多数人对参比卷烟提取物有毒理学反应，而口含烟提取物具有最小到中等程度的影响，尼古丁袋提取物在所有检测中几乎没有反应[64]。

4.7 人群效应及相关因素

对于烟草制品使用者来说，尼古丁袋可能被认为是传统卷烟、加热型烟草制品或无烟烟草制品危害较小的替代品，最好完全戒掉烟草和烟碱。然而，从不吸烟或不吸烟碱的人使用尼古丁袋会导致烟碱暴露，这可能会导致上瘾，甚至可能成为使用其他烟碱和烟草产品的途径。不幸的是，关于这些影响的数据有限。来自瑞典Match的数据显示，大多数Zyn使用者是前烟草使用者（43%），只有少数人从未使用过（4%）；大多数人每天都使用Zyn[22]。Zyn最吸引卷烟和无烟烟草双重使用者（76%）、无烟烟草使用者（52%）和吸烟者（36%），而从不吸烟和前烟草使用者表现出的兴趣要少得多，但仍有一定的兴趣（11%~12%）[22]。然而，新的烟碱产品可以迅速被青少年和年轻人接受，例如美国的Juul电子烟[3]。从不吸烟者，特别是年轻人，可以通过多种因素来刺激，如市场营销和设计，各种口味[3]。正如电子烟所报道的那样，发展为使用烟草制品的可能性尚不清楚，但烟碱水平足够高以维持成瘾，通常每袋烟碱≤ 50 mg/g，尽管某些市场上已经报道了烟碱含量高达120 mg/g的尼古丁袋[65, 66]。

尼古丁袋可能会破坏烟草控制政策，如禁止烟草制品（包括传统卷烟）中的调味剂。尼古丁袋可以在不允许吸烟的地方谨慎使用，并可能导致与传统卷烟的双重使用，这将破坏无烟政策的有益影响。阻断非吸烟规则规定的烟碱暴露（例如在工作场所、交通系统、餐馆和酒吧）有助于烟草使用者戒烟。因此，在不允许吸烟的地方持续服用烟碱会加剧成瘾，并降低戒烟的可能性。

另一个担忧是，尼古丁袋模糊了烟碱替代疗法和无烟烟草制品之间的区别[47]，因为一些制造商将这些产品作为烟碱替代疗法或停止使用无烟和吸烟烟草制品的工具进行推广。袋装广告明确或含蓄地承诺了它们在戒烟中的作用："为吸烟者而设计"(On!)，"我可以再次呼吸"(Zyn)和"永不回头"(Zyn)。

一项由Zonnic袋制造商和新西兰国家心脏基金会资助的研究表明，对于吸烟者来说，Zonnic袋在缓解成瘾方面与尼古丁口香糖一样有效，但主观上更有吸引力[67]。总的来说，尼古丁袋还不是经过验证的戒烟工具，目前尚不清楚它们的可用性将如何影响总体戒烟率；可能与已验证的戒烟工具竞争[47]。

4.8 监管和监管机制

根据电子烟情报[68]和WHO成员国的数据，尼古丁袋目前在30多个国家有售，并且市场将在未来几年扩大。在2020~2022年间，世界卫生组织分发了调查问卷，并从各个方面获取信息，以获取各国对尼古丁袋的经验、现有的监管机制和监管中发现的挑战。总共有71个国家提供了关于尼古丁袋的信息，并审查了124项国家法律。大多数提供信息的成员国报告说，尼古丁袋已在2018~2020年间进入他们的市场，许多国家报告说这些产品在本国的销售正成为一个问题。

已经采取了各种管制办法，包括将尼古丁袋作为消费品、毒品、医疗产品或药品、尼古丁袋（属于其本身的类别）、烟碱和烟草产品进行管制。这些分类导致在包括澳大利亚和俄罗斯在内的12个国家禁止使用尼古丁袋，在其他一些国家实施管制，并在其他国家实施现有的烟草控制条例。

虽然WHO审查确定了22个监管尼古丁袋的国家，但这些产品在161个国家似乎不受监管，尽管有一般消费者法适用。很少有烟草控制法律涵盖尼古丁袋，也很少有国家专门管制或禁止尼古丁袋。大多数管制尼古丁袋的国家都是通过非烟草控制法律进行管制的，例如关于药品、毒品、食品和一般消费者保护法。表4.1显示了通过各种方法管制尼古丁袋的示例。

表4.1　各国为监管尼古丁袋而采取的方法实例

监管方法	国家	法律或法规	描述
消费产品	奥地利、保加利亚、克罗地亚、塞浦路斯、多米尼加共和国、希腊、冰岛、卢森堡、马耳他、波兰、葡萄牙	消费者法适用	在奥地利，尼古丁袋被列为“消费产品”和“药品”。只要没有关于戒烟辅助的声明，它们就被归类为“消费产品”。在其他情况下，尼古丁袋被归类为“药品”
食品	德国、荷兰	商品法和欧盟178/2002号条例（EC）第14条、2002年1月28日的议会和理事会（荷兰）	在德国，尼古丁袋被禁止使用，因为它们含有烟碱，这是一种未经授权的新型食物成分。在荷兰，每袋含有≥ 0.035 mg烟碱的尼古丁袋不得再出售或交易，因为它们被归类为有害食品
毒品	文莱达鲁萨兰、爱尔兰	管理毒品法令	在文莱达鲁萨兰国，尼古丁袋被归类为“毒品”和“仿烟草制品”。根据《毒品法》，它们被列为“毒品”；进口和销售毒品需要有许可证。（见下文“仿烟草制品”说明）
药品或药剂制品	奥地利、加拿大、智利、芬兰、匈牙利、日本、马来西亚、南非	加拿大食品和药品法、芬兰药品法（第3节）	在奥地利，如果声明用于戒烟，尼古丁袋被列为“药物”。否则，它们就会被归类为“消费产品”。每剂量提供< 4 mg的尼古丁袋不受处方限制，被管制为“天然保健品”，受加拿大天然保健品条例的约束，在芬兰作为许可的自我药物产品。每剂量提供> 4 mg的药袋被认为是一种处方药，并符合食品和药物法规（加拿大）和药品法（芬兰）的要求。运送药物的药袋被认为是一类医疗设备。尼古丁袋尚未被授权在加拿大销售

续表

监管方法	国家	法律或法规	描述
尼古丁袋、含有烟碱的产品、无烟产品、烟草替代品、仿烟草制品	比利时、文莱达鲁萨兰、爱沙尼亚、新西兰、摩尔多瓦共和国	2005年烟草令（文莱达鲁萨兰国）、2020年无烟环境和受管制产品修正案（新西兰）、2015年修订的烟草和烟草产品法No. 278-XVI（摩尔多瓦共和国）	在比利时，尼古丁袋被归类为“类似烟草产品”。在文莱达鲁萨兰国，尼古丁袋被归类为“毒品”和“仿烟草制品”。尼古丁袋可能被认为是一种“仿烟草制品”，因此被禁止使用。尼古丁袋目前还没有在中国销售。参见“毒品”的注释。在爱沙尼亚，尼古丁袋被认为是“口含烟仿制产品”，并被作为“替代烟草产品”征税。详见表4.2。在新西兰，政府禁止进口销售、包装和分销口服烟碱产品（除非被批准为药品）。通过修改“烟草制品”的定义，避免对立法进行重大改变；相反，口服尼古丁袋被直接禁止，这符合新西兰法律对口含烟和咀嚼烟草的规定。关于摩尔多瓦共和国管制的更多细节见表4.2
烟草制品	美国	联邦法规守则第21篇第8卷	见表4.2

由于这些产品在许多市场上都相对较新，因此它们在一些国家不受监管。在一些国家，目前的烟草控制或其他法律没有适用该产品的措施。在其他国家，这些产品没有特别监管，但适用一般的消费者保护法。一些国家正在探索管控尼古丁袋的方法，比如将它们视为尼古丁口香糖并征收消费税。表4.2列出了在现有烟草控制法律下适用于尼古丁袋的定义及其法律解释。

表4.2　烟草控制法中适用于尼古丁袋的法律定义示例和法律解释

国家	相关法规或法律	相关定义	解释
爱沙尼亚	烟草法案（RT I 2005, 29, 210），于2018年修订	“与烟草制品相关的产品”被定义为“与烟草制品类似的模仿消费烟草制品的产品，以及用于替代烟草制品的制品，包括电子烟、用于吸烟的草药产品、替代水烟的不同材料和无烟草烟，无论这些产品的烟碱产量如何”	该法的规定适用于烟草制品和与烟草制品有关的产品。因此，尼古丁袋被认为是仿烟草制品，因此根据《烟草法》进行管制；禁止广告、向未成年人销售和销售点展示，烟袋作为替代烟草制品征税
摩尔多瓦共和国	2015年修订的烟草和烟草产品法No. 278-XVI	“含烟碱产品”的定义是“通过吸入、摄入或其他方式消费的任何产品，其中烟碱是在生产过程中添加的，或者是消费者在消费前或消费过程中自己添加的”；“烟草相关产品”被定义为“由植物制成的草本烟产品和含有烟碱的产品，包括电子烟”	根据该法律，含有烟碱的产品受到管制，第23a和第23e条特别适用于尼古丁袋，规定“a) 烟碱含量不超过每单位或产品2 mg”和“e) 产品不含第11条第 (3) 款规定的添加剂”。这包括消费者在消费前或消费期间添加的烟碱，以及符合特定定义的烟草相关产品

续表

国家	相关法规或法律	相关定义	解释
俄罗斯联邦/俄罗斯	2020年7月31日联邦法律No. 303-FZ《俄罗斯联邦关于保护公民健康免受食用烟碱产品后果的某些立法行为的修正案》	“含烟碱产品”被定义为“含有烟碱（包括合成烟碱）或其衍生物（包括烟碱盐）的产品，用于以吸吮、咀嚼、鼻吸或吸入的方式提供烟碱供人消费，包括加热型烟草制品，含有至少0.1 mg/mL液体烟碱的溶液、液体或凝胶，用于吸吮、咀嚼、鼻吸和吸入，以及不用于消费的含烟碱的液体、粉末或混合物（根据俄罗斯联邦法律注册的医疗产品、含天然烟碱的食品和烟草产品除外）”	该法律禁止用于咀嚼和吮吸的含有烟碱的产品的批发和零售贸易，有效地禁止了尼古丁袋类产品。这适用于任何形式的烟碱，包括合成烟碱
美国	联邦法规守则第21篇第8卷	烟草制品的定义为“任何为人类消费而制成或源自烟草的产品，包括烟草制品的任何成分、部分或附件（用于制造烟草制品的成分、部件或附件的烟草以外的原材料）”	该法律禁止用于咀嚼和吮吸的含有烟碱的产品的批发和零售贸易，有效地禁止了尼古丁袋类产品。这适用于任何形式的烟碱，包括合成烟碱

约翰·霍普金斯大学烟草控制研究所进行了一项补充但独立的研究，作为其两年一次的调查的一部分，该调查旨在收集 2021 年各国关于烟草和烟碱产品监管的信息[69]。WHO六大地区中，各种收入类别的67个国家提供了信息。政策扫描确定了34个管制尼古丁袋的国家，其中23个国家同时管制烟草来源的烟碱和合成烟碱，而其他11个国家只管制烟草来源的尼古丁袋。38个不管制合成尼古丁袋的国家的代表指出，其立法的措辞是管制合成烟碱的主要障碍。在33个报告尼古丁袋在其市场上销售的国家中，有20个国家实施了监管，而14个国家报告尼古丁袋未在其市场上销售，但仍有监管政策。

4.8.1　监管考虑

有兴趣对尼古丁袋采取监管行动的国家有两种可能的监管途径：禁止或监管。在禁止或管制这些产品时，世界卫生组织《烟草控制框架公约》（WHO FCTC）缔约方在制定或通过有关这些产品的政策时应考虑到其根据WHO FCTC承担的义务。各国还应考虑现有的相关国家法律（关于食品、消费者保护、药品和烟草）、贸易分类和根据产品成分或特征进行分类。由世界海关组织管理的协调商品描述和编码系统是许多国家用来对贸易产品进行分类的标准化系统[70]。这种分类可能会影响国内法律的适用，各国可能会考虑协调系统代码是否适用于尼古丁袋。

在监管尼古丁袋方面，缔约方可根据WHO FCTC第5 (2b)条，采取预防烟碱成瘾的措施。这规定，缔约方应根据其能力，采取并实施有效的立法、行政、管理和/或其他措施，并酌情与其他缔约方合作，制定适当的政策，以预防和减少烟草消费、烟碱成瘾和接触烟草烟气的情况。

各国也可根据WHO FCTC第2.1条禁止这些产品，该条款鼓励缔约方实施超出WHO FCTC要求以外的措施。如果一个国家选择监管而不是实施禁令，它也可以考虑禁止或限制可用于提高这些产品适口性（如调味剂）的成分，如WHO FCTC第9、10条部分实施指南第3.1.2.2段的建议[46]，以减少年轻人或从未使用过的人的摄入。

在监管尼古丁袋时，它们的分类是一个重要的考虑因素，因为它在很大程度上决定了一个产品的监管方式。在一些司法管辖区，如欧盟，烟碱被列为化学品；然而，尼古丁袋没有得到统一的监管，因为它们目前未包括在烟草产品指令中。各国还可以考虑国内法律是否适用于这些产品，包括消费品、食品和烟草控制法。

4.8.2 国家案例研究：荷兰

来自荷兰国家公共卫生与环境研究所（RIVM）的一份报告[14]描述了荷兰市场上销售的烟碱产品，并报告说尼古丁袋正变得越来越受欢迎。该报告称，烟碱会上瘾，特别是在高剂量时对健康有害，例如对神经系统有害，并可能导致心律失常。因此，RIVM建议荷兰卫生、福利和体育部通过实施更严格的法规和组织公共宣传运动来劝阻使用尼古丁袋。尼古丁袋目前属于《商品法》的管辖范围。RIVM考虑了哪些现有立法可以适用于没有烟草的烟碱产品的问题。根据卫生部的说法，这些产品目前不在《烟草法》的范围内，因为它们不含任何烟草。决策者可以考虑将不含烟草的烟碱产品纳入现行法律的监管范围，通过扩大烟草和相关产品的定义。鉴于尼古丁袋的有害和致瘾作用，应防止人们，特别是年轻人开始使用它们。2021年11月9日，荷兰卫生国务大臣宣布，他打算将不含烟草的烟碱产品纳入有关烟草和吸烟产品的法律，并特别禁止尼古丁袋[71]。在此之前，这些产品仍将受《大宗商品法》的管辖。RIVM还提议，每袋含有≥ 0.035 mg烟碱的尼古丁袋应被视为有害食品，此类产品在2021年11月的商品法下被禁止[72]。

4.9 讨　　论

尽管尼古丁袋在许多市场上相对较新，但烟草制造商似乎正在扩大他们的市场，并正在游说政府将尼古丁袋归类为非烟草制品并获得许可证。制造商还在寻求确保对这些产品实施比对传统烟草制品更宽松的法规。一个关键策略是混淆产品类别（即模糊不同产品类别之间的界限），以制造混乱，以便渗透全球市场、实现最大化利润并与监管机构在谈判桌上“占有一席之地”。监管机构有时缺乏关于这些产品造成的危害以及应对它们所带来的挑战的监管选项的信息。一些监管机构面临的挑战之一是一些制造商声称，由于产品不含烟草和/或产品中所含的烟碱不是来自烟草，因此不应根据烟草法对其进行监管。一些制造商还试图绕过卫生部，让其他部委注册产品，以逃避严格的规定。此外，这些产品被定位为传统产品的“低危害性”或“无烟”替代品，并以多种口味销售，这可能会破坏烟草控制政策，如禁止调味剂和无烟法律。

与传统卷烟相比，尼古丁袋中含有的成分和有毒物质要少得多，而且被宣传为“危害更小”。虽然尼古丁袋的风险可能比传统烟草制品小，但制造商不应做出此类声明，除非得到监管机构的证实和批准。政府可以利用其政策和监管框架，根据现有证据决定如何对民众进行教育。区分烟草和烟碱产品以及烟草衍生的烟碱和合成烟碱的法规为讨论某种产品是否“无烟草”提供了可能性，以及是否应该对其进行更宽松的监管。从公共卫生的角度来看，这种区分并不具有效益，因此，监管机构应考虑扩大其监管框架，将非治疗性烟碱产品纳入其中，无论它们是否含有烟草或烟碱是否来自烟草。

虽然关于使用率的数据有限，但现有证据表明，尼古丁袋的营销策略与传统烟草制品的营销策略类似。这些产品在外观上与传统的无烟烟草制品相似，如口含烟，它们含有烟碱，使用方式也类似。这些产品的吸引力（包括口味）表明它们可以通过提高适口性来维持使用。这是一个公共卫生问题，特别是对于年轻人和不吸烟者而言。其中一些产品的烟碱含量可能与传统烟草制品一样高或更高，这表明了对烟碱成瘾的合理担忧。尽管由于这些产品最近的推出，关于这些产品的数据有限，但鉴于它们与传统产品的相似性，有必要采取谨慎的方法。这些产品在网上和由烟草商销售，在许多国家，它们在很大程度上是不受控制或不受限制的，特别是在互联网上，包括欧洲向美国销售的高强度尼古丁袋。其中一些产品很难与传统的无烟烟草制品区分开来[8]。鉴于各国大力推销和使用吸引年轻人的口味，我们鼓励各国酌情保护其现有政策或酌情制定新政策。此外，它们还应扩大其监管要求，以覆盖在世界各地的几个市场上出现的各种烟草和烟碱产品。

我们对国家法律和调查结果的初步分析表明，由于制造商利用了监管空隙，尼古丁袋在多个司法管辖区不受监管或没有专门监管。然而，其他国家此前已使其法规和法律“面向未来”且具有弹性，以确保尼古丁袋受到现有法律的监管。一些国家最近更新了他们的法律，而一些国家仍然使用涵盖传统产品的定义。该行业可能会利用后者，采用战略“获得一席之地”，并将自己作为减少烟草使用的解决方案的一部分，尽管助长了烟碱的广泛使用。然而，一些国家仍将尼古丁袋指定为烟草制品，其他国家可能会考虑采取类似的行动。有意禁止或监管尼古丁袋的世界卫生组织《烟草控制框架公约》缔约方可以利用WHO FCTC的某些条款来保护其公民。一些世界卫生组织欧洲地区国家已对这些产品进行了立法修订，但遭到了烟草制造商的反对。正如《欧盟运作条约》所要求的那样，当务之急是统一对新型烟草制品的监管，以确保对健康的有力保护[8]。规范年轻人对这些产品的获得和推广尤其重要。

4.10 研究空白、优先事项和问题

目前，关于尼古丁袋的信息有限，包括其滥用可能性、伤害、使用者概况和人口影响。此外，由于这些产品在市场上停留的时间很短，目前还没有关于它们的长期依赖性的信息。需要更多的数据，最好是独立于烟草行业的研究：

- 流行率和使用者概况，包括烟草使用状况；
- 尼古丁袋是否可以帮助烟草使用者戒烟；
- 这些产品是否作为卷烟或其他烟碱和烟草产品同时使用（双重使用）；
- 监测产品使用情况，以确保尼古丁袋不会促进非吸烟者，特别是年轻人的烟碱成瘾；
- 这些产品有可能成为传统烟草制品使用和成瘾的入门途径，特别是对年轻人；
- 有提高吸引人的潜力，如风味概况，以及营销等因素对感知和使用的影响；
- 烟碱、调味剂、其他添加剂和污染物的精确含量；
- 烟碱和尼古丁袋中其他物质对健康的短期和长期影响；
- 完全从使用烟草制品转向尼古丁袋对暴露和健康的影响；
- 吸烟者、无烟烟草使用者、从未和曾经使用尼古丁袋者的实际结果。

4.11 关于产品监管和信息传播的政策建议

决策者应采取在许多司法管辖区成功应用于烟草和相关产品的共同监管原则，以便：

- 最小化产品的吸引力和年轻人的接受度；
- 提高产品的安全性；
- 减少错误的健康信念。

根据烟草制品的法律定义或其他相关法律中的定义，各国可以探索利用现有的烟草控制或其他相关法律来管制尼古丁袋。任何决定都应符合该国的国内监管背景，并应确保最大限度地保护其公民的健康，特别是儿童和青少年的健康。

以下是对决策者的建议，特别是对保护年轻人和非使用者的建议：

- 建立或扩大对产品和使用者的监测，包括他们的人口统计数据；使用其他烟草和相关产品；品牌和类型；尼古丁袋中使用的口味，以评估流行率和使用者概况。

- 管制尼古丁袋的所有营销形式，并采取一切必要的行动，以尽量减少年轻人的获取、吸引和接触。
- 告知公众这些尼古丁袋中烟碱相关的毒性和成瘾风险。
- 要求在尼古丁袋的包装上印上健康警示，例如烟碱的影响，这可能包括对使用者的影响，对孕妇及胎儿发育的有害影响，以及对年轻人大脑发育的有害影响，包括对学习的影响。
- 禁止制造商声称与健康有关，包括其作为戒烟产品的潜在功效，除非产品获得监管机构的许可和批准。
- 设定烟碱的限值，以降低产品的成瘾性和无意摄入的危害。
- 保护现有政策，并酌情制定新政策，以扩大监管要求，以覆盖世界各地几个市场上出现的各种烟草和烟碱产品。
- 以与外观、含量和用途相似的产品相同的方式管制尼古丁袋。
- 确保尼古丁袋不被归类为药品，除非它们被证明是烟碱替代疗法，并经过适当的国家监管机构的严格药品注册以获得许可。
- 规范尼古丁袋，以防止所有形式的营销，并采取所有其他必要行动，以减少年轻人的获取、吸引和接触。
- 保护烟草控制活动免受与尼古丁袋有关的所有商业和其他既得利益集团的影响，包括烟草行业的直接和间接利益，并禁止一切形式的营销和促销活动。
- 全面执行世界卫生组织《烟草控制框架公约》第5.3条，以保护政策不受烟草和相关行业的不正当影响。

4.12 结　论

尼古丁袋最近在全球许多市场上都有销售。它们含有足够的烟碱来引发和维持烟碱成瘾，并有许多吸引人的特性，如诱人的味道、包装以及隐蔽的使用方式。与传统烟草制品相比，它们含有的有毒物质更少，因此使用者接触到的有害和潜在有害的成分更少；然而，不建议使用非治疗用途的烟碱和烟草产品以最大限度地保护健康，因为戒烟的好处几乎是立即显现的。使用尼古丁袋会导致接触有毒的烟碱，这可能导致烟碱成瘾，并随后导致使用其他烟碱和烟草产品。在一些司法管辖区，尼古丁袋没有受到监管或没有专门监管，而其他国家已经制定了“面向未来”且具有弹性的法规和法律，因此尼古丁袋受现有法律的监管。其他国家保留的定义只指传统产品。

参考文献

[1] St Claire S, Fayokun R, Commar A, Schotte K, Prasad VM. The World Health Organization's World No Tobacco Day 2020 campaign exposes tobacco and related industry tactics to manipulate children and young people and hook a new generation of users. J Adolesc Health. 2020;67:334-7. doi:10.1016/j.jadohealth.2020.06.026.

[2] WHO study group on tobacco product regulation. Report on the scientific basis of tobacco product regulation: eighth report of a WHO study group (WHO Technical Report Series, No. 1029). Geneva: World Health Organization;2021. (https://www.who.int/publications/i/item/9789240022720).

[3] Robichaud MO, Seidenberg AB, Byron MJ. Tobacco companies introduce "tobacco-free" nicotine pouches. Tob Control. 2019;29(E1):e145-6. doi:10.1136/tobaccocontrol-2019-055321.

[4] The rise of products using synthetic nicotine. Washington DC: Campaign for Tobacco-Free Kids; 2022 (https://www.tobaccofreekids.org/assets/factsheets/0420.pdf).

[5] Stanfill S, Tran H, Tyx R, Fernandez C, Zhu W, Marynak K et al. Characterization of total and unprotonated (free) nicotine content of nicotine pouch products. Nicotine Tob Res. 2021;23:1590-6. doi:10.1093/ntr/ntab030.

[6] Jordt SE. Synthetic nicotine has arrived. Tob Control. 2023;32(e1):e113-7. doi:10.1136/tobaccocontrol-2021-056626.

[7] Requirements for products made with non-tobacco nicotine take effect April 14. Silver Spring (MD): Food and Drug Administration. 2022 (https://www.fda.gov/tobacco-products/ctp-newsroom/requirements-products-made-non-tobacco-nicotine-take-effect-april-14).

[8] Salokannel M, Ollila E. Snus and snus-like nicotine products moving across Nordic borders: Can laws protect young people? Nordic Stud Alcohol Drugs. 2021;38(6):540-4. doi:10.1177/145507252199570.

[9] Market sizes. London: Euromonitor International; 2022.

[10] Chapman M, Okoth E, Törnkvist A, Margottini L, Irfan A, Cheema U. New products, old tricks? Concerns big tobacco is targeting youngsters. London: The Bureau of Investigative Journalism, 21 February 2021 (https://www.thebureauinvestigates.com/stories/2021-02-21/new products-old-tricks-concerns-big-tobacco-is-targeting-youngsters).

[11] Pouches & gums. SRITA collection of pouch advertisements. Palo Alto (CA): Stanford Research into the Impact of Tobacco Advertising; 2023. (https://tobacco.stanford.edu/pouches_ gums/).

[12] Burki TK. Petition to ban nicotine pouches in Kenya. Lancet Oncol. 2021;22:756. doi:10.1016/S1470-2045(21)00267-9.

[13] Put the used pouch in the garbage. Stockholm: Swedish Match, 21 November 2016 (https://web.archive.org/web/20210617145948/https://www.swedishmatch.com/Media/Pressreleases-and-news/News/put-the-used-pouch-in-the-garbage/).

[14] Pauwels CGGM, Bakker-'t Hart IME, Hegger I, Bil W, Bos PMJ, Talhout R. Nicotineproducten zonder tabak voor recreatief gebruik [Nicotine products without tobacco for recreational use]. In. Bilthoven: National Institute for Public Health and the Environment; 2021 (https://open.

overheid.nl/documenten/ronl-fa15b269-76a3-466d-a513-fe457e3cf6f4/pdf).
[15] GrantSnus. Berlin: Kordula UAB. (https://grantsnus.com/).
[16] Lunell E, Fagerström K, Hughes J, Pendrill R. Pharmacokinetic comparison of a novel non-tobacco-based nicotine pouch (ZYN) with conventional, tobacco-based Swedish snus and American moist snuff. Nicotine Tob Res. 2020;22:1757-63. doi:10.1093/ntr/ntaa068.
[17] Lawler TS, Stanfill SB, Tran HT, Lee GE, Chen PX, Kimbrell JB et al. Chemical analysis of snus products from the United States and northern Europe. PLoS One. 2020;15:e0227837. doi:10.1371/ journal.pone.0227837.
[18] Delnevo CD, Hrywna M, Miller Lo EJ, Wackowski OA. Examining market trends in smokeless tobacco sales in the United States: 2011–2019. Nicotine Tob Res. 2021;23:1420-4. doi:10.1093/ntr/ntaa239.
[19] Marynak KL, Wang X, Borowiecki M, Kim Y, Tynan MA, Emery S. Nicotine pouch unit sales in the US, 2016-2020. JAMA. 2021;326:566-8. doi:10.1001/jama.2021.10366.
[20] Market report. Barcelona: Tobacco Intelligence; 2022.
[21] Market development. Smokefree products in the US. Stockholm: 瑞典火柴; 2021 (https://web.archive.org/web/20210617151128/https://www.swedishmatch.com/Our-business/smokefree/Market-development/smokefree-products-in-the-us/).
[22] Plurphanswat N, Hughes JR, Fagerstrom K, Rodu B. Initial information on a novel nicotine product. Am J Addict. 2020;29:279-86. doi:10.1111/ajad.13020.
[23] Q&A. Amsterdam: Snussie.com; 2019 (https://web.archive.org/web/20210118194301/https://www.snussie.com/en/service/qanda/).
[24] How many nicotine pouches per day? London: Haypp.com; 2020 (https://www.haypp.com/uk/nicopedia/how-many-nicotine-pouches-per-day/).
[25] Modern and traditional oral products. London: British American Tobacco; 2022 (https://web.archive.org/web/20210614154017/https://www.bat.com/snus).
[26] Brand portfolio. Bristol: Imperial Brands; undated (https://web.archive.org/web/20210617202521/https://www.imperialbrandsplc.com/about-us/brand-portfolio.html).
[27] Tobacco-free innovations ushers in new era SKRUF. Bristol: Imperial Brands; 2018 (https://web.archive.org/web/20191118113541/https:/www.imperialbrandsplc.com/sustainabil-ity/case-studies/reduced harm-ngps/tobacco-free-innovation-ushers-in-new-era-skruf.html).
[28] JTI Sweden and Nordic snus. Stockholm: Japan Tobacco International; undated (https://web.archive.org/web/20210617153749/https://www.jti.com/europe/sweden).
[29] Kôp all white och nikotinfri portion [Buy all white and nicotine-free portion]. Göteborg: Microzero AB; undated (https://web.archive.org/web/20210303173503/https://nixs.se/).
[30] FTC report finds annual cigarette salesincreased for the first time in 20 years. Washington DC: Federal Trade Commission; 2021 (https://www.ftc.gov/news-events/press-releases/2021/10/ftc-report-finds-annual-cigarette-sales-increased-first-time-20).
[31] Havermans A, Pennings JLA, Hegger I, Elling JM, de Vries H, Pauwels CGGM et al. Awareness, use and perceptions of cigarillos, heated tobacco products and nicotine pouches: A survey among Dutch adolescents and adults. Drug Alcohol Depend. 2021;229(Pt_B):109136. doi:10.1016/j.drugalcdep.2021.109136.
[32] Pillitteri JL, Shiffman S, Sembower MA, Polster MR, Curtin GM. Assessing comprehension and

perceptions of modified-risk information for snus among adult current cigarette smokers, former tobacco users, and never tobacco users. Addict Behav Rep. 2020;11:100254. doi:10.1016/j.abrep.2020.100254.

[33] Wackowski OA, Rashid M, Greene KL, Lewis MJ, O'Connor RJ. Smokers' and young adult non-smokers' perceptions and perceived impact of snus and e-cigarette modified risk messages. Int J Environ Res Public Health. 2020;17(18):6807. doi:10.3390/ijerph17186807.

[34] Everything about nicotine pouches and the difference between the tobacco pouch snus. London: Haypp.com; undated (https://web.archive.org/web/20210617203157/https://www.haypp.com/eu/whats-a-nicotine-pouch).

[35] Nicotine pouches. Stockholm: Zyn.com (https://web.archive.org/web/20210616130922/https://www.zyn.com/international/en/).

[36] Talbot EM, Giovenco DP, Grana R, Hrywna M, Ganz O. Cross-promotion of nicotine pouches by leading cigarette brands. Tob Control. 2021. doi:10.1136/tobaccocontrol-2021-056899.

[37] Czaplicki L, Patel M, Rahman B, Yoon S, Schillo B, Rose SW. Oral nicotine marketing claims in direct-mail advertising. Tob Control. 2021;31(5):663-66. doi:10.1136/tobaccocontrol-2020-056446.

[38] Dryft. Moorpark (CA): Kretek International Inc.; 2023 (https://www.kretek.com/company/brands/dryft.)

[39] The history of the tobacco pouch and nicotine pouches. London: Haypp; 2021 (https://web.archive.org/web/20210617203006/https://www.haypp.com/eu/the-history-of-nicotine-pouches).

[40] Nicotine pouches, a viable alternative to smoking? London: Euromonitor International; 2020 (https://www.euromonitor.com/nicotine-pouches-a-viable-alternative-to-smoking-/report).

[41] Li L, Borland R, Cummings KM, Gravely S, Quah ACK, Fong GT et al. Patterns of non-cigarette tobacco and nicotine use among current cigarette smokers and recent quitters: findings from the 2020 ITC Four Country Smoking and Vaping Survey. Nicotine Tob Res. 2021;23:1611-6. doi:10.1093/ntr/ntab040.

[42] Brose LS, McDermott MS, McNeill A. Heated tobacco products and nicotine pouches: a survey of people with experience of smoking and/or vaping in the UK. Int J Environ Res Public Health. 2021;18(16):8852. doi:10.3390/ijerph18168852.

[43] East KA, Reid JL, Rynard VL, Hammond D. Trends and patterns of tobacco and nicotine product use among youth in Canada, England, and the United States from 2017 to 2019. J Adolesc Health. 2021;69:447-56. doi:10.1016/j.jadohealth.2021.02.011.

[44] Gentzke AS, Wang TW, Cornelius M, Park-Lee E, Ren C, Sawdey MD et al. Tobacco product use and associated factors among middle and high school students-National Youth Tobacco Survey, United States, 2021. MMWR Surveill Summ. 2022;71(5):1-29. doi:10.15585/mmwr.ss7105a1.

[45] Hrywna M, Gonsalves NJ, Delnevo CD, Wackowski OA. Nicotine pouch product awareness, interest and ever use among US adults who smoke, 2021. Tob Control. 2022. doi:10.1136/tobaccocontrol-2021-057156.

[46] Partial guidelines for implementation of Articles 9 and 10 of the WHO Framework Convention on Tobacco Control. Geneva: World Health Organization; 2010 (https://www.who.int/fctc/guidelines/Guideliness_Articles_9_10_rev_240613.pdf?ua=1).

[47] Kostygina G, England L, Ling P. New product marketing blurs the line between nicotine re-

placement therapy and smokeless tobacco products. Am J Public Health. 2016;106:1219-22. doi:10.2105/AJPH.2016.303057.

[48] Aldeek F, McCutcheon N, Smith C, Miller JH, Danielson TL. Dissolution testing of nicotine release from OTDN pouches: product characterization and product-to-product comparison. Separations. 2021;8(1):7. doi:10.3390/separations8010007.

[49] Rensch J, Liu J, Wang J, Vansickel A, Edmiston J, Sarkar M. Nicotine pharmacokinetics and subjective response among adult smokers using different flavors of on!® nicotine pouches compared to combustible cigarettes. Psychopharmacology (Berl). 2021;238(11):3325-34. doi:10.1007/s00213-021-05948-y.

[50] Hukkanen J, Jacob P 3rd, Benowitz NL. Metabolism and disposition kinetics of nicotine. Pharmacol Rev. 2005;57:79-115. doi:10.1124/pr.57.1.3.

[51] Digard H, Proctor C, Kulasekaran A, Malmqvist U, Richter A. Determination of nicotine absorption from multiple tobacco products and nicotine gum. Nicotine Tob Res. 2013;15:255-61. doi:10.1093/ntr/nts123.

[52] Benowitz NL. Clinical pharmacology of nicotine: implications for understanding, preventing, and treating tobacco addiction. Clin Pharmacol Ther. 2008;83:531-41. doi:10.1038/clpt.2008.3.

[53] Pickworth WB, Rosenberry ZR, Gold W, Koszowski B. Nicotine absorption from smokeless tobacco modified to adjust pH. J Addict Res Ther. 2014;5:1000184. doi:10.4172/2155-6105.1000184.

[54] Palmer AM, Toll BA, Carpenter MJ, Donny EC, Hatsukami DK, Rojewski AM et al. Reappraising choice in addiction: novel conceptualizations and treatments for tobacco use disorder. Nicotine Tob Res. 2022;24(1):3-9. doi:10.1093/ntr/ntab148.

[55] A summary of data on the bioavailability of nicotine and other ingredients from the use of oral nicotine pouches and assessment of risk to users. London: HM Government, Committee on Toxicity of Chemicals in Food, Consumer Products and the Environment. 2021 (https://cot.food.gov.uk/sites/default/files/2021-05/TOX-2021-22%20Nicotine%20pouches.pdf).

[56] Price LR, Martinez J. Cardiovascular, carcinogenic and reproductive effects of nicotine exposure: A narrative review of the scientific literature. F1000Res. 2019;8:1586. doi:10.12688/f1000research.20062.2

[57] REACH (Registration, Evaluation, Authorisation and Restriction of Chemicals) (EC 1907/2006). Brussels: European Commission, Environment; 2006 (https://ec.europa.eu/environment/chemicals/reach/reach_en.htm).

[58] Ozga JE, Felicione NJ, Elswick D, Blank MD. Acute effects of snus in never-tobacco users: a pilot study. Am J Drug Alcohol Abuse. 2018;44:113-9. doi:10.1080/00952990.2016.1260581.

[59] Azzopardi D, Liu C, Murphy J. Chemical characterization of tobacco-free "modern" oral nicotine pouches and their position on the toxicant and risk continuums. Drug Chem Toxicol. 2022;45(5):2246-54. doi:10.1080/01480545.2021.1925691.

[60] Keogh A. Nicotine pouches. Br Dent J. 2021;230:61-2. doi:10.1038/s41415-021-2622-y.

[61] Sahni V. Blurred line on nicotine. Br Dent J. 2021;230:325. doi:10.1038/s41415-021-2855-9.

[62] Dalrymple A, Bean EJ, Badrock TC, Weidman RA, Thissen J, Coburn S et al. Enamel staining with e-cigarettes, tobacco heating products and modern oral nicotine products compared with cigarettes and snus: an in vitro study. Am J Dent. 2021;34:3-9. PMID:33544982.

[63] Chaffee BW, Couch ET, Vora MV, Holliday RS. Oral and periodontal implications of tobacco and nicotine products. Periodontology. 2000;87(1):241-53. doi:10.1111/prd.12395.

[64] Bishop E, East N, Bozhilova S, Santopietro S, Smart D, Taylor M et al. An approach for the extract generation and toxicological assessment of tobacco-free "modern" oral nicotine pouches. Food Chem Toxicol. 2020;145:111713. doi:10.1016/j.fct.2020.111713.

[65] Khouja JN, Suddell SF, Peters SE, Taylor AE, Munafò MR. Is e-cigarette use in non-smoking young adults associated with later smoking? A systematic review and meta-analysis. Tob Control. 2020;30(1):8-15. doi:10.1136/tobaccocontrol-2019-055433.

[66] Yoong SL, Hall A, Turon H, Stockings E, Leonard A, Grady A et al. Association between electronic nicotine delivery systems and electronic non-nicotine delivery systems with initiation of tobacco use in individuals aged < 20 years. A systematic review and meta-analysis. PLoS One. 2021;16(9):e0256044. doi:10.1371/journal.pone.0256044.

[67] Thornley S, McRobbie H, Lin RB, Bullen C, Hajek P, Laugesen M et al. A single-blind, randomized, crossover trial of the effects of a nicotine pouch on the relief of tobacco withdrawal symptoms and user satisfaction. Nicotine Tob Res. 2009;11:715-21. doi:10.1093/ntr/ntp054.

[68] ECigIntelligence.com.

[69] Duren M, Atella L, Welding K, Kennedy RD. Nicotine pouches: a summary of regulatory approaches across 67 countries. Tob Control. 2023. doi:10.1136/tc-2022-057734.

[70] Harmonized Commodity Description and Coding System. Brussels: World Customs Organization; 2023 (https://www.wcoomd.org/en/topics/nomenclature/overview.aspx#:~:text=What%20is%20the%20Harmonized%20System,World%20Customs%20Organization%20 (WCO)).

[71] Kamerbrief tabak en alcohol [Parliamentary brief tobacco and alcohol]. The Hague: State Secretary of Health, Ministry of Health, Well-being and Sport; 2021 (https://www.rijksoverheid. nl/documenten/kamerstukken/2021/11/09/verzamelbrief-tabak-en-alcohol).

[72] Beoordeling van het nicotinegehalte in nicotinezakjes waarbij de Acute reference Dose niet overschreden wordt [Assessment of the nicotine in nicotine pouches not exceeding the acute reference dose]. Bilthoven: National Institute for Public Health and the Environment; 2021 (https://www.rivm.nl/sites/default/files/2021-11/FO_nicotinezakjes%20tox_20211101_def_anon.pdf).

5. 电子烟碱传输系统和加热型烟草制品的暴露、效应和易感性的生物标志物及其优先次序评估

Irina Stepanov, McKnight Distinguished University Professor and Director, Institute for Global Cancer Prevention Research, University of Minnesota, Minneapolis (MN), USA

Stephen S. Hecht, PhD, Wallin Professor of Cancer Prevention, American Cancer Society Professor, Masonic Cancer Center, University of Minnesota, Minneapolis (MN), USA

摘　要

生物标志物已广泛应用于卷烟和其他传统烟草制品的研究，为吸烟者和非吸烟者（暴露于二手烟的人群）在有害暴露、生物效应以及疾病易感性方面提供了有价值的数据。本报告评估了已发表的文献，这些文献涉及在研究电子烟碱传输系统（ENDS）和加热型烟草制品（HTP）中使用此类生物标志物的情况，并对生物标志物在烟草控制中的潜在效用和局限性进行了分析。审查证据表明，从吸食传统卷烟转向专门使用 ENDS，与减少多种在吸烟引发疾病中起关键作用的有害成分和致癌物质的生物标志物水平相关。然而，在双重使用者（同时使用卷烟和 ENDS 的人群）中，这些生物标志物的水平往往更高，而这种双重使用的情况比完全转为使用 ENDS 更为普遍。此外，这种暴露变化对健康的影响尚不完全明确，而生物效应标志物表明，ENDS 对使用者，尤其是双重使用者，与完全不使用任何烟草和烟碱产品相比，存在一定的风险。对已发表文献的综述进一步强调，关于 HTP 使用的暴露和影响，目前缺乏独立且非行业资助的研究。本报告提出了一组针对烟草控制的优先生物标志物，明确了相关的研究空白，指出了行业独立研究的必要性，并建议了重点监管领域。

关键词：生物标志物，暴露，生物效应，毒性，电子烟，电子烟碱传输系统，加热型烟草制品，健康效应

5.1 背　景

本章的任务是针对21世纪出现的烟草和烟碱产品（特别是电子烟碱传输系统（ENDS）和加热型烟草制品（HTP）)，提供基于证据的建议，并提出政策选项，以实现相关决策（FCTC/COP8(22)）中概述的目标和措施。本章基于世界卫生组织烟草制品管制研究小组第九份技术报告的背景资料。

生物标志物是客观评估人体暴露于烟草制品、ENDS和HTP中的化学有害物质及致癌物，以及由此引发的与疾病相关的生物效应的强大工具。这种客观评估至关重要，因为仅通过产品分析无法充分预测成分的摄入量，而使用者的行为对摄入量有显著影响[1, 2]。此外，与烟草和烟碱产品相关的慢性疾病[如癌症、慢性阻塞性肺病（COPD）和心血管疾病]需要较长时间才能发展，虽然对这些疾病的监测对于长期烟草控制政策至关重要，但在新产品进入市场时，不适合用于作出监管决策。因此，生物标志物可以作为评估此类健康风险的替代指标。然而，生物标志物在监管中的应用仍然有限。

其中一个原因是，在资源有限的国家，必须有效优先分配资源，以减少烟草使用带来的公众健康危害。在许多情况下，测量生物标志物的活动优先级低于其他工作。此外，由于产品间的差异以及使用者行为的不同，区分生物标志物水平差异的挑战，导致 TobReg 在2008年得出结论，认为测量生物标志物并不是监测卷烟产品差异的适当监管策略[3]。

然而，在过去的15年中，大量新研究采用了新型生物标志物、新技术和新证据。此外，制造商持续向全球市场推出新型烟草和烟碱产品，其中一些产品的化学成分与传统产品（如卷烟）以及其他较新的产品显著不同。例如，HTP和最近出现的无烟碱烟袋在化学成分和使用方式上与电子烟有很大差异。因此，关于电子烟健康影响的知识无法直接指导对这类新兴产品的监管决策。因此，有必要在当前产品环境的背景下总结生物标志物研究的进展，以重新评估其在烟草控制中作为替代指标的潜在用途。

本章我们回顾了目前关于暴露生物标志物、生物效应标志物（包括与特定疾病相关的生物标志物）和易感性生物标志物的文献，这些生物标志物已被用于电子烟和其他ENDS和HTP的研究。具体来说，它包括：

- ENDS和HTP的暴露生物标志物；
- 各种相关疾病病理生理学的生物效应标志物；
- 易感性生物标志物；
- 讨论生物标志物的研究现状及其对烟草控制的影响，包括生物标志物的研究空白和局限性；

- 关于烟草控制可能的生物标志物优先次序的建议；
- 关于解决研究空白和优先事项的建议；
- 相关政策建议。

文献检索主要在PubMed数据库和SciFinder检索工具中进行，SciFinder检索工具从Medline和CAplus数据库检索数据。在数据库研究中获得的出版物引用的重要相关文章也包括在内。此外，我们还使用了美国疾病控制和预防中心、美国食品药品监督管理局的网站以及其他有关网站，这些网站载有与ENDS和HTPs有关的暴露和影响的信息。

5.2 暴露生物标志物

5.2.1 烟草和烟碱产品研究中常用暴露生物标志物的定义和概述

根据牛津词典定义，生物标志物是一种可测量的物质，其存在表明某些现象的发生，例如环境暴露。在这一广义定义中，暴露生物标志物指与特定暴露相关且能够可靠量化的实体。在本报告的背景下，暴露生物标志物可用于确认是否使用或接触了特定的烟草和烟碱产品，或在个体切换产品时指示特定化学化合物暴露的变化。本报告中讨论的生物标志物的结构及其来源如图5.1所示。

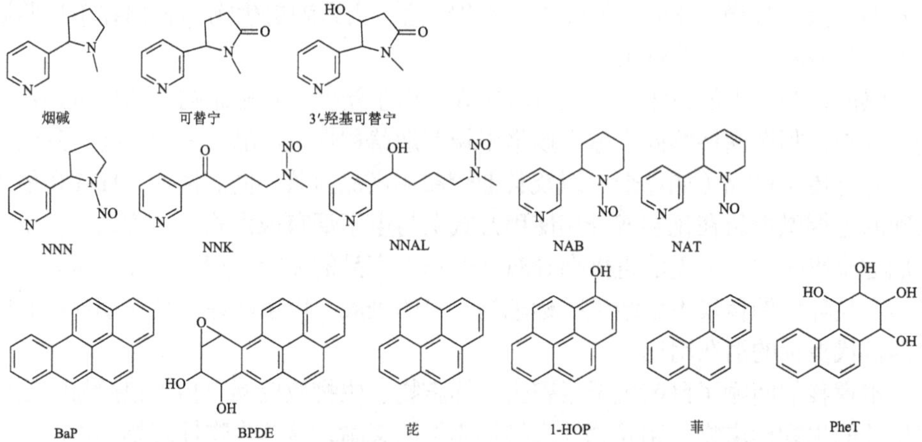

烟碱 可替宁 3'-羟基可替宁

NNN NNK NNAL NAB NAT

BaP BPDE 芘 1-HOP 菲 PheT

图5.1 本报告讨论的一些成分和生物标志物的结构

一氧化碳（CO）

一氧化碳是有机物不完全燃烧的产物。呼出的一氧化碳是一种有用且广泛应用的生物标志物，用于衡量涉及燃烧的所有烟草制品的使用情况，这些制品包括卷烟、雪茄、烟斗和水烟，也可能包括HTP，因为有证据表明使用这些产品时可能存在一定程度的燃烧。吸食大麻也会增加呼出CO的水平。在不涉及燃烧的情况下，使用ENDS、HTP或无烟烟草不会产生显著的一氧化碳量。一氧化碳暴露与血液中的碳氧血红蛋白（COHb）水平相关，但更常用的测量方法是呼出气体中的一氧化碳，因为这可以通过市售设备轻松测试。由于高水平环境污染等其他因素也会影响测量结果，因此提出了不同的呼出CO截止值以区分吸烟者和非吸烟者。建议以5~6 ppm的呼出气CO为分界点来区分吸烟者和非吸烟者[4]。一氧化碳通过迅速与血液中的血红蛋白结合，降低血红蛋白的携氧能力，从而可能引发多种健康问题。对于心血管疾病或肺部潜在疾病患者，这种影响尤为显著。雪茄抽吸是CO暴露的一个特别重要的来源[5]。

烟碱及其代谢物

对烟碱的成瘾是人们持续使用能够有效传递这种物质的产品的主要原因，尽管已知使用烟草制品对健康有害。所有烟草和烟碱产品都会递送烟碱，其药代动力学特性各不相同，从而使烟碱与大脑中的烟碱乙酰胆碱受体结合并释放多巴胺，产生愉悦感受[4, 6]。从吸入烟草烟气到释放多巴胺的时间仅需几秒钟，这有助于解释烟碱成瘾现象[4, 6]。由于烟碱在体内的半衰期仅约为2小时，因此其作为烟碱暴露的定量生物标志物的作用有限。烟碱的主要代谢物——可替宁，因其较长的半衰期（约16小时，范围8~30小时，取决于个体特征）已被广泛用于作为烟碱摄入的生物标志物。因此，可替宁可在血清、血浆、全血、唾液和尿液中量化，用于评估烟碱摄入情况。尽管可替宁是烟碱暴露的一个良好标志物，但其形成及进一步代谢所涉及的酶（包括CYP2A6和UGT2B10）的个体差异会影响其测量结果。因此，烟碱暴露的金标准生物标志物是“总烟碱当量”，包括尿液中的烟碱、可替宁和3′-羟基可替宁及其葡糖醛酸结合物。此生物标志物与尿液代谢物测量结果密切相关，不仅包括这些化合物，还包括几种次要的烟碱代谢物，如烟碱*N*-氧化物[4, 6]。

烟草特有亚硝胺及其代谢物

烟草特有亚硝胺是烟草在烘烤和加工过程中，由烟碱、降烟碱、假木贼碱、新烟草碱等烟草生物碱与烟草中的亚硝酸盐反应生成的一类致癌物[7-10]。所有含烟草的产品都含有烟草特有亚硝胺，包括 *N*-亚硝基降烟碱（NNN）、*N*-亚硝基假木贼碱、*N*-亚硝基新烟草碱和4-(甲基亚硝胺)-1-(3-吡啶基)-1-丁酮（NNK），以及一些次要产物[11]。如下文所述，加热型烟草制品（HTP）释放物中的烟草特有亚硝胺含量低于传统卷烟烟气，而在电子烟碱传输系统（ENDS）释放物中通常未检出或含量极低。顾名思义，这类致癌物是烟草制品所特有的，包括卷烟和无烟烟草制品[12]。NNN和NNK是强致癌物，在实验动物中可在相关组织（如口腔黏膜、食道和肺）诱发肿瘤[9]。通过长期低剂量暴露处理动物时也观察到肿瘤发生[13, 14]。研究明确表明，烟草使用者能够吸收 NNN 和 NNK [11]。因此，NNN和NNK 被广泛认为是使用无烟烟草制品或抽吸卷烟的人群中癌症的重要原因；国际癌症研究机构（IARC）已将其列为"对人体致癌"[11]。NNK 在实验动物和人体内代谢为 4-(甲基亚硝胺)-1-(3-吡啶基)-1-丁醇（NNAL），其致癌活性与 NNK 类似[9]。尿液中的 NNAL 已被广泛用作 NNK 暴露的生物标志物[10]。由于其烟草特异性和致癌活性，NNAL 成为评估烟草致癌物暴露和癌症风险的重要指标。对吸烟者的前瞻性流行病学研究表明，尿液中较高的 NNAL 水平与肺癌风险显著相关[15]。同样，尿液中的 NNN 已被用作 NNN 暴露和致癌性的生物标志物，在一项对吸烟者的前瞻性研究中，尿液中的 NNN 水平与食管癌的发病率显著相关[16]。因此，在使用烟草制品的人群中，NNAL 和 NNN 被认为是与癌症风险相关的潜在有用生物标志物，可用于预测癌症风险；然而，还需要进一步研究。在电子烟使用者的唾液中也检测到了 NNN，这是烟碱和/或其代谢物降烟碱在口腔内内源性形成的[17]。

多环芳烃（PAH）代谢物

多环芳烃与一氧化碳类似，是有机物不完全燃烧的产物。因此，在卷烟、雪茄、烟斗和水烟的烟气中都可发现多环芳烃混合物，而在不含烤烟的无烟烟草中的含量低得多，其存在部分归因于环境污染[18-20]。同样，ENDS和HTP释放物中的多环芳烃水平始终低于传统燃烧型烟草制品，或者根本检测不到[21]。自20世纪70年代以来，多环芳烃被认为在烟草烟气致癌中起重要作用，这一结论来自多种动物模型研究，分析了烟草烟气冷凝物中不同亚组分和单个化合物的致癌活性，包括苯并[*a*]芘（BaP）、䓛、甲基䓛、苯并芴和苯并[*a*]蒽等[22]。在烟草烟气中，至少部分鉴定出超过500种多环芳烃，而苯并[*a*]芘（BaP）作为代表性多环芳烃，被国际癌症研究机构（IARC）分类为"对人体致癌"[18, 23]。1-羟基芘（1-HOP）和羟基菲是非致癌性多环芳烃芘和菲的尿液代谢物，作为所有多环芳烃混合物的成分，已被广泛用作多环芳烃暴露的生物标志物[14]。基于人群的美国国家健康与营养调查（NHANES）显示，美国吸烟者的这些代谢物水平显著高于不吸烟者，后者的暴露来源于吸入污染空气或食用炭烤食物[24]。研究一致表明，吸烟是多环芳烃暴露的主要来源。

多环芳烃通过形成二醇环氧化物的代谢途径引发致癌作用[25]。这一关键代谢途径可通过尿液中苯并[*a*]芘四醇（BaP-tetraol）的分析来定量，苯并[*a*]芘四醇是苯并[*a*]芘二醇环氧化物代谢的最终产物。然而，更实际的方法是使用菲四醇（PheT），因为尿液中其浓度比苯并[*a*]芘四醇高出1000倍以上[10, 26]。在上海队列研究中，吸烟者体内的PheT水平与肺癌显著相关[15]。毫无疑问，多环芳烃增加了吸烟者患癌症的风险，但其在特定癌症病因中的相对作用，与本报告讨论的其他有害成分和致癌物相比，目前尚不明确。

挥发性有害成分和致癌物及其代谢物

烟草燃烧会产生大量挥发性有害成分和致癌物。这些包括国际癌症研究机构（IARC）第1类（对人体致癌）化合物（如甲醛、环氧乙烷、苯和1,3-丁二烯），第2A类（很可能对人体致癌）化合物（如丙烯醛、丙烯酰胺、二甲基甲酰胺和苯乙烯），以及第2B类（可能对人体致癌）化合物（如环氧丙烷、丙烯腈、巴豆醛和乙苯等）[27]。其他挥发性化合物，如甲基丙烯醛和甲基乙烯酮，也被证实具有与丙烯醛类似的有害作用[28-30]。丙烯醛被认为是烟草烟气中毒性最强的化合物之一，其非癌症危害指数在常见烟气成分中最高[31]。丙烯醛及相关化合物与慢性阻塞性肺病（COPD）的病因密切相关[32]。这些化合物或其代谢物（主要为巯基尿酸）普遍存在于人类的血液或尿液中，这可能源于内源性过程、炎症反应以及环境或膳食暴露。然而，与非吸烟者相比，吸烟（包括卷烟、雪茄、烟斗、水烟和大麻）会显著提高这些化合物的水平[33-46]。氰乙基巯基尿酸（CEMA）是丙烯腈的代谢物，它不是一种内源性化合物，除烟草烟气外，在一般环境中很少大量存在。CEMA是一种特别有用的生物标志物，可区分吸烟者和非吸烟者。研究表明，尿液中CEMA水平的分界点为27 pmol/mL时，其区分吸烟者和非吸烟者的敏感性和特异性均超过99%[47]。电子烟碱传输系统（ENDS）和加热型烟草制品（HTP）也会产生挥发性有害成分和致癌物，但其含量通常远低于传统卷烟[48, 49]。

金属

据报道，烟草中含有多种金属，包括砷、铍、镉、六价铬、钴、铅、镍和放射性钋[20]。在卷烟烟气中，颗粒物中镉和铅的平均浓度最高，分别为（40.2±5.4）ng/支和（11.0±1.1）ng/支（ISO抽吸方案）[50]。这些结果与其他研究一致[20]。NHANES研究显示，吸烟者的血液和尿液中的镉水平以及血铅水平显著高于非吸烟者[51]。德国人群的研究中也发现了类似结果[52]。镉及其化合物对人类具有致癌性，可导致肺癌，可能还与肾癌和前列腺癌相关[53]。铅对神经、肾脏、心血管、血液、免疫、生殖和发育系统均有毒害作用[54]。

一些研究报告显示，电子烟气溶胶中存在金属成分。例如，铬和铅在电子烟气溶胶中可被可靠测量[55, 56]。其他研究报告显示镉、铜、镍、锰、铝和锡的存在，并表明产品设计对金属含量有重要影响[57, 58]。由于镉和铅在人体内的停留时间较长，因此它们可以作为累积暴露的长期标志物[59]。

5.2.2 暴露生物标志物在ENDS和HTP研究中的应用

一氧化碳（CO）

随机临床试验和横断面研究等已对呼出CO或血液COHb进行了定量分析，结果显示，ENDS使用者的水平显著低于吸烟者，大多数研究未发现专门使用ENDS的人的CO或COHb水平升高（参见文献[21]和[60]的综述）。例如，Oliveri等[61]报告称，专门使用ENDS的成年人的COHb浓度比吸烟者低47%。Hatsukami等[62]观察到，当吸烟者在8周内改为使用ENDS时，呼出CO显著减少60%，尽管并非所有研究参与者都完全改用ENDS。McRobbie等[63]报告称，吸烟者在4周内改用ENDS后，呼出CO显著减少80%。Czoli等[64]发现，当受试者从7天的卷烟和ENDS混合使用改为7天的专用ENDS时，呼出CO显著减少41%。O'Connell等[65]发现，吸烟者在5天内改为使用ENDS后，呼出CO显著减少88%~89%，而COHb减少84%~86%。Cravo等[66]报告称，受试者从传统卷烟转为ENDS后，呼出CO和血液中的COHb含量迅速下降。在使用ENDS 1周后，呼出CO从20.3 ppm降至7.4 ppm，并在第2周至研究结束（12周）期间维持在7.6~9.0 ppm。COHb在使用ENDS 1周后从6.79%降至4.06%~4.37%，并一直维持到研究结束。Morris等[67]报告称，受试者从传统卷烟改用ENDS 9~14天后，COHb减少了79%。

Akiyama和Sherwood[60]（烟草行业研究人员）的一篇综述提供了吸烟后以及在30项加热型烟草制品（HTP）临床试验中的生物标志物对比结果，干预期的中位时间为8天。在大多数研究中，1周内呼出CO减少了80%~90%，而COHb减少了50%~90%。这些结果与HTP在加热过程中燃烧显著减少相符。例如，在一项研究中，受试者从传统卷烟改为HTP后，6~7天内呼出CO减少约80%，达到接近戒烟后的水平[68]。在薄荷醇HTP与薄荷醇卷烟的比较中，从吸烟转换为薄荷醇HTP的5天内，平均COHb降低62%，与戒烟5天后的结果相似[69]。

烟碱及其代谢物

Akiyama和Sherwood[60]以及Scherer等[21]回顾了随机临床试验和横断面研究中吸烟者与ENDS使用者烟碱及其代谢物生物标志物的水平。一些研究显示，ENDS使用者尿液中总烟碱水平低于吸烟者，而另一些研究则未发现差异。例如，Round等[70]开展了一项随机、平行组临床研究，分析了吸烟者在改用ENDS产品5天后的情况。24小时尿样中总烟碱水平下降了38.3%（$P < 0.05$），血浆可替宁和烟碱水平同样显著下降。Shahab等[71]进行的一项横断面研究则发现，吸烟者和ENDS使用者尿液中的总烟碱水平没有显著差异。理论上，吸烟者和ENDS使用者的总烟碱水平应该相似，因为这两种产品都旨在高效提供烟碱，并可能存在一

定程度的自我调节。然而，ENDS产品特性和使用模式的差异可能导致不同结果。值得注意的是，使用非盐基液体的ENDS，其中大部分烟碱以非质子化形式存在，主要通过口腔吸收，导致烟碱药代动力学减慢，滥用风险可能低于传统卷烟。然而，目前市场上许多销售的ENDS产品含有盐形式的烟碱，这使得ENDS气溶胶更容易吸入，并导致烟碱的吸收速度加快。

在Akiyama和Sherwood[60]的综述中，大多数研究发现HTP使用者的总烟碱水平与吸烟者相似，差异不超过20%。例如，在一项三组平行研究中，160名日本成年吸烟者被随机分配至薄荷醇HTP组（n=78）或薄荷醇卷烟组（n=42），分别在密闭环境中使用5天和流动环境中使用85天。两种情况下，HTP和传统卷烟使用者的总烟碱水平均无显著差异[72]。尽管不同HTP的特性差异显著，但这些差异可能与烟碱释放机制相关。

烟草特有亚硝胺及其代谢物

在吸烟者改用电子烟碱传输系统（ENDS）或加热型烟草制品（HTP）的所有随机临床试验中，尿液中的NNAL水平均显著降低。横断面研究（包括烟草与健康人口评估，PATH）也显示，ENDS使用者的NNAL水平显著低于吸烟者[60, 73, 74]。在ENDS使用者的尿液中几乎检测不到NNAL，因为NNAL是NNK的代谢物，而NNK仅存在于含烟草的产品中。偶尔检测到的低水平NNAL可能部分归因于之前使用烟草制品（由于NNAL的长半衰期）或暴露于二手烟草烟气[75-77]。对于NNN，ENDS使用者尿液中的水平要么极低，要么未被检测到[60, 78]。Bustamante等[17]提供的证据显示，NNN可在ENDS使用者的唾液中检测到（14.6 pg/mL±23.1 pg/mL），并得出结论认为，NNN为内源性生成，因为ENDS液体中未检测到超出痕量的NNN。Scherer等[79]的研究发现，ENDS或HTP使用者的尿液或唾液中与烟草特异性亚硝胺相关的代谢物水平并未显著高于非烟草产品使用者。

PAH代谢物

一项随机临床试验显示，吸烟者改用ENDS 5天后，尿液中的1-HOP和3-羟基苯并[*a*]芘分别显著减少63.5%和63.8%，而芴和萘的代谢物也显著降低[70]。当使用含薄荷醇的产品时，结果类似。另一项为期5天的试验中，1-HOP水平减少了70.5%[65]。在吸烟者改用ENDS的8周试验中，PheT水平显著降低了20%[62]。与吸烟者的三项研究相比，ENDS使用者尿液中的1-HOP水平显著减少了57%~61%[80]。根据Akiyama和Sherwood[60]的综述，多项随机临床试验表明，从传统卷烟转向HTP后，1-HOP水平有所降低，通常减幅超过60%，尽管有些试验报告的降幅为15%~30%。类似的结果在PATH研究中也得到了观察[74]。这些结果表明，无论使用ENDS还是HTP，使用者暴露于燃烧产物的水平都显著减少。

挥发性有害成分和致癌物及其代谢物

多项随机临床试验一致发现，吸烟者改用ENDS后，暴露于挥发性有害物质（如丙烯醛、丙烯酰胺、丙烯腈、苯、1,3-丁二烯、巴豆醛和环氧乙烷）的生物标志物显著减少[60]。在一项研究中，吸烟者被随机分配为完全改用电子烟碱传输系统（ENDS）8周，尿液中丙烯酰胺、丙烯醛、丙烯腈和巴豆醛的代谢物水平分别下降了32%、47%、66%和47%[62]。在另一项5天的随机研究中，丙烯醛（70.5%）、丙烯腈（85.9%）、苯（89.7%）、1,3-丁二烯（55.5%）、巴豆醛（77.5%）和环氧乙烷（62.3%）的巯基尿酸水平显著下降[71]。横断面研究（包括PATH研究[60, 74]）以及最近一项限制卷烟抽吸者或ENDS使用者3天的研究[81]也得出了类似的结果。

一些研究表明，ENDS使用者尿液中暴露于挥发性物质（如丙烯腈、丙烯醛、巴豆醛和环氧丙烷）的生物标志物水平高于不使用任何烟草和烟碱产品的人[82]。一项为期4~6个月的研究发现，ENDS使用者尿液中的3-羟丙基巯基尿酸（丙烯醛的主要代谢物）水平显著高于非使用者[82]。

由行业研究人员进行的几项随机临床试验显示，吸烟者改用HTP后，挥发性成分的巯基尿酸水平大幅下降。这些结果在所有已发表的行业试验中表现一致[60]。在一项临床研究中，每组10名受试者在限制环境中仅使用指定产品（卷烟、ENDS、HTP、口服烟草、烟碱替代疗法或不使用任何烟草和烟碱产品）3天。结果显示，与使用其他非卷烟产品的受试者相比，HTP使用者与丙烯醛、丙烯酰胺和巴豆醛相关的巯基尿酸水平略有升高[81]。

金属

如上所述，镉和铅是吸烟者暴露于毒性金属中含量最高的金属。PATH研究第一轮数据（2013~2014年）显示，ENDS使用者的尿液中镉和铅的含量显著高于从未使用任何烟草制品或ENDS的人群，分别高出23%和19%[73]。作者指出，金属暴露生物标志物的长半衰期可能是混淆因素，因为部分ENDS使用者可能是前吸烟者或通过其他途径暴露于金属。Prokopowicz等[83]报道称，吸烟者转用ENDS后，血液中的镉含量显著降低，而血铅含量没有显著变化。吸烟者的这两种生物标志物水平明显高于不吸烟者。一项关于尿液中元素的横断面研究，包括铬、镍、钴、银、铟、锰、钡、锶、钒和锑，以及镉和铅，显示ENDS使用者与不吸烟者的这些元素水平没有差异[84]。另有回顾性研究发现，ENDS使用者与不使用ENDS的人群相比，其生物标志物（如铅、铬、镍、硒和锶）的研究结果存在不一致性[85]。

目前尚无关于HTP使用者金属暴露生物标志物的报告。在PubMed中搜索“加热型烟草制品中的金属暴露”以及类似的谷歌搜索，未发现相关文章或专著。

唾液中的丙二醇作为一种新的ENDS生物标志物

丙二醇是电子烟气溶胶的主要成分。最近开发了一种检测唾液中丙二醇的方法[1]，结果显示ENDS使用者唾液中丙二醇的平均浓度约为不吸烟者的100倍，是吸烟者的30倍。因此，唾液中的丙二醇可作为一种新的生物标志物，用于验证ENDS的使用。

暴露生物标志物与双重使用者

许多采用ENDS的吸烟者仍继续吸传统卷烟（被称为“双重使用者”），并呈现出不同程度的替代[86]。研究表明，与完全吸烟者相比，双重使用者通常具有相似或更高水平的暴露生物标志物，而要实现显著的暴露减少，需要完全转用ENDS[87-90]。在最近的一项研究中，Anic等[90]使用了PATH研究中2475名成年人的生物标志物数据。这些成年人在第一轮（2013~2014年）为吸烟者，在第二轮（2015年）转为单一使用ENDS、双重使用或戒烟。结果显示，成为卷烟与ENDS双重使用者的吸烟者，其大多数评估的生物标志物水平未显著降低。表5.1列出了该研究的数据示例，展示了继续吸烟、成为双重使用者或停止任何烟草或烟碱使用的吸烟者的生物标志物水平。

表5.1 PATH研究中第二轮参与者的生物标志物水平（按产品使用状态划分）

生物标志物（来源）	产品使用状况			
	仅抽吸卷烟	双重使用和ENDS使用	仅使用ENDS	不使用烟草或烟碱
TNE, μmol/g肌酐（烟碱）	31.2 [28.0, 34.8]	38.5 [30.3, 48.9]	9.1 [3.6, 22.9]	0.1 [0.0, 0.1]
NNAL, ng/g肌酐（NNK）	218.1 [199.2, 238.8]	231.9 [187.0, 287.5]	12.5 [5.7, 27.3]	5.0 [3.6, 7.0]
1-HOP, ng/g肌酐（芘）	316.8 [298.3, 336.4]	308.1 [277.4, 342.2]	113.4 [93.0, 138.4]	167.9 [148.9, 189.4]
CEMA, μg/g肌酐（丙烯腈）	131.7 [121.1, 143.2]	128.1 [104.7, 156.6]	8.6 [4.9, 14.9]	3.9 [2.9, 5.3]
3HPMA, μg/g肌酐（丙烯醛）	1342.2 [1247.8,1443.7]	1531.6 [1321.6,1774.9]	303.8 [228.7, 403.7]	299.9 [255.8, 351.5]
4HBMA, μg/g肌酐（1,3-丁二烯）	31.8 [29.6, 34.2]	33.9 [28.9, 39.8]	5.1 [3.9,6.8]	5.4 [4.6, 6.4]

资料来源：Anic等[90]

注：每个单元格显示几何平均值和[95% CI]。这些分析的所有参与者在第一轮都是仅抽吸卷烟者

1 Tang MK, Carmella SG, 2022 年未发表的数据

5.3 效应生物标志物（危害或疾病）

5.3.1 烟草和烟碱产品研究中常用效应生物标志物的定义和概述

文献中对效应生物标志物有不同的定义[91]。这些生物标志物通常被称为“潜在危害生物标志物”，其定义为“由于暴露而产生影响的测量，包括早期生物效应，形态、结构或功能的改变，以及与危害相一致的临床症状（包括临床前变化）”[92-94]。应当指出，这一定义包括：连续的生物效应及一系列相关疾病。在烟草和烟碱产品的使用背景下，与之相关的主要健康后果包括癌症、心血管疾病和呼吸系统疾病。时间也是一个重要变量。急性变化与长期变化的解释可能取决于生物标志物。烟草行业研究人员提出了一个专门针对“燃烧型”烟草制品使用的定义，其中明确指出，效应生物标志物是“抽吸烟草制品后，生物样本中显著的、客观的、可测量的变化……这种变化在一定比例的吸烟者中发生，并且在戒烟后是可逆的”[95]。该定义包含了可逆性这一概念，这与研究传统烟草制品（如卷烟）使用者转向具有不同有害成分含量（如ENDS）的烟草或烟碱制品时生物效应的潜在变化有关。生物标志物可逆性的后果应在健康影响变化的研究中进一步调查。

DNA加合物

DNA加合物（通常称为加合物）是在细胞代谢吸入有害成分或致癌物质过程中，由某些有机或无机中间体与DNA反应产生的；它们也可能由一些内源性化合物的中间体与DNA反应形成。DNA碱基或磷酸盐的加合物是致癌过程的核心，因为它们可能引起DNA错误编码，导致许多与癌症相关的关键生长控制基因发生突变。

尽管细胞具有DNA修复系统来修复损伤，但当修复系统效率低下或容易出错时，加合基可能被错误解读，并由DNA聚合酶插入错误的碱基，导致DNA突变。这一结果可能在关键的生长控制基因中产生永久性突变，从而诱发癌症。在烟草致癌物代谢过程中，各种组织的DNA中产生了许多突变[96-101]。使用多种方法（包括^{32}P后标记法、免疫分析法和质谱法）进行的大量研究，已检测到吸烟者和非吸烟者各种组织中特定类型的DNA加合物[102-108]。许多研究显示，吸烟者组织中某些DNA加合物的水平高于非吸烟者，但解释较为复杂，因为某些情况下受试者数量较少或方法缺乏适当验证。

细胞因子、趋化因子和反应蛋白

炎症和氧化损伤是吸烟引起的各种疾病[包括癌症、心血管疾病和慢性阻塞性肺病（COPD）]的重要因素。这些过程涉及淋巴细胞、巨噬细胞和中性粒细胞在受压组织中的浸润，以及促炎症和抗炎症细胞因子等因子的分泌。这些生物标志物通常通过血浆或血清测量。例如，白细胞介素6（IL-6）和C反应蛋白（CRP）在吸烟者中明显高于非吸烟者[107]。这些标志物的水平与多种相关疾病直接相关。例如，CRP、BCA-1/CXCL13、MDC/CCL22 和 IL-1RA 与肺癌风险相关[108]。在心血管疾病（CVD）中，氧化LDL是脂质谱的代表性指标[109]。CRP、IL-6、纤维蛋白原和可溶性ICAM-1是血栓形成和内皮功能障碍的指标，它们在动脉粥样硬化、血管收缩和冠心病的发生和进展中起着基础性作用[110-114]。COPD与IL-8、TNF-α、IL-6 和 RANTES 的升高相关，这些标志物反映了巨噬细胞和T细胞的主导作用，与气流阻塞和肺气肿的严重程度相关，并在细胞凋亡中发挥主要作用，导致肺部破坏[115-124]。COPD还以纤维蛋白原升高为特征，这与肺功能降低有关[18-121, 125-131]。

尿前列腺素代谢物

前列腺素E2代谢物（PGEM）和(*Z*)-7-[(1*R*,2*R*,3*R*,5*S*)-3,5-二羟基-2-[(*E*,3*S*)-3-甲基辛-1-烯基]环戊基]庚-5-烯酸（8-*iso*-$PGF_{2\alpha}$）分别被认为是炎症和氧化损伤的生物标志物。PGEM是前列腺素E2的代谢物，而8-*iso*-$PGF_{2\alpha}$是脂质过氧化的产物[132]。8-*iso*-$PGF_{2\alpha}$也可通过血液检测。炎症和氧化损伤与吸烟明显相关，并通过促进协同癌变或肿瘤促进的机制，增强卷烟烟气致癌物的活性，从而在癌症诱导中起着重要作用[102]。它们在心血管疾病和COPD中也有明确作用[133]。上海队列研究发现，在调整吸烟强度、持续时间和其他可能的混杂因素后，吸烟者尿中8-*iso*-$PGF_{2\alpha}$水平与肺癌风险之间存在独立关联[134]。在有吸烟史的人中也观察到了显著关联，但在从未吸烟者中未观察到这种关联，这表明烟草烟气致癌物与氧化损伤之间可能存在相互作用[134]。戒烟后，尿中的8-*iso*-$PGF_{2\alpha}$下降的速度比暴露生物标志物慢[134]。PATH研究发现，8-*iso*-$PGF_{2\alpha}$的平均水平需要6个多月才能恢复到非吸烟者的水平，而另一项研究报告戒烟12周后下降27%[135, 136]。

5.3.2 效应生物标志物在ENDS和HTP研究中的应用

DNA加合物

关于电子烟碱传输系统（ENDS）对口腔细胞DNA损伤的研究很少发表[137]。在一些体外研究中，通过将培养的口腔细胞暴露于电子烟烟气或烟液中得到了不同的结果。一些研究表明可能存在DNA损伤，而其他研究则没有观察到类似情况。

一项临床研究检测了丙烯醛-DNA加合物 (8*R*/*S*)-3-(2′-脱氧核糖-1′-基)-5,6,7,8-四氢-8-羟基嘧啶[1,2-*a*]嘌呤-10(3*H*)-酮（γ-OH-Acr-dGuo）在电子烟使用者与不使用任何烟草制品者的口腔细胞中的水平。每组20人，每月访问诊所一次，持续3个月。研究发现，在电子烟使用者中，γ-OH-Acr-dGuo的水平显著高于非使用者，为其9倍，但低于吸烟者[106,138]。这些结果表明，与非使用者相比，电子烟使用者在口腔黏膜中形成了特定的DNA加合物，暗示可能存在致癌作用。

在另一项关于DNA缺失位点（即嘌呤或嘧啶缺失）的研究中，发现使用电子烟者的DNA损伤水平显著低于非吸烟者和吸烟者，分别减少了45%和42%。这项研究基于一次临床观察数据（每组30~35名受试者）得出。DNA缺失位点与电子烟使用或吸烟之间的直接关系尚不明确[139]。

目前尚无研究探讨加热型烟草制品（HTP）对DNA损伤的影响。

细胞因子、趋化因子和反应蛋白

表5.2总结了烟草和烟碱产品使用研究中常用的一些血液和尿液生物效应标志物的数据，涉及不同产品类型使用人群的几何均值比[135,140]。需要注意的是，这些生物标志物并不特定于某一种烟草和烟碱产品的暴露，可能受到既往吸烟引起的潜在亚临床疾病的影响。

表5.2 烟草和烟碱产品使用研究中常见的生物效应标志物的几何均值比（GMR）和范围（按产品使用情况划分）

生物标志物	基质	指标	ENDS使用者与卷烟抽吸者的GMR	ENDS使用者与非吸烟者的GMR	人群[a]
IL-6	血清	炎症	0.84（0.71~0.98）	0.98（0.82~1.18）	PATH
hs-CRP	血清或血浆	炎症，心血管风险	0.73（0.57~0.93）	0.86（0.66~1.11）	PATH
Fibrinogen	血浆	炎症、凝血、心血管风险	0.96（0.92~1.01）	0.99（0.94~1.04）	PATH
sICAM	血清	炎症，心血管风险	0.82（0.75~0.89）	1.02（0.95~1.1）	PATH
LDL	血浆	心血管疾病的风险	0.52（0.24, 1.14）	0.60（0.31, 1.16）	NHANES[b]
HDL-C	血浆	心血管疾病的风险	1.00（0.50, 2.00）	1.82（0.95, 3.49）[c]	NHANES[b]
TGL	血浆	心血管疾病的风险	0.26（0.06, 1.02）	0.42（0.12, 1.51）	NHANES[b]
8-*iso*-$PGF_{2\alpha}$	尿	氧化应激	0.75（0.68~0.83）	1.10（0.98~1.22）	PATH

a. 美国队列。NHANES: 国家健康与营养调查；PATH: 烟草与健康人口评估

b. 数据为没有吸烟史的仅使用ENDS者

c. 低水平的高密度脂蛋白胆固醇（HDL-C）与/较高的心血管风险相关

ENDS使用者 根据PATH研究第一轮的数据分析结果显示，转向仅使用电子烟（ENDS）的吸烟者在IL-6、hs-CRP和sICAM-1水平上显著低于仅使用传统卷烟的吸烟者，并且与完全戒烟的吸烟者水平相当[135]。然而，ENDS使用者与卷烟抽吸者的纤维蛋白原水平相似（GMR, 0.96；95% CI, 0.92~1.01）。分析还表明，目前单

纯ENDS使用者在这些生物标志物的水平上，无论使用频率如何，均无显著差异，且与戒烟时间无关。在两次国家健康与营养调查（2015~2016年和2017~2018年）中，对8688名成年人的HDL-C、低密度脂蛋白胆固醇、甘油三酯和空腹血糖数据的研究发现，仅使用ENDS对这些指标没有统计学意义上的显著影响[140]。尽管普遍报道仅使用ENDS者在细胞因子和其他循环炎症生物标志物水平上低于吸烟者，但研究结果在不同研究、生物标志物或设备类型之间并不一致[135, 141-144]。明确ENDS使用者中这些生物标志物的数据至关重要，因为许多体外、体内和人体研究表明，ENDS气溶胶可能诱发炎症并导致呼吸和心血管问题[145-148]。例如，Mohammadi等[149]的一项研究评估了慢性ENDS使用者、慢性吸烟者和非使用者的内皮功能，通过检测受试者血清对培养的内皮细胞一氧化氮（NO）和过氧化氢释放及细胞通透性的影响。结果显示，ENDS使用者的血清对内皮细胞的作用与吸烟者相似，包括减少血管内皮生长因子诱导的NO分泌、释放过氧化氢、增加细胞通透性，以及引起炎症、血栓形成和细胞黏附循环生物标志物的变化。这些结果表明，使用ENDS可能导致内皮功能改变。一项在印度进行的唾液炎症生物标志物研究表明，ENDS使用者的唾液CRP、TNF-α和IL-1β水平显著高于非使用者，且与吸烟者相似[150]。

HTP使用者　关于HTP使用者血液中炎症生物标志物的数据主要来自烟草行业主导的研究。例如，菲利普 · 莫里斯国际公司（PMI）发布了若干关于此类生物标志物的报告，这些研究的参与者均是从抽吸卷烟转向使用HTP的个体[151]。在一项对日本吸烟者的研究中，这些吸烟者在6天内转为使用HTP，研究测量了血清中间细胞16k Da蛋白（CC16）的水平，这是一种反映肺上皮细胞损伤的指标；结果未观察到变化[152]。另一项PMI研究在316名波兰吸烟者中进行，这些参与者被随机分为HTP组或继续吸烟组，研究为期1个月，并评估了一系列与心血管风险相关的生物标志物[153]。HTP组的HDL-C水平有所增加（即改善），但观察到红细胞计数、血红蛋白和红细胞压积的下降。大多数其他生物标志物在转向使用HTP一个月后无显著变化。PMI还在美国开展了一项更长期的研究，将984名成年吸烟者随机分配为HTP组或继续吸烟组，研究持续6个月[154]。结果显示，与继续吸烟组相比，HTP组在四种生物标志物[HDL-C、白细胞计数、第1秒用力呼气量（FEV_1）和COHb]上均有所降低。在所有这些研究中，HTP组参与者的烟气成分暴露和尿致突变性均显著降低。

英美烟草公司研究人员报告了一项为期12个月的动态临床研究，研究中吸烟者被随机分配至HTP组、继续吸烟组和戒烟组[155]。与继续吸烟相比，使用HTP 6个月后，白细胞计数（降低）和呼出一氧化氮（FeNO，增加）出现统计学显著的积极变化。此外，11-dTXB2水平在HTP使用6个月后有所降低，但HTP组与继续吸烟组的差异未达统计学意义。此外，HTP使用对sICAM-1或HDL水平无实质性影响（仅提供了描述性统计数据）。在该研究的最新报告中，大多数生物标志物在12个月时的水平与6个月时相似[156]。尽管切换到HTP后，某些生物效应标志物显示出积极变化（表明其危害小于吸烟），但其结果仍不及戒烟者。

日本烟草公司研究人员进行了一项反映真实世界情况的市场后研究[157]，研究测量了一系列炎症标志物，包括HDL-C、甘油三酯、sICAM-1、白细胞计数、11-DHTXB2和2,3-d-TXB2（血小板活化的生物标志物）。结果显示，HTP使用者（平均使用1.2年）的部分生物标志物水平有所降低，但尿2,3-d-TXB2水平较非吸烟者更高。

前列腺素代谢物及其他相关尿液生物标志物

前列腺素代谢物　来自PATH研究的数据分析表明，目前仅使用ENDS产品的前吸烟者尿液中的8-*iso*-$PGF_{2\alpha}$水平与未使用ENDS产品的前吸烟者及从未使用过烟草的参与者相似[135]。然而，目前尚不清楚非前吸烟者的ENDS使用者是否也存在8-*iso*-$PGF_{2\alpha}$水平升高，而吸烟停止后尿液中8-*iso*-$PGF_{2\alpha}$水平下降缓慢可能加剧了这种不确定性[135, 136]。值得注意的是，在该研究中，卷烟和电子烟双重使用者的8-*iso*-$PGF_{2\alpha}$水平显著高于仅吸烟者（GMR, 1.09；95% CI, 1.03~1.15）。在卷烟抽吸者转为HTP使用者的研究中，尿液中异前列腺素（包括8-*iso*-$PGF_{2\alpha}$）的变化幅度很小且无统计学显著性[158]。当20名吸烟者在一周不使用任何烟草制品后改用ENDS或HTP时，血液中的8-*iso*-$PGF_{2\alpha}$水平显著升高[159]。

多项行业资助的临床试验探讨了HTP对尿液中8-*iso*-$PGF_{2\alpha}$水平的影响。例如，在一项研究中，将健康的成年吸烟者随机分配到薄荷味HTP组或戒烟组，在91天后发现HTP组的尿液8-*iso*-$PGF_{2\alpha}$水平下降了13%（$P < 0.05$），且与戒烟组相似[160]。在PMI对984名美国吸烟者的研究中，与继续吸烟者相比，转为使用HTP 6个月的参与者尿液中8-*iso*-$PGF_{2\alpha}$水平减少了6.8%[154]。在日本关于HTP真实世界使用的市场后研究中，HTP使用者的这种生物标志物水平略高于非使用者，但差异的统计学意义接近边缘（P=0.0646）[157]。

临床前变化和症状的指标

氧化应激的其他尿液生物标志物　在DNA中，鸟嘌呤是炎症诱导的自由基直接氧化的主要靶点。这种氧化的最常见产物是8-氧-7,8-二氢-2′-脱氧鸟苷，它会导致染色体畸变并诱发突变，被广泛用作氧化应激的生物标志物[115]。一些研究发现，ENDS使用者体内8-氧-7,8-二氢-2′-脱氧鸟苷水平显著高于非吸烟者[161, 162]。Sakamaki-Ching等[162]发现，ENDS使用者与卷烟抽吸者之间此生物标志物水平无差异。

直接评估使用ENDS和HTP引发慢性疾病结局需要耗时数年且涉及大量参与

者的前瞻性队列研究。然而，由于这在多数研究中难以实现，且不适合时间敏感的监管决策，研究通常使用前临床变化和症状作为烟草和烟碱产品使用者呼吸与心血管风险的替代测量标准。

心血管疾病的前临床指标与症状可能包括血压、心率、动脉硬化、血小板反应性及其他心血管结果等测量指标。而对于COPD等呼吸系统疾病，常测量的前临床指标与症状包括第1秒用力呼气量（FEV_1）、用力肺活量（FVC）、FEV_1/FVC、肺一氧化碳弥散量（DLCO）以及咳嗽、喘息和呼吸短促等症状[163-166]。心肺前临床指标的解释通常基于血脂水平、纤维蛋白原、D-二聚体和hs-CRP等参数。

癌症的前临床指标尚无可靠标准。关于ENDS使用与疾病结局之间关联的大规模人群流行病学研究逐渐增多，其中大多数发表于2020年之后。这些新兴数据将在未来评估新型烟草制品中暴露与生物效应标志物预测价值方面发挥关键作用。

ENDS使用者　几项研究表明，与不使用烟草和烟碱产品相比，使用电子烟碱传输系统（ENDS）会增加血压、心率、动脉僵硬度、血小板反应性及其他心血管健康风险[141, 147, 148, 167, 168]。此外，ENDS使用者还报告了呼吸道症状（如气流阻力增加、肺部脂质性巨噬细胞积聚）。根据PATH研究的第一轮和第二轮纵向数据，初始仅使用ENDS的个体在随访期间发生呼吸道症状的比例高于非使用者（分别为33.6%和21.7%）[168-170]。另一份基于PATH研究第一至第四轮数据的报告指出，与非使用者相比，ENDS使用者患呼吸系统疾病（慢性阻塞性肺疾病、肺气肿、慢性支气管炎和哮喘）的风险更高[171]。

然而，从吸烟转向ENDS可能改善某些健康指标和结局。一项系统回顾分析了6项基于人群的研究，这些研究的样本量从19475到161529不等，结果发现，与当前吸烟者相比，使用ENDS的前吸烟者患呼吸系统疾病（慢性阻塞性肺疾病、慢性支气管炎、肺气肿、哮喘和喘息）的风险较低，但在心血管疾病（中风、心肌梗死和冠心病）方面无显著变化[172]。一项针对水烟使用者的随机交叉研究表明，与传统烟草水烟相比，使用电子水烟会导致动脉僵硬度增加及炎症标志物水平升高[173]。前临床指标与症状（如慢性阻塞性肺疾病与心血管疾病）之间的差异可能是暴露和生物学反应复杂交互的结果。例如，戒烟后，肺癌风险的降低通常需要20~25年，并且永远无法达到从未吸烟者的水平[174]。相比之下，心血管疾病的风险在戒烟1~3年后即可降至与从未吸烟者相同的水平[174]；然而，即使是低水平暴露，例如每天少于三支烟，或非吸烟者暴露于二手烟，仍会显著增加心血管疾病风险[175-177]。

HTP使用者　关于加热型烟草制品（HTP）使用者健康影响指标的独立研究报告较少。在2018年韩国青少年健康风险行为调查中，研究人员对58336名12~18岁的学生进行了横断面研究，发现HTP使用与哮喘、过敏性鼻炎和特应性皮炎存在关联[178]。在日本，有两例急性嗜酸性粒细胞增多症与HTP使用相关的病例报告。

一例为20岁男性，其每天使用20支HTP棒，持续6个月，并在住院前两周将使用量加倍[179]。另一例为16岁患有支气管哮喘的男孩，他在使用HTP后立即出现咳嗽、呼吸急促和疲劳的症状，并在连续使用HTP两周后症状加重[180]。关于HTP使用者前临床指标的烟草行业报告大多限于呼吸功能测量（如第1秒用力呼气量，FEV_1）。这些研究表明，从吸烟转向HTP后，该指标要么无显著变化，要么略有改善[155, 157]。

研究ENDS和HTP的新型效应生物标志物

DNA甲基化数据 大量文献支持使用表观遗传修饰作为评估吸烟影响的一种方法[181, 182]。在唾液和支气管肺泡灌洗液中对DNA甲基化的研究发现，电子烟碱系统（ENDS）使用者的表观遗传特征与非使用者相似[183, 184]，且ENDS的使用未影响cg05575921位点，这是吸烟者中高度低甲基化的芳香烃受体阻遏物（AHRR）位点，也是吸烟状态的一个敏感且特异的标志物[185, 186]。然而，在ENDS使用者的白细胞中，报告了LINE-1重复元件的低甲基化以及DNA羟甲基化的整体丧失，这提示了系统性效应的存在[187]。

在日本的Tsuruoka代谢组队列研究中评估了加热型烟草制品（HTP）使用对DNA甲基化的影响。研究发现，在17个与吸烟相关的基因中，有10个基因在HTP使用者中显著低甲基化，与非吸烟者相比，GPR15表达显著上调，尽管AHRR的表达明显低于吸烟者。这些结果表明HTP的使用可能导致不同的DNA甲基化和转录组谱，其影响需要进一步研究[188]。

基因表达 吸烟及其他有害暴露引起的基因表达变化可能表明细胞代谢途径的紊乱，这些变化可以作为与特定健康结果（包括肺癌）相关的生物效应标志物[189-191]。对ENDS使用者、吸烟者和非吸烟者的横断面研究显示，这些人群间存在差异性基因表达。例如，Martin等[192]发现，吸烟者的鼻上皮有53个基因差异下调，而ENDS使用者的鼻上皮与非吸烟者相比，有358个基因差异下调。只有一个基因——生长反应基因1（EGR1）的上调在卷烟抽吸者和ENDS使用者中相同，其余过表达基因则是两种产品所特有的。在另一项横断面研究中，利用口腔细胞评估相同两组的基因表达，结果显示吸烟者的差异表达基因多于ENDS使用者[193]。卷烟抽吸者和ENDS使用者中最严重的调控紊乱通路与致癌通路相关。急性暴露于ENDS后的基因表达研究表明，口腔、血液和呼吸细胞在暴露后发生显著变化[194, 195]。这些发现的病理生理后果需要更多研究来明确。

口腔微生物群 口腔微生物群是口腔内化学物质暴露的复杂受体介质。口腔化学环境的变化会为某些细菌种群创造有利或不利的条件，从而改变口腔微生物群的组成和功能。口腔也是细菌进入呼吸道的门户[196-198]，越来越多的研究表明，口腔微生物群与多种慢性疾病（包括癌症、心血管疾病和慢性阻塞性肺疾病）之间存在关联[196, 199-207]。吸烟通过免疫抑制、促进生物膜形成、改变口腔氧张力和

pH值以及改变口腔化学环境等方式影响口腔微生物群[208-212]。烟草类型、使用频率和吸烟史被报道会影响这种变化的程度和性质[213, 214]。

近期研究表明，ENDS使用者的口腔微生物群特征明显不同于吸烟者和前吸烟者或从未吸烟者，包括分类组成的改变、微生物多样性增加、参与碳水化合物和氨基酸代谢的微生物途径丰度显著增加以及多样化的致病因子[215-217]。这些特征中有些是有利的（例如，多样性高于卷烟抽吸者），而另一些则提示炎症过程的存在。与卷烟抽吸者类似，韦永氏菌（*Veillonella*）在ENDS使用者的口腔细胞和唾液中的相对丰度显著高于非使用者[215, 216]。韦永氏菌与多种感染有关，包括口腔、肺部和心脏感染[218-220]，某些韦永氏菌种可能通过将硝酸盐还原为亚硝酸盐的能力，在内源性亚硝化过程中发挥作用[216, 221]。这可能是Bustamante等[17]发现ENDS使用者和卷烟抽吸者口腔中NNN水平相当的主要原因。

关于HTP使用者口腔微生物群的研究仅有一项[222]。该研究在乌克兰的65名14~18岁青少年中进行，这些青少年为HTP或ENDS使用者或不使用任何烟草制品的对照组。研究发现，HTP使用者的口腔微生物群组成与ENDS使用者不同。研究结果表明，这两种产品都会减少驻留菌斑菌群的数量，从而导致机会性短暂链球菌（如肺炎链球菌和化脓性链球菌）的出现。

5.4 易感性生物标志物

5.4.1 烟草和烟碱产品研究中易感性生物标志物的定义和概述

个体与群体之间对烟草和烟碱产品中毒性物质及致癌物的吸收和/或代谢的差异，可能导致对不良生物效应的敏感性差异并最终产生不同的健康结果。易感性生物标志物是一种预测性指标，用于反映驱动这些差异的个体特征（例如基因多态性）。在烟草控制领域，这些生物标志物有助于解释和预测同类产品使用者之间暴露标志物或生物效应水平的潜在群体差异。此外，这些标志物可用于识别易感人群，以便开展有针对性的戒烟干预。

烟碱代谢率　在研究烟草使用和疾病风险时，最常用的易感性生物标志物是两种烟碱代谢产物（反式3′-羟基可替宁与可替宁）之比，即烟碱代谢率（NMR）。该标志物反映了负责烟碱代谢的主要酶 CYP2A6 的活性，其活性主要由*CYP2A6*基因中功能性多态性的存在或缺失决定[223]。研究发现，不同个体间的NMR差异与吸烟行为及剂量[224]、戒烟能力[225]以及肺癌风险[6]有关。根据 PATH 研究第一轮数据的最新分析，美国日常烟草使用者的NMR，包括按年龄、性别和种族或民族划分的代表性值，现已公开[226]。

尿液中致癌物和有害成分的代谢物　如前所述，烟草烟气成分的尿液代谢物（如烟碱、烟草特有亚硝胺、多环芳烃和挥发性有机化合物）是公认的暴露生物标志物。此外，对上海队列研究和新加坡华人健康研究这两项大型前瞻性流行病学研究的数据分析表明，总烟碱当量、总 NNAL、总 NNN 和 PheT 与吸烟者的癌症风险显著升高独立相关（表 5.3）[10,15,227,228]。因此，这些代谢物也可以被视为癌症易感性的生物标志物。在这些研究中，总烟碱当量作为烟草烟气所有其他成分剂量的监测指标。尽管对总烟碱当量进行校正后，总 NNAL、总 NNN 和 PheT 的影响仍然显著，但对于 1,3-丁二烯、环氧乙烷、苯、丙烯醛和巴豆醛的巯基乙酸代谢物标志物，情况则不相同[229]。使用尿液中的致癌物和毒性物质生物标志物评估电子烟碱传输系统（ENDS）和加热型烟草制品（HTP）使用者风险的相关性需要进一步研究，因为这些产品的使用者似乎对相关母体化合物（NNK、NNN 和 PAH）的暴露有限。如前所述，Bustamante 等[17]提供了证据，表明 ENDS 使用者唾液中 NNN 水平有所升高。

表5.3　尿致癌物和有害代谢物与吸烟者肺癌风险的潜在关联

成分	生物标志物	比值比	研究人群	参考文献
烟碱	可替宁	0.85~3.52	上海，新加坡，美国	[229-231]
NNK	总NNAL	1.57~2.64	上海，新加坡，美国	[229-231]
PAH	PheT	1.23~2.34	上海，美国	[229,231]
挥发性有机化合物	巯基尿酸	0.97~1.20	上海	[228]

其他潜在的易感性生物标志物　上文所述的某些生物标志物，如DNA甲基化、基因表达和微生物组，也是重要的个体特征，可影响烟草成分的代谢和/或保护机制（如免疫反应、DNA修复）抵御其有害影响。因此，DNA甲基化、基因表达和微生物组可以作为烟草或烟碱产品易感性的生物标志物。

5.4.2　易感性生物标志物在ENDS和HTP研究中的应用

在对ENDS和HTP的研究中，易感性生物标志物的使用仍然有限。

烟碱代谢率

目前尚不清楚烟碱代谢率（NMR）是否能够预测电子烟碱传输系统（ENDS）或加热型烟草制品（HTP）的使用行为，或者是否能预测使用这些产品引起的任何健康结果。由于这些产品上市时间较短、种类繁多且不断演变，以及烟碱递送量的差异，这可能是缺乏数据的主要原因。一项对PATH研究（第一轮和第二轮）数据的分析发现，NMR与吸烟和ENDS使用转变之间存在显著的双向交互作用[232]。研究表明，NMR较高（即烟碱代谢更快）的女性戒用ENDS的可能性比NMR较低的女性低10倍。这些结果表明，NMR可能成为女性戒用ENDS的潜在生物标志物。

其他潜在的易感性生物标志物

尿液中致癌物和有害成分的代谢物 目前尚无一致证据表明，与未使用任何烟草和烟碱产品的人相比，ENDS或HTP使用者尿液中的NNK或PAH代谢物显著增加[60,233]。

DNA甲基化、基因表达和微生物组 在研究烟草和烟碱产品中使用这些生物标志物仍处于相对较新的阶段，应用范围有限。这些生物标志物与ENDS或HTP成分代谢之间的潜在关联，或与ENDS或HTP使用者生物学效应的潜在关联尚未被研究。

5.5 生物标志物测量的已建立、经验证的方法

针对大多数常用的暴露和生物效应标志物，特别是用于大型队列研究的标志物，已建立明确的方法。液相色谱（LC）或气相色谱（GC）联合质谱（MS）是这些测量的首选方法，具有高度的灵敏度和选择性。表5.4展示了一些方法及其分析参数的示例。对于循环中的生物效应标志物（如细胞因子和活性蛋白），通常采用免疫分析方法，并使用商业化试剂盒。这些试剂盒的性能由制造商验证其质量和特异性。

表5.4 一些暴露生物标志物和生物效应标志物的验证方法实例

生物标志物	方法描述	方法特点	参考文献
暴露生物标志物			
尿中的总烟碱当量（TNE）	LC-MS/MS分析烟碱、可替宁、3′-羟基可替宁及其葡萄糖醛酸盐经尿液酶处理（从其葡萄糖醛酸盐缀合物中释放这些生物标志物）和固相萃取后的含量	准确度：93%~96% 日内CV：4.2%~7.1% 日间CV：0.4%~5%	[234]
尿中的总NNAL	酶法处理尿液及两步提取后NNAL及其*O*-和*N*-葡萄糖醛酸酯的LC-MS/MS分析	准确度：94% 日内CV：3.0% 日间CV：5.7%	[235]
尿中的CEMA	96孔板混合模式阴离子交换纯化步骤后LC-MS/MS分析	准确度：98% 日内CV：6.4% 日间CV：6.6%	[236]
尿中的PheT	经*β*-葡萄糖醛酸酶和芳基硫酸酯酶处理，苯二乙烯基苯板96孔纯化后，GC-NICI-MS/MS分析	准确度：95% 日内CV：2.9% 日间CV：3.7%	[235]
生物效应标志物			
尿中的8-*iso*-$PGF_{2\alpha}$	单步纯化后LC-MS/MS分析	准确度：103% 日内CV：4.0% 日间CV：5.5%	[237]
口腔细胞中丙烯醛的DNA加合物	分离DNA，经水解和固相萃取后，用LC-MS/MS对加合物进行定量	准确度：96% 日内CV：1.6% 日间CV：3.4%	[238]

注：CV，变异系数；NICI，负离子化学电离

5.6 ENDS 和 HTP 生物标志物证据总结及其对公共卫生的影响

过去15年中，用于评估烟草和烟碱产品的生物标志物研究迅速增加，扩展了现有生物标志物的应用范围，提供了新的生物标志物，并揭示了使用者体内生物标志物水平的新证据。美国国家纵向PATH研究及其他大型纵向队列成为使用广泛生物标志物评估各种烟草制品使用者暴露和效应的重要平台。大量文献支持使用暴露生物标志物评估电子烟碱传输系统（ENDS），并新增了对口腔细胞DNA加合物及某些心肺生物效应标志物应用的支持。同样，这些暴露与效应生物标志物可能对评估加热型烟草制品（HTP）也有用；然而，目前关于HTP的生物标志物研究大多由烟草行业主导。关于易感性生物标志物NMR在评估ENDS或HTP使用者暴露与效应中的潜在作用的信息仍然有限。

5.6.1 现有数据总结及对公共卫生的影响

以下是对ENDS和HTP生物标志物数据的主要审查结论及其健康影响的概述。

从传统卷烟抽吸转为仅使用电子烟碱传输系统（ENDS）与减少暴露于多种有毒物质和致癌物，这些有毒物质和致癌物在吸烟引发的疾病中起关键作用。这一结论得到了大量文献的支持，包括基于PATH研究第一轮和第二轮的最新分析[74, 90]以及对使用ENDS进行戒烟试验的Cochrane系统评价的二次分析[239]。

对公共卫生的影响

- 这些减少暴露对健康影响的程度和性质尚未被充分理解。暴露生物标志物无法反映多个单一成分可能产生的联合效应，因此在预测疾病风险变化方面能力有限。例如，许多暴露标志物的水平（如挥发性有毒物质丙烯腈和丙烯醛的水平）在仅使用ENDS的使用者中高于未使用任何烟碱和烟草产品的人[88, 142, 170, 240-242]，而这些低水平暴露的影响尚未充分研究。此外，ENDS使用者可能接触某些有机磷酸酯阻燃剂（可能来自ENDS设备的污染物）[243]，且非靶向化学分析研究表明，ENDS气溶胶可能含有2000多种尚未被表征的化学成分[244]。
- 尽管对ENDS长期使用的影响知之甚少，但完全转向ENDS的吸烟者可能因减少接触许多烟草有害成分和致癌物而获益。

一组暴露生物标志物可用于确定或验证产品使用情况（表5.5）。5~6 ppm CO的截止点可以将燃烧型卷烟抽吸者与电子烟碱传输系统（ENDS）使用者和非使用者区分开来。ENDS使用者的尿液总烟碱当量（TNE，至少2000 pmol/mg肌酐）[242]和可替宁代谢物CEMA浓度较低（< 27 pmol/mL尿液），而非使用者的任何烟草和烟碱产品将有最小（基本上为零）的TNE [242]和CEMA（< 27 pmol/mL尿液）[245, 246]。如果参与者的烟碱使用状态不明确，可以测量NNAL（1~2 pmol/mL尿液）作为烟草特有的NNK的生物标志物，ENDS使用者中的NNK水平较低（0.023 pmol/mL尿液）[242, 247]。可以使用唾液中的丙二醇（ENDS使用者为3.5 μmol/mL，非使用者为0.004 μmol/mL，我们未发表的数据）来确认ENDS的使用。在使用这些临界值时，应考虑生物标志物的半衰期，特别是在研究近期从吸烟转为使用ENDS的情况时。此外，阿纳巴辛（一种次要烟草生物碱）不应出现在仅ENDS使用者中，少数研究已应用这一标志物进行分析。

表5.5　各种烟草和烟碱产品使用者的生物标志物的预期相对值

使用类别	尿生物标志物				唾液中的丙二醇
	CO	TNE	总NNAL	CEMA	
仅抽吸卷烟	高	高	高	高	低或ND
仅使用ENDS	低	高	低或ND	低或ND	高
ENDS和卷烟双重使用	变量	高	高	高	变量
不使用任何烟草和烟碱产品	低	低或ND	低或ND	低或ND	低或ND

注：ND，未检测到

公共卫生影响

- 使用提议的生物标志物组合对ENDS使用的暴露和效应进行研究非常重要。

ENDS和传统卷烟双重使用者在大多数暴露生物标志物方面并未出现明显的减少。吸烟量似乎是决定双重使用者暴露程度的主要因素。

公共卫生影响

- 双重使用者不太可能在生物效应方面比仅吸烟有所改善。
- 双重使用会令使用者暴露于与吸烟同样程度的烟草有毒物质和致癌物质，还会暴露于ENDS的释放物中。这种混合暴露的健康后果尚未明了；然而，一项系统评估表明，双重使用可能与仅抽吸卷烟相比涉及同样或显著更高的自我报告症状或疾病风险[248]。

口腔细胞中的DNA加合物是有用的致癌物剂量和生物效应标志物，可以用来比较电子烟碱传输系统（ENDS）与传统卷烟的效果。在可能的情况下，应更广泛地使用这类生物标志物。

公共卫生影响

- 如上所述，暴露生物标志物在预测疾病风险变化方面的能力有限。由于DNA加合物的形成是化学致癌过程中的关键步骤，这些生物标志物可能表明癌症风险。
- 此外，鉴于ENDS和ENNDS气溶胶中存在的醛类和其他炎症介质的活性，尿液暴露生物标志物可能无法完全捕捉气溶胶与人体组织（如口腔）直接接触处的重要暴露和生物效应。

关于电子烟碱传输系统（ENDS）使用后与心血管和呼吸效应相关的循环炎症细胞因子的研究结果并不一致。解释这些发现的一个挑战是，不同的研究使用了不同类型的生物标志物和生物标志物组合。

对公共卫生的影响

- 关于ENDS使用者这些生物标志物的明确数据是必要的，因为大量研究表明，ENDS可能是炎症暴露的一个来源，从而对使用者的呼吸和心血管造成影响。

具有临床前变化和症状指标的研究表明，与不使用任何产品相比，ENDS增加了呼吸和（潜在）心血管影响的风险。

对公共卫生的影响

- 虽然有报道称吸烟者改用ENDS后呼吸道症状有所改善，但不应鼓励前吸烟者长期使用ENDS。戒烟干预措施对于帮助使用者戒烟和不使用ENDS是有必要的。
- 从不吸烟的人使用ENDS可能会增加他们患病的风险。

对HTP使用者的生物标志物的独立研究严重缺乏。

对公共卫生的影响

- 尽管烟草行业的报告表明，吸烟者转向使用HTP后，暴露生物标志物和一些生物效应标志物显著减少，但需要独立的学术研究来证实这些发现。
- HTP使用者可能暴露于比ENDS更高水平的有害成分[249]；然而，目前还没有生物标志物的研究。

5.6.2 生物标志物的局限性

生物标志物的局限性主要集中在其特异性、稳定性以及测量的可行性方面。

- 许多暴露和生物效应标志物并不特定于某一烟草和烟碱产品，容易受到饮食、环境和职业暴露、健康状况及身体活动等因素的影响。因此，单一或组合生物标志物无法全面捕捉与烟草和烟碱产品相关的所有暴露或效应。

- 许多生物效应标志物并非针对一种疾病。例如，氧化应激和炎症作为癌症、心血管疾病及呼吸系统疾病的常见病理生理机制，其标志物难以区分个体对特定疾病的风险。
- 生物标志物的半衰期有限且多变。例如，呼出一氧化碳的半衰期仅约8小时，尿液中总NNAL的半衰期约为3.5周，而尿液中镉的半衰期可长达38年。在设计和解释研究时需考虑这种生物稳定性，特别是在吸烟历史或ENDS/HTP使用持续时间的背景下。
- 某些生物标志物的测量需要高度专业化的知识和仪器，这限制了其在更广泛研究中的应用。

5.6.3 研究空白

这篇关于ENDS和HTP使用研究中使用暴露、效应和易感性生物标志物的证据的综述指出了以下研究空白和优先事项：

- 对ENDS、HTP和其他新兴烟草和烟碱产品（如尼古丁袋）的独立（非烟草业）研究；
- 研究包括对产品释放物和生物标本进行非靶向分析，以确定ENDS和HTP特有的生物标志物，因为目前使用的大多数生物标志物都是基于暴露卷烟烟气的情况；
- 评估因使用不同设计和不同成分的ENDS和HTP而产生的生物标志物；
- 针对暴露于烟草和烟碱产品的生物效应标志物的研究，或者特定于个体健康影响（例如心血管疾病或呼吸系统疾病）；
- 通过横断面和纵向研究，评估和比较各国人群中ENDS和HTP暴露情况（包括与未使用任何产品的人群进行比较）；
- 对ENDS和HTP使用者的暴露、生物效应和临床疾病特异性表现进行系统研究，以更好地表征与各种生物标志物的关联以及生物标志物水平与特定疾病结果的关联；
- 更准确地描述和阐释同时使用两种及多种产品（尤其是不同吸烟程度下）所产生的影响的研究。

5.7 烟草控制生物标志物可能优先次序的建议

生物标志物是客观测量各种烟草和烟碱产品使用者的有害暴露及其与疾病病

理生理相关生物效应的重要工具。因此，暴露生物标志物和效应生物标志物在烟草控制中具有潜在应用价值。

暴露生物标志物　暴露生物标志物的优势在于它们考虑了产品使用模式对有害成分传递给使用者的影响。这些效应可能不会被实验室中标准化的、基于机器的产品测试所捕捉。因此，生物标志物可以用来更准确地描述产品的毒性和滥用潜力。根据所审查的证据，以下生物标志物组合被推荐用于评估电子烟碱传输系统（ENDS）和加热型烟草制品（HTP）使用者对与吸烟相关伤害有关的组分的暴露水平：尿液中的总烟碱当量（成瘾性）、NNAL（暴露于烟草衍生致癌物质）和CEMA（暴露于燃烧产物）。可以在这个组合中添加呼出一氧化碳和唾液中的丙二醇的测量；然而，呼出一氧化碳的半衰期很短，测量唾液中丙二醇的方法需要验证。本报告中审查的其他暴露生物标志物也可以使用；然而，所提出的有限组合可以提供关于关键有害成分类别暴露的足够信息，并允许按产品使用状态进行分类。

生物效应标志物　生物效应标志物解释了几种有害暴露之间的相互作用，以及可能无法通过产品测试或尿液暴露生物标志物捕获的独特暴露。根据所审查的证据，建议优先监测以下生物效应：

- 口腔细胞中的DNA加合物，由丙烯醛和其他挥发性有害成分（如甲醛）形成。在吸烟者、电子烟使用者和非烟草使用者之间，这些生物标志物的水平存在显著差异。由于DNA加合物的形成与癌症风险直接相关，监测这些标志物对于评估新型产品的潜在风险具有重要意义，不仅是相对于传统卷烟，还包括相对于不使用任何烟草和烟碱产品的情况。口腔细胞中的DNA加合物也可作为低水平但生物学相关的挥发性有害成分暴露的标志物。
- 临床前变化和症状指标。指标如血压、心率、动脉僵硬度、血小板反应性、呼吸症状（如咳嗽、哮喘和呼吸短促）以及呼吸功能测量（如FEV_1、FVC、FEV_1/FVC和DLCO）。这些指标通常在一般临床实践中评估，易于测量，建议将其纳入关于电子烟碱传输系统和加热型烟草制品使用的横断面和纵向研究中，作为呼吸和心血管风险的替代性指标。

尿异前列腺素和选定的细胞因子，在ENDS或HTP使用者与吸烟者的比较中获得了一致的结果，也可以使用。与非烟草和烟碱产品使用者相比，这些生物标志物水平在使用者组之间没有差异，而水平升高表明ENDS和HTP使用者中持续的系统性氧化应激和炎症。然而，对这些生物标志物的解释应包括认识到氧化应激和炎症并非暴露于这些产品所特有的。

易感性生物标志物　任何暴露于产品释放物中有害成分的人都可能面临风险（如成瘾或患病）。易感性生物标志物的价值在于识别特别容易受到有害影响的人群亚组。尽管对这些标志物的研究仍在发展，现有数据不足以支持其在烟草控制中的直接应用，但建议将烟碱代谢率（NMR）测量纳入监测ENDS和HTP使用者的TNE和总NNAL的方案和研究中。

对疾病风险进行直接流行病学评估 新兴的流行病学研究通常显示，ENDS的健康结果风险高于与卷烟烟气相关的暴露生物标志物水平所预期的风险。这部分原因可能是由于双重使用（即同时使用传统卷烟和ENDS）的高流行率，这可能没有在这些研究中被捕捉到，以及ENDS使用的独特生物学效应尚未得到很好的描述。因此，应记录直接监测因使用ENDS、HTP以及任何新型烟草制品而导致的健康效应。

5.8 解决研究空白和优先事项的建议

以下是针对独立研究以解决已确定的研究空白和优先事项的推荐策略：

- 开展研究，开发和使用暴露、效应和易感性生物标志物，以更好地描述ENDS、HTP和其他新兴产品对公共卫生的影响。
- 开展研究，比较使用HTP和ENDS的相对风险（暴露和生物效应），并与不使用烟草和烟碱产品进行比较。
- 建立横断面和纵向队列，实时监测ENDS和HTP使用者的暴露和生物效应以及其他新兴产品，向监管机构提供此类产品的相对风险和不断变化的产品特性的潜在影响。
- 对各类生物标志物和指标进行系统监测，以更好地了解暴露各类产品的差异如何转化为生物效应、临床前症状和疾病风险。
- 开展传播策略研究，使民众了解ENDS和HTP的暴露和影响；防止与卷烟同时使用；避免对产品风险的误解。

5.9 政 策 建 议

提出以下监管建议，供决策者、研究人员和公共卫生界审议：

- 在制定关于ENDS、HTP和其他新兴烟草和烟碱产品的政策决策时，要考虑基于生物标志物的发现（来自所有国家），依赖于独立于烟草或ENDS行业获得的数据，并考虑生物标志物的局限性。
- 优先考虑并支持独立研究，包括建立测量生物标志物和流行病学研究的能力，以解决与ENDS、HTP产品和其他新兴烟草和烟碱产品的公共卫生影响有关的研究空白和优先事项。
- 在具备必要能力的国家，监测ENDS、HTP和其他新兴烟草和烟碱产品使用者中推荐的生物标志物组。

- 向卫生保健专业人员和公众明确告知，目前没有证据表明使用HTP可以减少伤害。
- 鉴于新产品的推出速度之快以及科学研究在暴露和效应因素上的滞后，强烈鼓励成员国考虑要求制造商在允许这些产品在其国家市场销售之前提供以下信息：①选定有害化学物质的释放水平；②使用者中推荐生物标志物组合的水平。

参考文献

[1] Hecht SS, Murphy SE, Carmella SG, Li S, Jensen J, Le C et al. Similar uptake of lung carcinogens by smokers of regular light and ultra-light cigarettes. Cancer Epidemiol Biomarkers Prev. 2005;14:693-8. doi: 10.1158/1055-9965.EPI-04-0542.

[2] Harris JE, Thun MJ, Mondul AM, Calle EE. Cigarette tar yields in relation to mortality from lung cancer in the cancer prevention study II prospective cohort 1982–8. BMJ. 2004;328(7431):72-9. doi:10.1136/bmj.37936.585382.44.

[3] Burns DM, Dybing E, Gray N, Hecht S, Anderson C, Sanner T et al. Mandated lowering of toxicants in cigarette smoke: a description of the World Health Organization TobReg proposal. Tob Control. 2008;17(2):132-41. doi:10.1136/tc.2007.024158.

[4] Benowitz NL, Bernert JT, Foulds J, Hecht SS, Jacob P, Jarvis M. J et al. Biochemical verification of tobacco use and abstinence: 2019 update. Nicotine Tob. Res. 2020 22 (7) 1086-1097. doi:10.1093/ntr/ntz132.

[5] Mead AM, Geller AB, Teutsch SM, editors. Premium cigars: patterns of use, marketing, and health effects. Washington DC: National Academies Press; 2022. doi:10.17226/26421.

[6] Murphy SE. Biochemistry of nicotine metabolism and its relevance to lung cancer. J Biol Chem. 2021;296:100722. doi:10.1016/j.jbc.2021.100722.

[7] Hoffmann D, Hecht SS, Ornaf RM, Wynder EL. N′-Nitrosonornicotine in tobacco. Science. 1974;186:26-7. doi:10.1126/science.186.4160.265.

[8] Hoffmann D, Hecht SS. Nicotine-derived N-nitrosamines and tobacco related cancer: current status and future directions. Cancer Res. 1985;45:935-44. PMID:3882226.

[9] Hecht SS. Biochemistry biology and carcinogenicity of tobacco-specific N-nitrosamines. Chem Res Toxicol. 1998;11:559-603. doi:10.1021/tx980005y.

[10] Smokeless tobacco and some tobacco-specific N-nitrosamines (IARC Monographs on the Evaluation of Carcinogenic Risks to Humans, vol. 89). Lyon: International Agency for Research on Cancer; 2007 (https://publications.iarc.fr/Book-And-Report-Series/Iarc-Monographs-On-The-Identification-Of-Carcinogenic-Hazards-To-Humans/Smokeless-Tobacco-And-Some-Tobacco-specific-Em-N-Em-Nitrosamines-2007).

[11] Personal habits and indoor combustions. In: IARC Monographs on the Evaluation of Carcinogenic Risks to Humans, vol. 100E. Lyon: International Agency for Research on

Cancer; 2012:319-31 (https://publications.iarc.fr/Book-And-Report-Series/Iarc-Monographs-On-The-Identification-Of-Carcinogenic-Hazards-To-Humans/Personal-Habits-And-Indoor-Combustions-2012).

[12] Balbo S, James-Yi S, Johnson CS, O'Sullivan G, Stepanov I, Wang M et al. (*S*)-*N*′-Nitrosonornicotine a constituent of smokeless tobacco is a powerful oral cavity carcinogen in rats. Carcinogenesis. 2013;34:2178-83. doi:10.1093/carcin/bgt162.

[13] Balbo S, Johnson CS, Kovi RC, James-Yi SA, O'Sullivan MG, Wang M et al. Carcinogenicity and DNA adduct formation of 4-(methylnitrosamino)-1-(3-pyridyl)-1-butanone and enantiomersof its metabolite 4-(methylnitrosamino)-1-(3-pyridyl)-1-butanol in F-344 rats. Carcinogenesis. 2014;35(12):2798-806. doi:10.1093/carcin/bgu204.

[14] Hecht SS, Hatsukami DK Smokeless tobacco and cigarette smoking: chemical mechanisms and cancer prevention. Nature Rev. Cancer 22 143-155, 2022.

[15] Yuan JM, Butler LM, Stepanov I, Hecht SS. Urinary tobacco smoke-constituent biomarkers for assessing risk of lung cancer. Cancer Res. 2014;74(2):401-11. doi:10.1158/0008-5472.CAN-13-3178.

[16] Stepanov I, Sebero E, Wang R, Gao YT, Hecht SS, Yuan JM. Tobacco-specific N-nitrosamine exposures and cancer risk in the Shanghai Cohort Study: remarkable coherence with rat tumor sites. Int J Cancer. 2014;134(10):2278-83. doi:10.1002/ijc.28575.

[17] Bustamante G, Ma B, Yakovlev G, Yershova K, Le C, Jensen J et al. Presence of the carcinogen N′-nitrosonornicotine in saliva of e-cigarette users. Chem Res Toxicol. 2018;31(8):731-8. doi:10.1021/acs.chemrestox.8b00089.

[18] Snook ME, Severson RF, Arrendale RF, Higman HC, Chortyk OT. Multi-alkyated polynuclear aromatic hydrocarbons of tobacco smoke: separation and identification. Beitr Tabakforsch. 1978;9:222-47. doi:10.2478/cttr-2013-0452.

[19] Rodgman A, Perfetti T. The chemical components of tobacco and tobacco smoke. Boca Raton (FL): CRC Press; 2009 (https://www.routledge.com/The-Chemical-Components-of-Tobacco-and-Tobacco-Smoke/Rodgman-Perfetti/p/book/9781466515482).

[20] Li Y, Hecht SS. Carcinogenic components of tobacco and tobacco smoke: A 2022 update. Food Chem Toxicol. 2022;165:113179. doi:10.1016/j.fct.2022.113179.

[21] Scherer G, Pluym N, Scherer M. Intake and uptake of chemicals upon use of various tobacco/nicotine products: Can users be differentiated by single or combinations of biomarkers? Contrib Tob Nicotine Res. 2022;30:167-98. doi:10.2478/cttr-2021-0014.

[22] Hecht SS. Tobacco smoke carcinogens and lung cancer. In: Penning TM, editor. Chemical carcinogenesis – current cancer research. Cham: Springer; 2011:53-74.

[23] Some non-heterocyclic polycyclic aromatic hydrocarbons and some related exposures (IARC Monographs on the Evaluation of Carcinogenic Risks to Humans, vol. 92). Lyon: International Agency for Research on Cancer; 2010:35-818 (https://publications.iarc.fr/Book-And-Report-Series/Iarc-Monographs-On-The-Identification-Of-Carcinogenic-Hazards-To-Humans/Some-Non-heterocyclic-Polycyclic-Aromatic-Hydrocarbons-And-Some-Related-Exposures-2010).

[24] Jain RB. Contributions of dietary demographic disease lifestyle and other factors in explaining variabilities in concentrations of selected monohydroxylated polycyclic aromatic hydrocarbons

in urine: Data for US children adolescents and adults. Environ Pollut. 2020;266(1):115178. doi:10.1016/j.envpol.2020.115178.

[25] Conney AH. Induction of microsomal enzymes by foreign chemicals and carcinogenesis by-polycyclic aromatic hydrocarbons: GHA Clowes Memorial Lecture. Cancer Res. 1982;42:4875–917. PMID:6814745.

[26] Zhong Y, Carmella SG, Hochalter JB, Balbo S, Hecht SS. Analysis of r-t-8,9 c-10-tetrahydroxy-7,8,9,10-tetrahydrobenzo[a]pyrene in human urine: a biomarker for directly assessing carcinogenic polycyclic aromatic hydrocarbon exposure plus metabolic activation. Chem Res Toxicol. 2011;24:73-80. doi:10.1021/tx100287n.

[27] Agents classified by the IARC Monographs volumes 1-132. Lyon: International Agency for Research on Cancer; 2022 (https://monographs.iarc.who.int/agents-classified-by-the-iarc/, accessed 29 Novermber 2022).

[28] Larsen ST, Nielsen GD. Effects of methacrolein on the respiratory tract in mice. Toxicol Lett. 2000;114(1-3):197-202. doi:10.1016/s0378-4274(99)00300-8.

[29] Morgan DL, Price HC, O'Connor RW, Seely JC, Ward SM, Wilson RE et al. Upper respiratory tract toxicity of inhaled methylvinyl ketone in F344 rats and B6C3F1 mice. Toxicol Sci. 2000;58(1):182-94. doi:10.1093/toxsci/58.1.182.

[30] Acrolein, crotonaldehyde and arecoline (IARC Monographs on the Identification of Carcinogenic Hazards to Humans, vol. 128). Lyon: International Agency for Research on Cancer; 2021 (https://publications.iarc.fr/Book-And-Report-Series/Iarc-Monographs-On-The-Identification-Of-Carcinogenic-Hazards-To-Humans/Acrolein-Crotonaldehyde-And-Arecoline-2021).

[31] Haussmann HJ. Use of hazard indices for a theoretical evaluation of cigarette smoke composition. Chem Res Toxicol. 2012;25(4):794-810. doi:10.1021/tx200536w.

[32] Burcham PC. Acrolein and human disease: Untangling the knotty exposure scenarios accompanying several diverse disorders. Chem Res Toxicol. 2017;30(1):145-61. doi:10.1021/acs.chemrestox.6b00310.

[33] Mathias PI, B'Hymer C. Mercapturic acids: recent advances in their determination by liquid chromatography/mass spectrometry and their use in toxicant metabolism studies and in occupational and environmental exposure studies. Biomarkers. 2016;21(4):293-315. doi:10.3109/1354750X.2016.1141988.

[34] Chen M, Carmella SG, Li Y, Zhao Y, Hecht SS. Resolution and quantitation of mercapturic acids derived from crotonaldehyde, methacrolein and methyl vinyl ketone in the urine of smokers and nonsmokers. Chem Res Toxicol. 2020;33(2):669-77. doi:10.1021/acs.

[35] Scherer G, Urban M, Engl J, Hagedorn HW, Riedel K. Influence of smoking charcoal filter tipped cigarettes on various biomarkers of exposure. Inhal Toxicol. 2006;18(10):821-9. doi:10.1080/08958370600747945.

[36] Scherer G, Urban M, Hagedorn HW, Feng S, Kinser RD, Sarkar M et al. Determination of two mercapturic acids related to crotonaldehyde in human urine: influence of smoking. Hum Exp Toxicol. 2007;26(1):37-47. doi:10.1177/0960327107073829.

[37] Carmella SG, Chen M, Han S, Briggs A, Jensen J, Hatsukami DK et al. Effects of smoking cessation on eight urinary tobacco carcinogen and toxicant biomarkers. Chem Res Toxicol. 2009;22(4):734-41. doi:10.1021/tx800479s.

[38] Alwis KU, Blount BC, Britt AS, Patel D, Ashley DL. Simultaneous analysis of 28 urinary VOC metabolites using ultra high performance liquid chromatography coupled with electrospray ionization tandem mass spectrometry (UPLC-ESI/MSMS). Anal Chim Acta. 2012;750:152-60. doi:10.1016/j.aca.2012.04.009.

[39] Wei B, Alwis KU, Li Z, Wang L, Valentin-Blasini L, Sosnoff CS et al. Urinary concentrations of PAH and VOC metabolites in marijuana users. Environ Int. 2016;88:1-8. doi:10.1016/j.envint.2015.12.003.

[40] Pluym N, Gilch G, Scherer G, Scherer M. Analysis of 18 urinary mercapturic acids by two high-throughput multiplex-LC-MS/MS methods. Anal Bioanal Chem. 2015;407(18):5463-76. doi:10.1007/s00216-015-8719-x.

[41] Bagchi P, Geldner N, de Castro BR, De Jesus VR, Park SK, Blount BC. Crotonaldehyde exposure in US tobacco smokers and nonsmokers: NHANES 2005-2006 and 2011-2012. Environ Res. 2018;163:1-9. doi:10.1016/j.envres.2018.01.033.

[42] Hatsukami DK, Luo X, Jensen JA, al'Absi M, Allen SS, Carmella SG et al. Effect of immediate vs gradual reduction in nicotine content of cigarettes on biomarkers of smoke exposure: a randomized cinical trial. JAMA. 2018;320(9):880-91. doi:10.1001/jama.2018.11473.

[43] Frigerio G, Mercadante R, Polledri E, Missineo P, Campo L, Fustinoni S. An LC-MS/MS method to profile urinary mercapturic acids metabolites of electrophilic intermediates of occupational and environmental toxicants. J Chromatogr B Anal Technol Biomed Life Sci. 2019;1117:66-76. doi:10.1016/j.jchromb.2019.04.015.

[44] Kotapati S, Esades A, Matter B, Le C, Tretyakova N. High throughput HPLC-ESI--MS/MS methodology for mercapturic acid metabolites of 1,3-butadiene: biomarkers of exposure and bioactivation. Chem Biol Interact. 2015;241:23-31. doi:10.1016/j.cbi.2015.02.009.

[45] Ashley DL, Bonin MA, Cardinali FL, McCraw JM, Wooten JV. Measurement of volatile organic compounds in human blood. Environ Health Perspect. 1996;104(Suppl_5):871-7. doi:10.1289/ehp.96104s5871.

[46] Ashley DL, Bonin MA, Hamar B, McGeehin MA. Removing the smoking confounder from blood volatile organic compounds measurements. Environ Res. 1995;71(1):39-45. doi:10.1006/enrs.1995.1065.

[47] Luo X, Carmella SG, Chen M, Jensen JA, Wilkens LR, Le Marchand L et al. Urinary cyanoethyl mercapturic acid, a biomarker of the smoke toxicant acrylonitrile, clearly distinguishes smokers from non-smokers. Nicotine Tob Res. 2020;22:1744-7. doi:10.1093/ntr/ntaa080.

[48] Cancelada L, Sleiman M, Tang X, Russell ML, Montesinos VN, Litter MI et al. Heated tobacco products: volatile emissions and their predicted impact on indoor air quality. Environ Sci Technol. 2019;53(13):7866-76. doi:10.1021/acs.est.9b02544.

[49] Farsalinos KE, Yannovits N, Sarri T, Voudris V, Poulas K, Leischow SJ. Carbonyl emissions from a novel heated tobacco product (IQOS): comparison with an e-cigarette and a tobacco cigarette. Addiction. 2018;113(11):2099-106. doi:10.1111/add.14365.

[50] Pappas RS, Fresquez MR, Martone N, Watson CH. Toxic metal concentrations in mainstream smoke from cigarettes available in the USA. J Anal Toxicol. 2014;38(4):204-11. doi:10.1093/jat/bku013.

[51] Marano KM, Naufal ZS, Kathman SJ, Bodnar JA, Borgerding MF, Garner CD et al. Cadmium

exposure and tobacco consumption: biomarkers and risk assessment. Regul Toxicol Pharmacol. 2012;6 (2):243-52. doi:10.1016/j.yrtph.2012.07.008.

[52] Hoffmann K, Becker K, Friedrich C, Helm D, Krause C, Seifert B. The German Environmental Survey 1990/1992 (GerES II): cadmium in blood urine and hair of adults and children. J Expo Anal Environ. Epidemiol. 2000;10:126-35. doi:10.1038/sj.jea.7500081.

[53] Arsenic metals fibres and dusts. In: IARC Monographs on the Evaluation of Carcinogenic Risks to Humans, vol. 100C. Lyon: International Agency for Research on Cancer; 2012:121-45 (https://monographs.iarc.who.int/wp-content/uploads/2018/06/mono100C.pdf).

[54] Toxicological profile for lead. Atlanta (GA): Agency for Toxic Substances and Disease Registry; 2020 (https://www.atsdr.cdc.gov/toxprofiles/tp13.pdf).

[55] Belushkin M, Tafin Djoko D, Esposito M, Korneliou A, Jeannet C, Lazzerini M et al. Selected harmful and potentially harmful constituents levels in commercial e-cigarettes. Chem Res Toxicol. 2020;33(2):657-68. doi:10.1021/acs.chemrestox.9b00470.

[56] Margham J, McAdam K, Forster M, Liu C, Wright C, Mariner D et al. Chemical composition of aerosol from an e-cigarette: a quantitative comparison with cigarette smoke. Chem Res Toxicol. 2016;29(10):1662-78. doi:10.1021/acs.chemrestox.6b00188.

[57] Eshraghian EA, Al-Delaimy WK. A review of constituents identified in e-cigarette liquids and aerosols. Tob Prev Cessat. 2021;7:10. doi:10.18332/tpc/131111.

[58] Gaur S, Agnihotri R. Health effects of trace metals in electronic cigarette aerosols – a systematic review. Biol Trace Elem Res. 2019;188(2):295-315. doi:10.1007/s12011-018-1423-x.

[59] Paschal DC, Burt V, Caudill SP, Gunter EW, Pirkle JL, Sampson EJ et al. Exposure of the US population aged 6 years and older to cadmium: 1988–1994. Arch Environ Contam Toxicol. 2000;38:377-83. doi:10.1007/s002449910050.

[60] Akiyama Y, Sherwood N. Systematic review of biomarker findings from clinical studies of electronic cigarettes and heated tobacco products. Toxicol Rep. 2021;8:282-94. doi:10.1016/j.toxrep.2021.01.014.

[61] Oliveri D, Liang Q, Sarkar M. Real-world evidence of differences in biomarkers of exposure to select harmful and potentially harmful constituents and biomarkers of potential harm between adult e-vapor users and adult cigarette smokers. Nicotine Tob Res. 2020;22(7):1114-22. doi:10.1093/ntr/ntz185.

[62] Hatsukami DK, Meier E, Lindgren BR, Anderson A, Reisinger SA, Norton KJ et al. A randomized clinical trial examining the effects of instructions for electronic cigarette use on smoking-related behaviors and biomarkers of exposure. Nicotine Tob. Res. 2020;22(9):1524-32. doi:10.1093/ntr/ntz233.

[63] McRobbie H, Phillips A, Goniewicz ML, Smith KM, Knight-West O, Przulj D et al. Effects of switching to electronic cigarettes with and without concurrent smoking on exposure to nicotine carbon monoxide and acrolein. Cancer Prev Res. 2015;8(9):873-8. doi:10.1158/1940-6207.CAPR-15-0058.

[64] Czoli CD, Fong GT, Goniewicz ML, Hammond D. Biomarkers of exposure among "dual users" of tobacco cigarettes and electronic cigarettes in Canada. Nicotine Tob Res. 2019;21(9):1259-66. doi:10.1093/ntr/nty174.

[65] O'Connell G, Graff DW, D'Ruiz CD. Reductions in biomarkers of exposure (BoE) to harmful

or potentially harmful constituents (HPHCs) following partial or complete substitution of cigarettes with electronic cigarettes in adult smokers. Toxicol Mech Meth. 2016;26(6):443-54. doi:10.1080/15376516.2016.1196282.

[66] Cravo AS, Bush J, Sharma G, Savioz R, Martin C, Craige S et al. A randomised parallel group study to evaluate the safety profile of an electronic vapour product over 12 weeks. Regul Toxicol Pharmacol. 2016;81)Suppl–1):S1-14. doi:10.1016/j.yrtph.2016.10.003.

[67] Morris P, McDermott S, Chapman F, Verron T, Cahours X, Stevenson M et al. Reductions in biomarkers of exposure to selected harmful and potentially harmful constituents following exclusive and partial switching from combustible cigarettes to myblu electronic nicotine delivery systems (ENDS). Intern Emerg Med. 2022;17(2):397-410. doi:10.1007/s11739-021-02813-w.

[68] Gale N, McEwan M, Eldridge AC, Fearon IM, Sherwood N, Bowen E et al. Changes in biomarkers of exposure on switching from a conventional cigarette to tobacco heating products: a randomized controlled study in healthy Japanese subjects. Nicotine Tob Res. 2019;21(9):1220-7. doi:10.1093/ntr/nty104.

[69] Haziza C, de La Bourdonnaye G, Donelli A, Poux V, Skiada D, Weitkunat R et al. Reduction in exposure to selected harmful and potentially harmful constituents approaching those observed upon smoking abstinence in smokers switching to the menthol tobacco heating system 2.2 for 3 months (Part 1). Nicotine Tob Res. 2020;22(4):539-48. doi:10.1093/ntr/ntz013.

[70] Round EK, Chen P, Taylor AK, Schmidt E. Biomarkers of tobacco exposure decrease after smokers switch to an e-cigarette or nicotine gum. Nicotine Tob. Res. 2019;21(9):1239-47. doi:10.1093/ntr/nty140.

[71] Shahab L, Goniewicz ML, Blount BC, Brown J, McNeill A, Alwis KU et al. Nicotine, carcinogen and toxin exposure in long-term e-cigarette and nicotine replacement therapy users: a cross-sectional study. Ann Intern Med. 2017;166(6):390-400. doi:10.7326/M16-1107.

[72] Ludicke F, Picavet P, Baker G, Haziza C, Poux V, Lama N et al. Effects of switching to the tobacco heating system 2.2 menthol smoking abstinence or continued cigarette smoking on biomarkers of exposure: a randomized controlled open-label multicenter study in sequential confinement and ambulatory settings (Part 1). Nicotine Tob Res. 2018;20(2):161-72. doi:10.1093/ntr/ntx028.

[73] Goniewicz ML, Smith DM, Edwards KC, Blount BC, Caldwell KL, Feng J et al. Comparison of nicotine and toxicant exposure in users of electronic cigarettes and combustible cigarettes. JAMA Netw Open. 2018;1(8):e185937. doi:10.1001/jamanetworkopen.2018.5937.

[74] Dai H, Benowitz NL, Achutan C, Farazi PA, Degarege A, Khan AS. Exposure to toxicants associated with use and transitions between cigarettes, e-cigarettes and no tobacco. JAMA Netw Open. 2022;5(2):e2147891. doi:10.1001/jamanetworkopen.2021.47891.

[75] Hecht SS, Carmella SG, Chen M, Koch JFD, Miller AT, Murphy SE et al. Quantitation of urinary metabolites of a tobacco-specific lung carcinogen after smoking cessation. Cancer Res. 1999;59:590-96. PMID:9973205.

[76] Goniewicz ML, Havel CM, Peng MW, Jacob P III, Dempsey D, Yu L et al. Elimination kinetics of the tobacco-specific biomarker and lung carcinogen 4-(methylnitrosamino)-1-(3-pyridyl)-1-butanol. Cancer Epidemiol Biomarkers Prev. 2009;18(12):3421-5. doi:10.1158/1055-9965.EPI-09-0874.

[77] Hecht SS, Carmella SG, Murphy SE, Akerkar S, Brunnemann KD, Hoffmann D. A tobac-

co-specific lung carcinogen in the urine of men exposed to cigarette smoke. N Engl J Med. 1993;329:1543-6. doi:10.1056/NEJM199311183292105.

[78] Kotandeniya D, Carmella SG, Pillsbury ME, Hecht SS. Combined analysis of *N′*-nitrosonornicotine and 4-(methylnitrosamino)-1-(3-pyridyl)-1-butanol in the urine of cigarette smokers and e-cigarette users. J Chromatogr B Analyt Technol Biomed Life Sci. 2015;1007:121-6. doi:10.1016/j.jchromb.2015.10.012.

[79] Scherer G, Scherer M, Mutze J, Hauke T, Pluym N. Assessment of the exposure to tobacco-specific nitrosamines and minor tobacco alkaloids in users of various tobacco/nicotine products. Chem Res Toxicol. 2022;35(4):684-93. doi:10.1021/acs.chemrestox.2c00020.

[80] Hecht SS, Carmella SG, Kotandeniya D, Pillsbury ME, Chen M, Ransom BW et al. Evaluation of toxicant and carcinogen metabolites in the urine of e-cigarette users versus cigarette smokers. Nicotine Tob Res. 2015;17(6):704-9. doi:10.1093/ntr/ntu218.

[81] Scherer G, Pluym N, Scherer M. Comparison of urinary mercapturic acid excretions in users of various tobacco/nicotine products. Drug Test Anal. 2022. doi:10.1002/dta.3372.

[82] Chen M, Carmella SG, Lindgren BR, Luo X, Ikuemonisan J, Niesen B et al. Increased levels of the acrolein metabolite 3-hydroxypropyl mercapturic acid in the urine of e-cigarette users. Chem Res Toxicol. 2022. doi:10.1021/acs.chemrestox.2c00145.

[83] Prokopowicz A, Sobczak A, Szula-Chraplewska M, Ochota P, Kosmider L. Exposure to cadmium and lead in cigarette smokers who switched to electronic cigarettes. Nicotine Tob Res. 2019;21(9):1198-205. doi:10.1093/ntr/nty161.

[84] Prokopowicz A, Sobczak A, Szdzuj J, Grygoyc K, Kosmider L. Metal concentration assessment in the urine of cigarette smokers who switched to electronic cigarettes: a pilot study. Int J Environ Res Public Health. 2020;17(6):1877. doi:10.3390/ijerph17061877.

[85] Hiler M, Weidner AS, Hull LC, Kurti AN, Mishina EV. Systemic biomarkers of exposure associated with ENDS use: A scoping review. Tob Control. 2021. doi:10.1136/tobaccocontrol-2021-056896.

[86] Baig SA, Giovenco DP. Behavioral heterogeneity among cigarette and e-cigarette dual-users and associations with future tobacco use: findings from the Population Assessment of Tobacco and Health Study. Addict Behav. 2020;104:106263. doi:10.1016/j.addbeh.2019.106263.

[87] Smith DM, Shahab L, Blount BC, Gawron M, Kosminder L, Sobczak A et al. Differences in exposure to nicotine, tobacco-specific nitrosamines, and volatile organic compounds among electronic cigarette users, tobacco smokers, and dual users from three countries. Toxics. 2020;8(4):88. doi:10.3390/toxics8040088.

[88] Keith RJ, Fetterman JL, Orimoloye OA, Dardari Z, Lorkiewicz PK, Hamburg NM et al. Characterization of volatile organic compound metabolites in cigarette smokers, electronic nicotine device users, dual users, and nonusers of tobacco. Nicotine Tob Res. 2020;22(2):264-72. doi:10.1093/ntr/ntz021.

[89] Piper ME, Baker TB, Benowitz NL, Kobinsky KH, Jorenby DE. Dual users compared to smokers: demographics, dependence and biomarkers. Nicotine Tob Res. 2019;21(9):1279-84. doi:10.1093/ntr/nty231.

[90] Anic GM, Rostron BL, Hammad HT, van Bemmel DM, Del Valle-Pinero AY, Christensen CH et al. Changes in biomarkers of tobacco exposure among cigarette smokers transitioning to ENDS

use: the Population Assessment of Tobacco and Health Study 2013-2015. Int J Environ Res Public Health. 2022;19(3):1462. doi:10.3390/ijerph19031462.

[91] Scherer G. Suitability of biomarkers of biological effects (BOBEs) for assessing the likelihood of reducing the tobacco related disease risk by new and innovative tobacco products: a literature review. Regul Toxicol Pharmacol. 2018;94:203-33. doi:10.1016/j.yrtph.2018.02.002.

[92] Chang CM, Cheng YC, Cho M, Mishina E, Del Valle-Pinero AY, van Bemmel D et al. Biomarkers of potential harm: summary of an FDA-sponsored public workshop. Nicotine Tob Res. 2019;21(1):3-13. doi:10.1093/ntr/ntx273.

[93] Chang JT, Vivar JC, Tam J, Hammad HT, Christensen CH, van Bemmel DM et al. Biomarkers of potential harm among adult cigarette and smokeless tobacco users in the PATH Study wave 1 (2013–2014): a cross-sectional analysis. Cancer Epidemiol Biomarkers Prev. 2021;30(7):1320-7. doi:10.1158/1055-9965.EPI-20-1544.

[94] Stratton K, Shetty P, Wallace R, Bondurant S. Clearing the smoke: assessing the science base for tobacco harm reduction. Washington DC: Institute of Medicine; 2001. doi:10.17226/10029.

[95] Gregg EO, Fisher AL, Lowe F, McEwan M, Massey ED. An approach to the validation of biomarkers of harm for use in a tobacco context. Regul Toxicol Pharmacol. 2006;44(3):262-7. doi:10.1016/j.yrtph.2005.12.006.

[96] Leemans CR, Snijders PJF, Brakenhoff RH. The molecular landscape of head and neck cancer. Nat Rev Cancer. 2018;18(5):269-82. doi:10.1038/nrc.2018.11.

[97] Hecht SS. Tobacco smoke carcinogens and lung cancer. J Natl Cancer Inst. 1999;91:1194-210.

[98] Basu AK. DNA damage, mutagenesis and cancer. Int J Mol Sci. 2018;19(4):970. doi:10.3390/ijms19040970.

[99] Hwa Yun B, Guo J, Bellamri M, Turesky RJ. DNA adducts: formation biological effects and new biospecimens for mass spectrometric measurements in humans. Mass Spectrom Rev. 2020;39(1-2):55-82. doi:10.1002/mas.21570.

[100] Geacintov NE, Broyde S. Repair-resistant DNA lesions. Chem Res Toxicol. 2017;30(8)1517-48. doi:10.1021/acs.chemrestox.7b00128.

[101] Ma B, Stepanov I, Hecht SS. Recent studies on DNA adducts resulting from human exposure to tobacco smoke. Toxics. 2019;7(1):16. doi:10.3390/toxics7010016.

[102] Tobacco smoke and involuntary smoking (IARC Monographs on the Evaluation of Carcinogenic Risks to Humans, vol. 83). Lyon: International Agency for Research on Cancer; 2004:35-102 (https://publications.iarc.fr/Book-And-Report-Series/Iarc-Monographs-On-The-Identification-Of-Carcinogenic-Hazards-To-Humans/Tobacco-Smoke-And-Involuntary-Smoking-2004).

[103] Phillips DH, Venitt S. DNA and protein adducts in human tissues resulting from exposure to tobacco smoke. Int J Cancer. 2012;131(12):2733-53. doi:10.1002/ijc.27827.

[104] Boysen G, Hecht SS. Analysis of DNA and protein adducts of benzo[a]pyrene in human tissues using structure-specific methods. Mutat Res. 2003;543:17-30. doi:10.1016/s1383-5742(02)00068-6.

[105] Hecht SS. Oral cell DNA adducts as potential biomarkers for lung cancer susceptibility in cigarette smokers. Chem Res Toxicol. 2017;30(1):367-75. doi:10.1021/acs.chemrestox.6b00372.

[106] Paiano V, Maertens L, Guidolin V, Yang J, Balbo S, Hecht SS. Quantitative liquid chromatography–nanoelectrospray ionization–high-resolution tandem mass spectrometry analysis of acro-

lein–DNA adducts and etheno-DNA adducts in oral cells from cigarette smokers and nonsmokers. Chem Res Toxicol. 2020;33(8):2197-207. doi:10.1021/acs.chemrestox.

[107] Frost-Pineda K, Liang Q, Liu J, Rimmer L, Jin Y, Feng S et al. Biomarkers of potential harm among adult smokers and nonsmokers in the total exposure study. Nicotine Tob Res. 2011;13(3):182-93. doi:10.1093/ntr/ntq235.

[108] Shiels MS, Pfeiffer RM, Hildesheim A, Engels EA, Kemp TJ, Park JH et al. Circulating inflammation markers and prospective risk for lung cancer. J Natl Cancer Inst. 2013;105:1871-80. doi:10.1093/jnci/djt309.

[109] Holvoet P, Jenny NS, Schreiner PJ, Tracy RP, Jacobs DR, Multi-Ethnic Study of Atherosclerosis. The relationship between oxidized LDL and other cardiovascular risk factors and subclinical CVD in different ethnic groups: the Multi-Ethnic Study of Atherosclerosis (MESA). Atherosclerosis. 2007;194(1):245-52. doi:10.1016/j.atherosclerosis.2006.08.002.

[110] Libby P. Inflammation during the life cycle of the atherosclerotic plaque. Cardiovasc Res. 2021;117(13):2525-36. doi:10.1093/cvr/cvab303.

[111] Mendall MA, Patel P, Asante M, Ballam L, Morris J, Strachan DP et al. Relation of serum cytokine concentrations to cardiovascular risk factors and coronary heart disease. Heart. 1997;78(3):273-7. doi:10.1136/hrt.78.3.273.

[112] Tuut M, Hense HW. Smoking other risk factors and fibrinogen levels. Evidence of effect modification. Ann Epidemiol. 2001;11(4):232-8. doi:10.1016/s1047-2797(00)00226-x.

[113] Luc G, Arveiler D, Evans A, Amouyel P, Ferrieres J, Bard JM et al. Circulating soluble adhesion molecules ICAM-1 and VCAM-1 and incident coronary heart disease: the PRIME Study. Atherosclerosis. 2003;170(1):169-76. doi:10.1016/s0021-9150(03)00280-6.

[114] Tracy RP, Psaty BM, Macy E, Bovill EG, Cushman M, Cornell ES et al. Lifetime smoking exposure affects the association of C-reactive protein with cardiovascular disease risk factors and subclinical disease in healthy elderly subjects. Arterioscler Thromb Vasc Biol. 1997;17(10):2167-76. doi:10.1161/01.atv.17.10.2167.

[115] Finkelstein R, Fraser RS, Ghezzo H, Cosio MG. Alveolar inflammation and its relation to emphysema in smokers. Am J Respir Crit Care Med. 1995;152(5:1):1666-72. doi:10.1164/ajrccm.152.5.7582312.

[116] Sullivan AK, Simonian PL, Falta MT, Mitchell JD, Cosgrove GP, Brown K et al. Oligoclonal CD4+ T cells in the lungs of patients with severe emphysema. Am J Respir Crit Care Med. 2005;172(5):590-6. doi:10.1164/rccm.200410-1332OC.

[117] Majo J, Ghezzo H, Cosio MG. Lymphocyte population and apoptosis in the lungs of smokers and their relation to emphysema. Eur Respir J. 2001;17(5):946-53. doi:10.1183/.

[118] Moy ML, Teylan M, Weston NA, Gagnon DR, Danilack VA, Garshick E. Daily step count is associated with plasma C-reactive protein and IL-6 in a US cohort with COPD. Chest. 2014;145(3):54250. doi:10.1378/chest.13-1052.

[119] Celli BR, Locantore N, Yates J, Tal-Singer R, Miller BE, Bakke P et al. Inflammatory biomarkers improve clinical prediction of mortality in chronic obstructive pulmonary disease. Am J Respir Crit Care Med. 2012;185(10):1065-72. doi:10.1164/rccm.201110-1792OC.

[120] Dickens JA, Miller BE, Edwards LD, Silverman EK, Lomas DA, Tal-Singer R et al. COPD association and repeatability of blood biomarkers in the ECLIPSE cohort. Respir Res. 2011;12:146.

doi:10.1186/1465-9921-12-146.

[121] Agusti A, Edwards LD, Rennard SI, MacNee W, Tal-Singer R, Miller BE et al. Persistent systemic inflammation is associated with poor clinical outcomes in COPD: a novel phenotype. PLoS One. 2012;7(5):e37483. doi:10.1371/journal.pone.0037483.

[122] Pinto-Plata V, Casanova C, Mullerova H, de Torres JP, Corado H, Varo N et al. Inflammatory and repair serum biomarker pattern: association to clinical outcomes in COPD. Respir Res. 2012;13:71. doi:10.1186/1465-9921-13-71.

[123] Tsai JJ, Liao EC, Hsu JY, Lee WJ, Lai YK. The differences of eosinophil- and neutrophil-related inflammation in elderly allergic and non-allergic chronic obstructive pulmonary disease. J Asthma. 2010;47(9):1040-4. doi:10.1080/02770903.2010.491145.

[124] Man SF, Xuekui Z, Vessey R, Walker T, Lee K, Park D et al. The effects of inhaled and oral corticosteroids on serum inflammatory biomarkers in COPD: an exploratory study. Ther Adv Respir Dis. 2009;3(2):73-80. doi:10.1177/1753465809336697.

[125] Thomsen M, Ingebrigtsen TS, Marott JL, Dahl M, Lange P, Vestbo J et al. Inflammatory biomarkers and exacerbations in chronic obstructive pulmonary disease. JAMA. 2013;309(22):2353-61. doi:10.1001/jama.2013.5732.

[126] van Durme YM, Verhamme KM, Aarnoudse AJ, Van Pottelberge GR, Hofman A, Witteman JC et al. C-reactive protein levels haplotypes and the risk of incident chronic obstructive pulmonary disease. Am J Respir Crit Care Med. 2009;179(5):375-82. doi:10.1164/rccm.200810-1540OC.

[127] Thomsen M, Dahl M, Lange P, Vestbo J, Nordestgaard BG. Inflammatory biomarkers and comorbidities in chronic obstructive pulmonary disease. Am J Respir Crit Care Med. 2012;186(10):982-8. doi:10.1164/rccm.201206-1113OC.

[128] Higashimoto Y, Iwata T, Okada M, Satoh H, Fukuda K, Tohda Y. Serum biomarkers as predictors of lung function decline in chronic obstructive pulmonary disease. Respir Med. 2009;103(8):1231-8. doi:10.1016/j.rmed.2009.01.021.

[129] Mannino DM, Valvi D, Mullerova H, Tal-Singer R. Fibrinogen COPD and mortality in a nationally representative US cohort. COPD. 2012;9(4):359-66. doi:10.3109/15412555.

[130] Cockayne DA, Cheng DT, Waschki B, Sridhar S, Ravindran P, Hilton H et al. Systemic biomarkers of neutrophilic inflammation tissue injury and repair in COPD patients with differing levels of disease severity. PLoS One. 2012;7(6):e38629. doi:10.1371/journal.

[131] Dahl M, Tybjaerg-Hansen A, Vestbo J, Lange P, Nordestgaard BG. Elevated plasma fibrinogen associated with reduced pulmonary function and increased risk of chronic obstructive pulmonary disease. Am J Respir Crit Care Med. 2001;164(6):1008-11. doi:10.1164/ajrccm.

[132] Milne GL, Yin H, Hardy KD, Davies SS, Roberts LJ. Isoprostane generation and function. Chem Rev. 2011;111(10):5973-96. doi:10.1021/cr200160h.

[133] How tobacco smoke causes disease: the biology and behavioral basis for smoking-attributable disease: a report of the Surgeon General. Atlanta (GA): Centers for Disease Control and Prevention, Office on Smoking and Health; 2010 (https://www.ncbi.nlm.nih.gov/books/NBK53017/).

[134] Yuan JM, Carmella SG, Wang R, Tan YT, Adams-Haduch J, Gao YT et al. Relationship of the oxidative damage biomarker 8-epi-prostaglandin F2 alpha to risk of lung cancer development in the Shanghai Cohort Study. Carcinogenesis. 2018;39(7):948-54. doi:10.1093/carcin/bgy060.

[135] Christensen CH, Chang JT, Rostron BL, Hammad HT, van Bemmel DM, Del Valle-Pinero

AY et al. Biomarkers of inflammation and oxidative stress among adult former smoker current e-cigarette users-results from wave 1 PATH Study. Cancer Epidemiol Biomarkers Prev. 2021;30(10):1947-55. doi:10.1158/1055-9965.EPI-21-0140.

[136] McElroy JP, Carmella SG, Heskin AK, Tang MK, Murphy SE, Reisinger SA et al. Effects of cessation of cigarette smoking on eicosanoid biomarkers of inflammation and oxidative damage. PLoS One. 2019;14(6):e0218386. doi:0.1371/journal.pone.0218386.

[137] Guo J, Hecht SS. DNA damage in human oral cells induced by use of e-cigarettes. Drug Test Anal. 2022. doi:10.1002/dta.3375.

[138] Cheng G, Guo J, Carmella SG, Lindgren B, Ikuemonisan J, Niesen B et al. Increased acrolein-DNA adducts in buccal brushings of e-cigarette users. Carcinogenesis. 2022;43(5):437-44. doi:10.1093/carcin/bgac026.

[139] Guo J, Ikuemonisan J, Hatsukami DK, Hecht SS. Liquid chromatography–nanoelectrospray ionization–high-resolution tandem mass spectrometry analysis of apurinic/apyrimidinic sites in oral cell DNA of cigarette smokers e-cigarette users and nonsmokers. Chem Res Toxicol. 2021;34(12):2540-8. doi:10.1021/acs.chemrestox.1c00308.

[140] Okafor CN, Okafor N, Kaliszewski C, Wang L. Association between electronic cigarette and combustible cigarette use with cardiometabolic risk biomarkers among US adults. Ann Epidemiol. 2022;71:44-50. doi:10.1016/j.annepidem.2022.02.002.

[141] Benowitz NL, St Helen G, Nardone N, Addo N, Zhang JJ, Harvanko AM et al. Twenty-fourhour cardiovascular effects of electronic cigarettes compared with cigarette smoking in dual users. J Am Heart Assoc. 2020;9(23):e017317. doi:10.1161/JAHA.120.017317.

[142] Perez MF, Mead EL, Atuegwu NC, Mortensen EM, Goniewicz M, Oncken C. Biomarkers of toxicant exposure and inflammation among women of reproductive age who use electronic or conventional cigarettes. J Womens Health (Larchmt). 2021;30(4):539-50. doi:10.1089/jwh.2019.8075.

[143] Majid S, Keith RJ, Fetterman JL, Weisbrod RM, Nystoriak J, Wilson T et al. Lipid profiles in users of combustible and electronic cigarettes. Vasc Med. 2021;26(5):483-8. doi:10.1177/1358863X211009313.

[144] Rao P, Liu J, Springer ML. JUUL and combusted cigarettes comparably impair endothelial function. Tob Regul Sci. 2020;6(1):30-7. doi:10.18001/TRS.6.1.4.

[145] Chaumont M, van de Borne P, Bernard A, Van Muylem A, Deprez G, Ullmo J et al. Fourth generation e-cigarette vaping induces transient lung inflammation and gas exchange disturbances: results from two randomized clinical trials. Am J Physiol Lung Cell Mol Physiol. 2019;316(5):L705-19. doi:10.1152/ajplung.00492.2018.

[146] Muthumalage T, Lamb T, Friedman MR, Rahman I. E-cigarette flavored pods induce inflammation epithelial barrier dysfunction and DNA damage in lung epithelial cells and monocytes. Sci Rep. 2019;9(1):19035.

[147] Moheimani RS, Bhetraratana M, Yin F, Peters KM, Gornbein J, Araujo JA et al. Increased cardiac sympathetic activity and oxidative stress in habitual electronic cigarette users: implications for cardiovascular risk. JAMA Cardiol. 2017;2(3):278-84.

[148] Metzen D, M'Pembele R, Zako S, Mourikis P, Helten C, Zikeli D et al. Platelet reactivity is higher in e-cigarette vaping as compared to traditional smoking. Int J Cardiol. 2021.343:146-8.

doi:10.1016/j.ijcard.2021.09.005.

[149] Mohammadi L, Han DD, Xu F, Huang A, Derakhshandeh R, Rao P et al. Chronic e-cigarette use impairs endothelial function on the physiological and cellular levels. Arterioscler Thromb Vasc Biol. 2022;42(11):1333-50.

[150] Verma A, Anand K, Bhargava M, Kolluri A, Kumar M, Palve DH. Comparative evaluation of salivary biomarker levels in e-cigarette smokers and conventional smokers. J Pharm Bioallied Sci. 2021;13(Suppl_2):S1642-5.

[151] Patskan G, Reininghaus W. Toxicological evaluation of an electrically heated cigarette. Part 1: Overview of technical concepts and summary of findings. J Appl Toxicol. 2003;23(5):323-8.

[152] Tricker AR, Kanada S, Takada K, Martin Leroy C, Lindner D, Schorp MK et al. Reduced exposure evaluation of an electrically heated cigarette smoking system. Part 6: 6-day randomized clinical trial of a menthol cigarette in Japan. Regul Toxicol Pharmacol. 2012;64(2-Suppl):S64-73.

[153] Martin Leroy C, Jarus-Dziedzic K, Ancerewicz J, Lindner D, Kulesza A, Magnette J. Reduced exposure evaluation of an electrically heated cigarette smoking system. Part 7: a onemonth randomized ambulatory controlled clinical study in Poland. Regul Toxicol Pharmacol. 2012;64(2-Suppl):S74-84.

[154] Lüdicke F, Ansari SM, Lama N, Blanc N, Bosilkovska M, Donelli A et al. Effects of switching to a heat-not-burn tobacco product on biologically relevant biomarkers to assess a candidate modified risk tobacco product: a randomized trial. Cancer Epidemiol Biomarkers Prev. 2019;28(11):1934-43.

[155] Gale N, McEwan M, Camacho OM, Hardie G, Proctor CJ, Murphy J. Changes in biomarkers after 180 days of tobacco heating product use: a randomised trial. Intern Emerg Med. 2021;16(8):2201-12.

[156] Gale N, McEwan M, Hardie G, Proctor CJ, Murphy J. Changes in biomarkers of exposure and biomarkers of potential harm after 360 days in smokers who either continue to smoke switch to a tobacco heating product or quit smoking. Intern Emerg Med. 2022;17(7):2017-30.

[157] Sakaguchi C, Nagata Y, Kikuchi A, Takeshige Y, Minami N. Differences in levels of biomarkers of potential harm among users of a heat-not-burn tobacco product, cigarette smokers, and never-smokers in Japan: a post-marketing observational study. Nicotine Tob Res. 2021;23(7):1143-52.

[158] Ogden MW, Marano KM, Jones BA, Morgan WT, Stiles MF. Switching from usual brand cigarettes to a tobacco-heating cigarette or snus: Part 3. Biomarkers of biological effect. Biomarkers. 2015;20(6-7):404-10.

[159] Biondi Zoccai G, Carnevale R, Vitali M, Tritapepe L, Martinelli O, Macrina F et al. A randomized trial comparing the acute coronary systemic and environmental effects of electronic vaping cigarettes versus heat-not-burn cigarettes in smokers of combustible cigarettes undergoing invasive coronary assessment: rationale and design of the SUR-VAPES 3 trial. Minerva Cardioangiol. 2020;68(6):548-55.

[160] Haziza C, de La Bourdonnaye G, Donelli A, Skiada D, Poux V, Weitkunat R et al. Favorable changes in biomarkers of potential harm to reduce the adverse health effects of smoking in smokers switching to the menthol tobacco heating system 2.2 for 3 months (Part 2). Nicotine Tob Res. 2020;22(4):549-59.

[161] Singh KP, Lawyer G, Muthumalage T, Maremanda KP, Khan NA, McDonough SR et al. Systemic biomarkers in electronic cigarette users: implications for noninvasive assessment of vaping-associated pulmonary injuries. ERJ Open Res. 2019;5(4):00182-2019. doi:10.1183/23120541.00182-2019.

[162] Sakamaki-Ching S, Williams M, Hua M, Li J, Bates SM, Robinson AN et al. Correlation between biomarkers of exposure effect and potential harm in the urine of electronic cigarette users. BMJ Open Respir Res. 2020;7(1). doi:10.1136/bmjresp-2019-000452.

[163] Leidy NK, Rennard SI, Schmier J, Jones MK, Goldman M. The breathlessness cough and sputum scale: the development of empirically based guidelines for interpretation. Chest. 2003;124(6):2182-91. doi:10.1378/chest.124.6.2182. doi:10.1038/npjpcrm.2016.83.

[164] Leidy NK, Schmier JK, Jones MK, Lloyd J, Rocchiccioli K. Evaluating symptoms in chronic obstructive pulmonary disease: validation of the Breathlessness Cough and Sputum Scale. Respir Med. 2003;97(Suppl_A):S59-70. PMID:12564612.

[165] Miller MR, Hankinson J, Brusasco V, Burgos F, Casaburi R, Coates A et al. Standardisation of spirometry. Eur Respir J. 2005;26(2):319-38. doi:10.1183/09031936.05.00034805.

[166] Miller MR, Crapo R, Hankinson J, Brusasco V, Burgos F, Casaburi R et al. General considerations for lung function testing. Eur Respir J. 2005;26(1):153-61. doi:10.1183/09031936.05.00034505.

[167] Martinez-Morata I, Sanchez TR, Shimbo D, Navas-Acien A. Electronic cigarette use and blood pressure endpoints: a systematic review. Curr Hypertens Rep. 2020;23(1):2. doi:10.1007/s11906-020-01119-0.

[168] Keith R, Bhatnagar A. Cardiorespiratory and immunologic effects of electronic cigarettes. Curr Addict Rep. 2021;8(2:336-46. doi:10.1007/s40429-021-00359-7.

[169] Shields PG, Song MA, Freudenheim JL, Brasky TM, McElroy JP, Reisinger SA et al. Lipid laden macrophages and electronic cigarettes in healthy adults. EBioMedicine. 2020;60:102982. doi:10.1016/j.ebiom.2020.102982.

[170] Dai H, Khan AS. A longitudinal study of exposure to tobacco-related toxicants and subsequent respiratory symptoms among US adults with varying e-cigarette use status. Nicotine Tob Res. 2020;22(Suppl-1):S61-9. doi:10.1093/ntr/ntaa180.

[171] Xie W, Kathuria H, Galiatsatos P, Blaha MJ, Hamburg NM, Robertson RM et al. Association of electronic cigarette use with incident respiratory conditions among US adults from 2013 to 2018. JAMA Netw Open. 2020;3(11):e2020816. doi:10.1001/jamanetworkopen.2020.20816.

[172] Goniewicz ML, Miller CR, Sutanto E, Li D. How effective are electronic cigarettes for reducing respiratory and cardiovascular risk in smokers? A systematic review. Harm Reduct J. 2020;17(1):91. doi:10.1186/s12954-020-00440-w.

[173] Rezk-Hanna M, Gupta R, Nettle CO, Dobrin D, Cheng CW, Means A et al. Differential effects of electronic hookah vaping and traditional combustible hookah smoking on oxidation inflammation and arterial stiffness. Chest. 2022;161(1):208-18. doi:10.1016/j.chest.2021.

[174] The health consequences of smoking – 50 years of progress: a report of the Surgeon General. Rockville (MD): United States Department of Health and Human Services; 2014. PMID:24455788.

[175] Luoto R, Uutela A, Puska P. Occasional smoking increases total and cardiovascular mortality

among men. Nicotine Tob Res. 2000;2(2):133-9. doi:10.1080/713688127.

[176] Bjartveit K, Tverdal A. Health consequences of smoking 1–4 cigarettes per day. Tob Control. 2005;14(5):315-20. doi:10.1136/tc.2005.011932.

[177] Pope CA 3rd, Burnett RT, Krewski D, Jerrett M, Shi Y, Calle EE et al. Cardiovascular mortality and exposure to airborne fine particulate matter and cigarette smoke: shape of the exposure–response relationship. Circulation. 2009;120(11):941-8. doi:10.1161/CIRCULATIONAHA.

[178] Lee A, Lee SY, Lee KS. The use of heated tobacco products is associated with asthma, allergic rhinitis and atopic dermatitis in Korean adolescents. Sci Rep. 2019;9(1):17699. doi:10.1038/s41598-019-54102-4.

[179] Kamada T, Yamashita Y, Tomioka H. Acute eosinophilic pneumonia following heat-not-burn cigarette smoking. Respirol Case Rep. 2016;4(6):e00190. doi:10.1002/rcr2.190.

[180] Aokage T, Tsukahara K, Fukuda Y, Tokioka F, Taniguchi A, Naito H et al. Heat-not-burn cigarettes induce fulminant acute eosinophilic pneumonia requiring extracorporeal membrane oxygenation. Respir Med Case Rep. 2019;26:87-90. doi:10.1016/j.rmcr.2018.12.002.

[181] Liu Y, Sanoff HK, Cho H, Burd CE, Torrice C, Ibrahim JG et al. Expression of p16(INK4a) in peripheral blood T-cells is a biomarker of human aging. Aging Cell. 2009;8(4):439-48. doi:10.1111/j.1474-9726.2009.00489.x.

[182] Lu AT, Quach A, Wilson JG, Reiner AP, Aviv A, Raj K et al. DNA methylation GrimAge strongly predicts lifespan and healthspan. Aging (Albany NY). 2019;11(2):303-27. doi:10.18632/aging.101684.

[183] Richmond RC, Sillero-Rejon C, Khouja JN, Prince C, Board A, Sharp G et al. Investigating the DNA methylation profile of e-cigarette use. Clin Epigenetics. 2021;13(1):183. doi:10.1186/s13148-021-01174-7.

[184] Song MA, Freudenheim JL, Brasky TM, Mathe EA, McElroy JP, Nickerson QA et al. Biomarkers of exposure and effect in the lungs of smokers, nonsmokers and electronic cigarette users. Cancer Epidemiol Biomarkers Prev. 2020;29(2):443-51. doi:10.1158/1055-9965.EPI-19-1245.

[185] Philibert R, Hollenbeck N, Andersen E, Osborn T, Gerrard M, Gibbons FX et al. A quantitative epigenetic approach for the assessment of cigarette consumption. Front Psychol. 2015;6:656. doi:10.3389/fpsyg.2015.00656.

[186] Andersen A, Reimer R, Dawes K, Becker A, Hutchens N, Miller S et al. DNA methylation differentiates smoking from vaping and non-combustible tobacco use. Epigenetics. 2022;17(2):178-90. doi:10.1080/15592294.2021.1890875.

[187] Caliri AW, Caceres A, Tommasi S, Besaratinia A. Hypomethylation of LINE-1 repeat elements and global loss of DNA hydroxymethylation in vapers and smokers. Epigenetics. 2020;15(8):816-29. doi:10.1080/15592294.2020.1724401.

[188] Ohmomo H, Harada S, Komaki S, Ono K, Sutoh Y, Otomo R et al. DNA methylation abnormalities and altered whole transcriptome profiles after switching from combustible tobacco smoking to heated tobacco products. Cancer Epidemiol Biomarkers Prev. 2022;31(1):269-79. doi:10.1158/1055-9965.EPI-21-0444.

[189] Spira A, Beane J, Shah V, Liu G, Schembri F, Yang X et al. Effects of cigarette smoke on the human airway epithelial cell transcriptome. Proc Natl Acad Sci U S A. 2004;101(27):10143-8. doi:10.1073/pnas.0401422101.

[190] Beane J, Sebastiani P, Liu G, Brody JS, Lenburg ME, Spira A. Reversible and permanent effects of tobacco smoke exposure on airway epithelial gene expression. Genome Biol. 2007;8(9):R201. doi:10.1186/gb-2007-8-9-r201.

[191] Pavel AB, Campbell JD, Liu G, Elashoff D, Dubinett S, Smith K et al. Alterations in bronchial airway miRNA expression for lung cancer detection. Cancer Prev Res (Phila). 2017;10(11):651-9. doi:10.1158/1940-6207.CAPR-17-0098.

[192] Martin EM, Clapp PW, Rebuli ME, Pawlak EA, Glista-Baker E, Benowitz NL et al. E-cigarette use results in suppression of immune and inflammatory-response genes in nasal epithelial cells similar to cigarette smoke. Am J Physiol Lung Cell Mol Physiol. 2016;311(1):L135-44. doi:10.1152/ajplung.00170.2016.

[193] Tommasi S, Caliri AW, Caceres A, Moreno DE, Li M, Chen Y et al. Deregulation of biologically significant genes and associated molecular pathways in the oral epithelium of electronic cigarette users. Int J Mol Sci. 2019;20(3):738. doi:10.3390/ijms20030738.

[194] Hamad SH, Brinkman MC, Tsai YH, Mellouk N, Cross K, Jaspers I et al. Pilot study to detect genes involved in DNA damage and cancer in humans: potential biomarkers of exposure to e-cigarette aerosols. Genes (Basel). 2021;12(3):448. doi:10.3390/genes12030448.

[195] Staudt MR, Salit J, Kaner RJ, Hollmann C, Crystal RG. Altered lung biology of healthy never smokers following acute inhalation of e-cigarettes. Respir Res. 2018;19(1):78. doi:10.1186/s12931-018-0778-z.

[196] Gaeckle NT, Pragman AA, Pendleton KM, Baldomero AK, Criner GJ. The oral–lung axis: the impact of oral health on lung health. Respir Care. 2020;65(8):1211-20. doi:10.4187/respcare.07332.

[197] Dickson RP, Huffnagle GB. The lung microbiome: new principles for respiratory bacteriology in health and disease. PLoS Pathog. 2015;11(7):e1004923. doi:10.1371/journal.ppat.1004923.

[198] Mathieu E, Escribano-Vazquez U, Descamps D, Cherbuy C, Langella P, Riffault S et al. Paradigms of lung microbiota functions in health and disease particularly in asthma. Front Physiol. 2018;9:1168. doi:10.3389/fphys.2018.01168.

[199] Hajishengallis G, Darveau RP, Curtis MA. The keystone-pathogen hypothesis. Nat Rev Microbiol. 2012;10(10):717-25. doi:10.1038/nrmicro2873.200.

[200] Shi J, Yang Y, Xie H, Wang X, Wu J, Long J et al. Association of oral microbiota with lung cancer risk in a low-income population in the Southeastern USA. Cancer Causes Control. 2021;32(12):1423-32. doi:10.1007/s10552-021-01490-6.

[201] Hosgood HD, Cai Q, Hua X, Long J, Shi J, Wan Y et al. Variation in oral microbiome is associated with future risk of lung cancer among never-smokers. Thorax. 2021;76(3):256-63. doi:10.1136/thoraxjnl-2020-215542.

[202] Hayes RB, Ahn J, Fan X, Peters BA, Ma Y, Yang L et al. Association of oral microbiome with risk for incident head and neck squamous cell cancer. JAMA Oncol. 2018;4(3)358-65. doi:10.1001/jamaoncol.2017.4777.

[203] Mammen MJ, Scannapieco FA, Sethi S. Oral–lung microbiome interactions in lung diseases. Periodontology. 2000;83(1):234-41. doi:10.1111/prd.12301.

[204] Plachokova AS, Andreu-Sanchez S, Noz MP, Fu J, Riksen NP. Oral microbiome in relation to periodontitis severity and systemic inflammation. Int J Mol Sci. 2021;22(11):5876. doi:10.3390/

ijms22115876.

[205] Schulz S, Reichert S, Grollmitz J, Friebe L, Kohnert M, Hofmann B et al. The role of Saccharibacteria (TM7) in the subgingival microbiome as a predictor for secondary cardiovascular events. Int J Cardiol. 2021;331:255-61. doi:10.1177/0022034519831671.

[206] Kato-Kogoe N, Sakaguchi S, Kamiya K, Omori M, Gu YH, Ito Y et al. Characterization of salivary microbiota in patients with atherosclerotic cardiovascular disease: a case–control study. J Atheroscler Thromb. 2022;29(3):403-21. doi:10.5551/jat.60608.

[207] Goh CE, Trinh P, Colombo PC, Genkinger JM, Mathema B, Uhlemann AC et al. Association between nitrate-reducing oral bacteria and cardiometabolic outcomes: results from ORIGINS. J Am Heart Assoc. 2019;8(23):e013324. doi:10.1161/JAHA.119.013324.208.

[208] Wu J, Peters BA, Dominianni C, Zhang Y, Pei Z, Yang L et al. Cigarette smoking and the oral microbiome in a large study of American adults. ISME J. 2016;10(10):2435-46. doi:10.1038/ismej.2016.37.

[209] Droemann D, Goldmann T, Tiedje T, Zabel P, Dalhoff K, Schaaf B. Toll-like receptor 2 expression is decreased on alveolar macrophages in cigarette smokers and COPD patients. Respir Res. 2005;6:68. doi:10.1186/1465-9921-6-68.

[210] Kulkarni R, Antala S, Wang A, Amaral FE, Rampersaud R, Larussa SJ et al. Cigarette smoke increases Staphylococcus aureus biofilm formation via oxidative stress. Infect Immun. 2012;80(11):3804-11. doi:10.1128/IAI.00689-12.

[211] Ganesan SM, Joshi V, Fellows M, Dabdoub SM, Nagaraja HN, O'Donnell B et al. A tale of two risks: smoking diabetes and the subgingival microbiome. ISME J. 2017;11(9):2075-89. doi:10.1038/ismej.2017.73.

[212] Mueller DC, Piller M, Niessner R, Scherer M, Scherer G. Untargeted metabolomic profiling in saliva of smokers and nonsmokers by a validated GC-TOF-MS method. J Proteome Res. 2014;13(3):1602-13. doi:10.1021/pr401099r.

[213] Valles Y, Inman CK, Peters BA, Ali R, Wareth LA, Abdulle A et al. Types of tobacco consumption and the oral microbiome in the United Arab Emirates Healthy Future (UAEHFS) pilot study. Sci Rep. 2018;8(1):11327. doi:10.1038/s41598-018-29730-x.

[214] Gopinath D, Wie CC, Banerjee M, Thangavelu L, Kumar RP, Nallaswamy D et al. Compositional profile of mucosal bacteriome of smokers and smokeless tobacco users. Clin Oral Investig. 2022;26(2):1647-56. doi:10.1007/s00784-021-04137-7.

[215] Chopyk J, Bojanowski CM, Shin J, Moshensky A, Fuentes AL, Bonde SS et al. Compositional differences in the oral microbiome of e-cigarette users. Front Microbiol. 2021;12:599664. doi:10.3389/fmicb.2021.599664.

[216] Pushalkar S, Paul B, Li Q, Yang J, Vasconcelos R, Makwana S et al. Electronic cigarette aerosol modulates the oral microbiome and increases risk of infection. iScience. 2020;23(3):100884. doi:10.1016/j.isci.2020.100884.

[217] Ganesan SM, Dabdoub SM, Nagaraja HN, Scott ML, Pamulapati S, Berman ML et al. Adverse effects of electronic cigarettes on the disease-naive oral microbiome. Sci Adv. 2020;6(22):eaaz0108. doi:10.1126/sciadv.aaz0108.

[218] Goolam Mahomed T, Peters RPH, Allam M, Ismail A, Mtshali S, Goolam Mahomed A et al. Lung microbiome of stable and exacerbated COPD patients in Tshwane South Africa. Sci Rep.

2021;11(1):19758. doi:10.1038/s41598-021-99127-w.

[219] Schulz-Weidner N, Weigel M, Turujlija F, Komma K, Mengel JP, Schlenz MA et al. Microbiome analysis of carious lesions in pre-school children with early childhood caries and congenital heart disease. Microorganisms. 2021;9(9):1904. doi:10.3390/microorganisms.

[220] Rocas IN, Siqueira JF Jr. Culture-independent detection of Eikenella corrodens and Veillonella parvula in primary endodontic infections. J Endod. 2006;32(6):509-12. doi:10.1016/j.joen.2005.

[221] Kato I, Vasquez AA, Moyerbrailean G, Land S, Sun J, Lin HS et al. Oral microbiome and history of smoking and colorectal cancer. J Epidemiol Res. 2016;2(2):92-101. doi:10.5430/jer.v2n2p92.

[222] Tishchenko OV, Kryvenko LS, Gargina VV. Influence of smoking heating up tobacco products and e-cigarettes on the microbiota of dental plaque. Pol Merkur Lekarski. 2022;50(295):16-20. PMID:35278292.

[223] Benowitz NL, Swan GE, Jacob P 3rd, Lessov-Schlaggar CN, Tyndale RF. CYP2A6 genotype and the metabolism and disposition kinetics of nicotine. Clin Pharmacol Ther. 2006;80(5):457-67. doi:10.1016/j.clpt.2006.08.011.

[224] Benowitz NL, Hukkanen J, Jacob P 3rd. Nicotine chemistry metabolism kinetics and biomarkers. Handb Exp Pharmacol. 200;192:29-60. doi:10.1007/978-3-540-69248-5_2.

[225] Lerman C, Schnoll RA, Hawk LW Jr, Cinciripini P, George TP, Wileyto EP et al. Use of the nicotine metabolite ratio as a genetically informed biomarker of response to nicotine patch or varenicline for smoking cessation: a randomised double-blind placebo-controlled trial. Lancet Respir Med. 2015;3(2):131-8. doi:10.1016/S2213-2600(14)70294-2.

[226] Sosnoff CS, Caron K, Akins JR, Dortch K, Hunter RE, Pine BN et al. Serum concentrations of cotinine and trans-3′-hydroxycotinine in US adults: results from wave 1 (2013-2014) of the Population Assessment of Tobacco and Health Study. Nicotine Tob Res. 2022;24(5):736-44. doi:10.1093/ntr/ntab240.

[227] Yuan JM, Nelson HH, Carmella SG, Wang R, Kuriger-Laber J, Jin A et al. CYP2A6 genetic polymorphisms and biomarkers of tobacco smoke constituents in relation to risk of lung cancer in the Singapore Chinese Health Study. Carcinogenesis. 2017;38(4):411-8. doi:10.1093/carcin/bgx012.

[228] Yuan JM, Gao YT, Wang R, Chen M, Carmella SG, Hecht SS. Urinary levels of volatile organic carcinogen and toxicant biomarkers in relation to lung cancer development in smokers. Carcinogenesis. 2012;33:804-9. doi:10.1093/carcin/bgs026.

[229] Yuan JM, Gao YT, Murphy SE, Carmella SG, Wang R, Zhong Y et al. Urinary levels of cigarette smoke constituent metabolites are prospectively associated with lung cancer development in smokers. Cancer Res. 2011;71(21):6749-57. doi:10.1158/0008-5472.CAN-11-0209.

[230] Yuan JM, Koh WP, Murphy SE, Fan Y, Wang R, Carmella SG et al. Urinary levels of tobacco-specific nitrosamine metabolites in relation to lung cancer development in two prospective cohorts of cigarette smokers. Cancer Res. 2009;69:2990-5. doi:10.1158/0008-5472.CAN-08-4330.

[231] Church TR, Anderson KE, Caporaso NE, Geisser MS, Le CT, Zhang Y et al. A prospectively measured serum biomarker for a tobacco-specific carcinogen and lung cancer in smokers. Cancer Epidemiol Biomarkers Prev. 2009;18(1):260-6. doi:10.1158/1055-9965.EPI-08-0718.

[232] Verplaetse TL, Peltier MR, Roberts W, Moore KE, Pittman BP, McKee SA. Associations between nicotine metabolite ratio and gender with transitions in cigarette smoking status and e-cigarette

use: findings across waves 1 and 2 of the Population Assessment of Tobacco and Health (PATH) study. Nicotine Tob Res. 2020;22(8):1316-21. doi:10.1093/ntr/ntaa022.

[233] Znyk M, Jurewicz J, Kaleta D. Exposure to heated tobacco products and adverse health effects, a systematic review. Int J Environ Res Public Health. 2021;18(12):6651. doi:10.3390/ijerph18126651.

[234] Murphy SE, Park SS, Thompson EF, Wilkens LR, Patel Y, Stram DO et al. Nicotine *N*-glucuronidation relative to *N*-oxidation and C-oxidation and UGT2B10 genotype in five ethnic/racial groups. Carcinogenesis. 2014;35(11):2526-33. doi:10.1093/carcin/bgu191.

[235] Carmella SG, Ming X, Olvera N, Brookmeyer C, Yoder A, Hecht SS. High throughput liquid and gas chromatography–tandem mass spectrometry assays for tobacco-specific nitrosamine and polycyclic aromatic hydrocarbon metabolites associated with lung cancer in smokers. Chem Res Toxicol. 2013;26(8):1209-17. doi:10.1021/tx400121n.

[236] Chen M, Carmella SG, Sipe C, Jensen J, Luo X, Le CT et al. Longitudinal stability in cigarette smokers of urinary biomarkers of exposure to the toxicants acrylonitrile and acrolein. PLoS One. 2019;14(1):e0210104. doi.:10.1371/journal.pone.0210104.

[237] Yan W, Byrd GD, Ogden MW. Quantitation of isoprostane isomers in human urine from smokers and nonsmokers by LC-MS/MS. J Lipid Res. 2007;48(7):1607-17. doi:10.1194/jlr.M700097-JLR200.

[238] Chen HJ, Lin WP. Quantitative analysis of multiple exocyclic DNA adducts in human salivary DNA by stable isotope dilution nanoflow liquid chromatography–nanospray ionization tandem mass spectrometry. Anal Chem. 2011;83(22):8543-51. doi:10.1021/ac201874d.

[239] Hartmann-Boyce J, Butler AR, Theodoulou A, Onakpoya IJ, Hajek P, Bullen C et al. Biomarkers of potential harm in people switching from smoking tobacco to exclusive e-cigarette use dual use or abstinence: secondary analysis of Cochrane systematic review of trials of e-cigarettes for smoking cessation. Addiction. 2023;118(3):539-45. doi:10.1111/add.16063.

[240] De Jesus VR, Bhandari D, Zhang L, Reese C, Capella K, Tevis D et al. Urinary biomarkers of exposure to volatile organic compounds from the Population Assessment of Tobacco and Health Study wave 1 (2013-2014). Int J Environ Res Public Health. 2020;17(15):5408. doi:10.3390/ijerph17155408.

[241] Majeed B, Linder D, Eissenberg T, Tarasenko Y, Smith D, Ashley D. Cluster analysis of urinary tobacco biomarkers among US adults: Population Assessment of Tobacco and Health (PATH) biomarker study (2013-2014). Prev Med. 2020;140:106218. doi:10.1016/j.ypmed.2020.106218.

[242] Goniewicz ML, Smith DM, Edwards KC, Blount BC, Caldwell KL, Feng J et al. Comparison of nicotine and toxicant exposure in users of electronic cigarettes and combustible cigarettes. JAMA Netw Open. 2018;1(8):e185937. doi:10.1001/jamanetworkopen.2018.5937

[243] Wei B, Goniewicz ML, O'Connor RJ, Travers MJ, Hyland AJ. Urinary metabolite levels of flame retardants in electronic cigarette users: a study using the data from NHANES 2013-2014. Int J Environ Res Public Health. 2018;15(2):201. doi:10.3390/ijerph15020201.

[244] Tehrani MW, Newmeyer MN, Rule AM, Prasse C. Characterizing the chemical landscape in commercial e-cigarette liquids and aerosols by liquid chromatography–high-resolution mass spectrometry. Chem Res Toxicol. 2021;34(10):2216-26. doi:10.1021/acs.chemrestox.1c00253.

[245] Smith DM, Christensen C, van Bemmel D, Borek N, Ambrose B, Erives G et al. Exposure to

nicotine and toxicants among dual users of tobacco cigarettes and e-cigarettes: Population Assessment of Tobacco and Health (PATH) study 2013-2014. Nicotine Tob Res. 2021;23(5):790-7. doi:10.1093/ntr/ntaa252.

[246] Luo X, Carmella SG, Chen M, Jensen JA, Wilkens LR, Le Marchand L et al. Urinary cyanoethyl mercapturic acid a biomarker of the smoke toxicant acrylonitrile clearly distinguishes smokers from nonsmokers. Nicotine Tob Res. 2020;22(10):1744-7. doi:10.1093/ntr/ntaa080.

[247] Schick SF, Blount BC, Jacob PR, Saliba NA, Bernert JT, El Hellani A et al. Biomarkers of exposure to new and emerging tobacco delivery products. Am J Physiol Lung Cell Mol Physiol. 2017;313(3):L425-52. doi: 10.1152/ajplung.00343.2016.

[248] Pisinger C, Rasmussen SKB. The health effects of real-world dual use of electronic and conventional cigarettes versus the health effects of exclusive smoking of conventional cigarettes: a systematic review. Int J Environ Res Public Health. 2022;19(20):13687. doi:10.3390/ijerph192013687.

[249] Rudasingwa G, Kim Y, Lee C, Lee J, Kim S, Kim S. Comparison of nicotine dependence and biomarker levels among traditional cigarette, heat-not-burn cigarette, and liquid e-cigarette users: results from the Think Study. Int J Environ Res Public Health. 2021;18(9):4777. doi:10.3390/ijerph18094777.

6. 烟草和非治疗性烟碱产品的互联网营销及相关监管考量[1]

Becky Freeman, Prevention Research Collaboration, School of Public Health, Faculty of Medicine and Health, University of Sydney, Sydney, Australia

Pamela Ling, MD, Department of Medicine and Center for Tobacco Control Research and Education, University of California, San Francisco (CA), USA

Stella Aguinaga Bialous, DrPH, School of Nursing and Center for Tobacco Control Research and Education, University of California, San Francisco (CA), USA

摘　要

对烟草广告、促销和赞助（TAPS）的禁令是全面控制烟草法律的基石。世界卫生组织《烟草控制框架公约》（WHO FCTC）的通过推动了全球在实施TAPS禁令方面的进展。然而，TAPS禁令的执行在很大程度上依赖于娱乐和数字化产业生产商以及在线平台的自我管理。随着媒体环境的不断演变，TAPS法规必须与时俱进，尤其是在跨境TAPS的监测和报告上，需要通过在线数字媒体平台进行。此外，TAPS法律还必须及时应对非治疗性烟碱设备和新型烟草制品的快速变化。此类产品包括电子烟碱传输系统（ENDS）、电子非烟碱传输系统（ENNDS）、个人蒸发器、加热型烟草制品、烟碱盐、类烟碱替代疗法的其他产品，以及使用与烟草制品相同释放装置或销售渠道的各种维生素和大麻产品。由于制造商利用烟草制品定义的漏洞，或处于监管不明的灰色地带，许多产品仍未受到有效监管。因此，必须制定前瞻性政策，来应对烟草、烟碱及其相关产品，以及不断演变的网络和数字媒体营销环境的变化。

关键词：烟草广告；社交媒体；在线数字媒体；营销；法规

1　本章借鉴了作者近期发表的著作中的部分内容[1, 2]。此外，还借鉴了 WHO 跨界烟草广告和促销专家咨询委员会主席 Becky Freeman 的报告和发现[3]，以及 WHO FCTC 第 13 条工作组关于烟草广告、促销和赞助的报告和发现[4]

6.1 背　　景

在世界大部分地区，直接烟草广告早已在传统大众媒体（如电视、广播和印刷品）以及广告牌等渠道被禁止。然而，关于烟草广告、促销和赞助（TAPS）的法律必须不断进步，以应对烟草行业无休止推广产品、维持现有客户、吸引戒烟者和新用户的手段[1]。当只有某些形式的TAPS受到监管时，行业会将促销活动和预算转向不受监管的促销方式[5]。如果不更新以涵盖促销手段和产品创新，所谓的全面和进步的TAPS禁令很快就会过时。数字媒体环境的快速发展，尤其是在线社交媒体的爆炸性增长和主导地位，也使烟草行业能够开发和利用新的促销形式[6]。广泛规避TAPS法律与新媒体和新产品形式[2]的结合并非不可解决：尽管具有挑战性，但可以通过新的政策方法进一步限制TAPS及其暴露风险。

WHO FCTC 对烟草广告和促销的定义是“任何形式的商业传播、建议或行动，其目的、效果或可能效果是直接或间接促进烟草制品或烟草使用”[7]。鼓励缔约方考虑将ENDS和ENNDS纳入任何TAPS监管办法。这些定义是有意泛化的，以确保涵盖烟草行业推销其产品的各种方式。WHO FCTC各方必须确保其TAPS法律是全面的，除非有任何宪法障碍，禁止一切形式的TAPS。本章重点关注烟草和其他非治疗性烟碱释放产品制造商如何使用在线数字媒体共享平台（尤其是社交媒体）来营销其产品，尽管我们承认TAPS不仅限于在线环境。此外，世界卫生组织烟草制品管制研究小组的第八份报告[8]包括“新型烟草制品的全球营销和推广及其影响”。本章不是重复，而是在这项工作的基础上补充和建立。

WHO FCTC第13条实施指南规定，在电影、在线视频和电脑游戏等娱乐媒体中描绘烟草是一种TAPS形式[9]。无论烟草行业是否参与了内容的创作或资助，这些娱乐媒体形式的商业本质决定了它们所包含的烟草描述是TAPS。许多娱乐媒体内容是通过个人互联网设备（如智能手机）上的社交媒体和流媒体平台来访问的。这种类型的内容也可以在一个国家创建、上传或传播，然后在另一个国家观看和共享。

跨境数字媒体消费为烟草行业规避TAPS禁令提供了更多渠道。如果在线信息没有相关的商业链接，某些形式的支持烟草的内容可能被视为“合法表达”。例如，如果用户在社交媒体上发布自己使用烟草产品的图片时没有获得任何财务或其他利益（如免费产品），则不会被视为进行“商业传播”。尽管社交媒体平台上的大部分内容本质上不是商业性的，但平台本身依赖商业内容实现盈利。商业付费内容通常根据用户的人口统计数据和兴趣被推送到他们的社交媒体订阅源中。此外，商业实体也会在其社交媒体平台账户上发布所谓的“有机”内容，即无需支付费用的帖子，旨在吸引用户[10]。这类“有机”帖子虽然未付费，但很容易被归类为商业传播的一部分，因为它们通常是制造商战略营销计划的一环。下文将对这些概念进行说明。

6.2 网络和社交媒体营销对烟草和 ENDS 使用的影响

接触社交媒体中的烟草内容以及与其互动对青少年的ENDS和烟草使用模式有重大影响。由于青少年花费大量时间接触在线内容，社交媒体可能是一个重要的干预场所，适合传播反烟草或烟草预防信息[11]。

一项关于接触社交媒体烟草内容与终生烟草使用、过去30天烟草使用以及从不吸烟者对烟草使用易感性之间关系的系统调研和荟萃分析发现，接触社交媒体烟草内容的参与者比未接触者相比，终生烟草使用的概率更高（OR，2.18；95% CI，1.54；3.08；$I^2 = 94\%$），过去30天烟草使用的概率更高（OR，2.19；95% CI，1.79；2.67；$I^2 = 84\%$），以及从不吸烟者对烟草使用的易感性更高（OR，2.08；95% CI，1.65；2.63；$I^2 = 73\%$）。亚组分析显示，烟草促销、主动参与、被动参与、终生接触烟草内容、在两个以上平台上接触烟草内容，以及青少年和年轻人接触烟草内容，都存在类似的关联[12]。

6.2.1 年轻人使用ENDS

烟草制品广告长期以来已被确立为导致年轻人开始使用烟草制品的一个原因[13]。目前ENDS和其他烟碱产品的广告大多基于与过去推广传统烟草制品相似的方法和主题[14]。

年轻人接触烟草和ENDS的营销，并参与支持烟草和ENDS的信息，增加了他们使用ENDS的可能性[15]。即使在对基线电子烟使用和电子烟使用导致的接触信息和参与的反馈环路进行调整后，ENDS使用可能性的增加仍然存在。相比之下，接触反烟草和反ENDS信息降低了他们使用电子烟的可能性。这些发现不仅强调了监管社交媒体上烟草和烟碱产品促销和营销信息的重要性，还表明社交媒体可以作为一个经济高效的平台，用于传播反烟草、反ENDS的运动信息，以预防年轻人使用ENDS产品[15]。

6.2.2 网络和数字媒体中TAPS示例

多个包含数字媒体TAPS示例的在线资料库涵盖了烟草和烟碱产品。斯坦福大学的烟草广告影响研究的网站拥有大量（60000个例子）的TAPS示例，涵盖所有形式的烟草和烟碱产品，不仅包括传统卷烟，还包括ENDS、尼古丁袋、水烟、无烟烟草和雪茄[16]。

在线媒体上的直接付费烟草广告是最容易识别、监测和执行的在线TAPS形式。然而，可能很难区分直接、有偿的烟草促销和没有商业联系的内容[17]。例如，2018年的一项调查发现，烟草公司正在吸引有流量的社交媒体人，通过他们高收视率的社交媒体资料推广烟草制品。媒体人发布的内容没有透露他们在做烟草广告，也没有透露他们收到了烟草行业的奖励，在其个人资料和页面上发布烟草品牌和形象。媒体人也广泛参与电子烟和其他新型烟碱设备的推广[18]。

除了娱乐和在线媒体中的TAPS和烟草宣传之外，烟草行业的公司宣传活动也是支持烟草的一个重要信息来源。这些促销活动避开了TAPS法律和TAPS定义。例子包括企业社会责任宣传[19]、行业资助的"基金会"运动[20]、行业资助的科学研究、政治游说活动和促销活动，包括新闻媒体上的付费社论（广告）[21]，以及公司品牌账户和员工在社交媒体上的无偿帖子[22]。这些企业传播通常属于"合法表达"的豁免范围[7]，但与直接广告有相同的媒体平台和目的。监管机构应考虑在公司不恰当地使用"合法表达"豁免时进行干预，例如当其在大众媒体渠道（如报纸广告）中的公关内容包含几乎不加掩饰的产品促销时。

1. 数字媒体共享平台

- 通过付费广告直接推广产品。这类直接促销通常会标明"付费赞助"、"付费合作伙伴"或"广告"等字样。
- 影响者推广。烟草行业及其利益相关者会鼓励或赞助个人在网上发布与产品或品牌相关的内容。社交媒体影响者拥有数千甚至数百万的粉丝，他们会获得品牌的补偿，并接受来自影响者营销公司的指导，了解如何在最佳时间发布内容以获得最大曝光，并避免发布看起来像刻意摆拍的广告内容。策略还可能包括组织品牌赞助的派对和竞赛，并鼓励参与者在其社交媒体上发帖。影响者可能会被指示通过与品牌相关和不相关但具有巨大收视率的标签（例如#爱，#艺术，#时尚）来放大他们的促销社交媒体帖子。

示例：TakeAPart对烟草影响者营销进行了深入报道，包括发布的图像以及与不同品牌和产品相关的信息。

- 消费者分享自己的烟草使用经历并进行商业推广。使用烟草制品的消费者可能会分享描述烟草使用的内容，也可能会直接评论宣传烟草消费或推荐特定品牌或产品的内容。视情况而定，这可能构成合法的言论，例如无偿的个人通信。为烟草行业利益工作的其他各方可以选择通过付费的数字媒体传播平台扩大这些内容的传播范围，将这些个人合法的言论帖子转化为商业促销。
- 活动推广。活动的参与者或团队由烟草公司和社交媒体赞助，视听共享平台播放该活动及相关图像。就赛车等重大体育赛事而言，其影响范围可以是全球性的，因为这些赛事广泛传播，包括在传统媒体上。

示例：参考文献[22]。

- 公司和活动促销。烟草公司或那些致力于促进其利益的公司，推广企业或运动品牌，而不是烟草制品品牌，并经营推广企业或运动品牌的社交媒体账户[23]。公司的促销活动和行动往往将烟草公司描绘成创新的践行者和对社会负责的行为主体，并将新型烟草制品宣传为传统卷烟的“危害较小的替代品”，尽管往往缺乏独立的科学证据来支持这种说法。

示例：菲利普·莫里斯国际公司的“不吸烟”运动[24]。

- 菲利普·莫里斯国际公司和奥驰亚公司的宣传运动：“无烟未来”和“超越烟雾”运动揭露了“一家真正关心健康的烟草公司”的虚伪性[25]。
- 商业内容中包含的烟草使用警示，被认为不是合法的表现形式。虽然社交媒体上的大部分内容本质上不是商业性的，但商业内容吸引了大量的用户流量（例如音乐视频、短片、网络系列），并且与内容创建者相关联，内容创建者从用户流量和用户购买推荐或评论的产品中产生收入。例如，音乐视频被广泛观看和分享，视听分享网站上的流行内容也是全球接触烟草预防的主要来源之一。

示例：Cranwell等[26]。

- 产品集成。烟草公司，或那些致力于促进其利益的公司，与制片人、制作公司和编剧合作，建立涉及其产品的故事情节，并将其无缝整合到其作品中。
- 赞助新闻或“信息娱乐”内容。烟草行业或那些致力于促进其利益的人，为新闻或时事记者或编辑提供设施访问，推销故事想法，或赞助相关或不相关主题的新闻报道。

示例：Meade [21]。

- 设备广告推广和赞助。广告或促销用于消费烟草制品的一个或多个装置可以直接或间接地广告或促销烟草制品本身。
- 烟草公司及其利益相关方运营社交媒体账户和网站，其内容可跨境传播。这些网站不仅常用于合法言论表达，还被用于推广企业品牌、推销特定产品或以提供消费者信息为幌子传播品牌信息，或作为所谓的企业社会责任活动。烟草行业利用社交网络和公司网站将自己重塑为一个现代化、社会责任感强且可持续的行业，试图与其产品所带来的危害划清界限。许多跨国烟草公司正在利用付费的整版“公共关系”公告，在长期以来禁止在其版面上刊登烟草广告的著名报纸和杂志上恢复品牌宣传。

示例：无烟世界基金会[27]和Freeman等[28]。

2. 电影、电视和流媒体内容是烟草展示的重要来源

对年轻人有吸引力的内容，如真人秀电视节目，被发现包含大量的烟草画面。

示例：Barker等[29]

在线电影数据库https://smokefreemedia.ucsf.edu/sfmmedia提供了电影院吸烟的发生率。

3. 流媒体电视节目

随着传统电视观众的减少以及在线流媒体和付费订阅的日益增加，流媒体内容愈发成为烟草促销的来源。从全球来看，在高收入和低收入国家中，年轻人（18~34岁）比35岁及以上的人更有可能使用互联网和智能手机。在流行的流媒体内容中，烟草展示比传统的广播或有线电视节目更为普遍。许多国家长期以来一直禁止通过“大众媒体”，通常定义为广播媒体（电视和电台），进行烟草广告。如今，社交媒体已成为一个强大的新型大众媒体分销渠道，尤其倾向于年轻观众。

示例：参考文献[30]提供了一份详细的报告和分析，表明一家全球流媒体巨头的节目比广播节目描绘了更多的吸烟图像。另见Barker等[31]。

4. 视频和电脑游戏

包装和在线视频游戏在年轻人中很受欢迎，但几乎没有任何控制措施来保护或防止用户接触游戏中嵌入的烟草展示，或游戏内、应用内购买烟草制品的内容。年龄限制可能未考虑到烟草使用，且年轻玩家很容易规避。

示例：参考文献[32]描述了以烟草使用为特征的游戏。

5. 智能手机应用程序

一些智能手机应用程序（APP）显示卷烟品牌的图像或类似于现有品牌的图像。支持吸烟的应用程序包括卡通游戏和模拟高质量吸烟体验的机会、免费应用程序或促进烟草制品销售的应用程序以及新型烟草制品，包括为消费此类产品而设计的设备。

示例：BinDhim等[33]。

6.2.3 社交媒体平台的烟草广告政策

主流社交媒体平台，包括Facebook[34]、Instagram[35]和Twitter[36]，都已采取禁止付费烟草广告的政策。然而，这些政策并不适用于由烟草行业赞助的政治或企业信息广告[37]，也不限制烟草公司使用标签来吸引社交媒体帖子的关注，也不阻止烟草公司在这些平台上运营免费的“有机”账户，这些账户成为品牌广告的热

门传播途径。例如，PMI运营着一个拥有100多万粉丝的Facebook页面[38]。谷歌还制定了关于危险产品或服务的广告政策，禁止烟草或任何含有烟草的产品，烟草制品的组成部分，直接促进或推动烟草消费的产品和服务，以及用于模拟吸烟的产品[39]。然而，对于烟草零售商的谷歌搜索，仍提供本地化的结果和销售网点的直接链接。

2020年，一项针对青少年用户众多的社交媒体平台TikTok上的电子烟和相关使用帖子的研究显示，大多数帖子对电子烟和其使用持正面态度[40]。2022年，TikTok更新了社区指南，声称禁止提供购买、销售、交易或招揽毒品或其他受管制物质、酒精或烟草制品（包括电子烟）、无烟烟草制品或传统烟草制品、合成烟碱产品以及其他ENDS的内容。新政策进一步规定，描述成年人使用烟草制品或提及受管制物质的内容不符合推荐条件。请注意，建议、描绘、模仿或促进未成年人拥有或消费酒精饮料、烟草或毒品的内容是禁止的。根据社区指南，也禁止针对未成年人提供关于如何购买、销售或交易酒精、烟草或管制药物的指导的内容[41]。

2021年5月对美国11个受年轻人欢迎的网站上的烟草制品推广和销售相关的社交媒体政策进行了评估[42]。11个网站中有9个禁止烟草制品的“付费广告”；然而，只有3个网站禁止宣传烟草的“赞助内容”。六个平台限制“销售烟草制品”的内容，三个平台声称“禁止未成年人访问”宣传或销售烟草制品的内容。虽然大多数平台政策禁止付费烟草广告，但很少有平台政策涉及赞助或影响力内容等不太直接的策略，也很少有平台政策通过年龄门槛来防止年轻人访问。

这些自愿性政策是否能有效减少公众接触TAPS，尚无充分证据。快速发展的媒体环境，加上对社交媒体平台的监管松散，尤其是过度依赖平台自我监管[43]，导致要将全面的TAPS禁令真正覆盖至线上媒体变得更加复杂。所有社交媒体平台必须禁止包括付费广告和自然内容在内的所有形式的烟草、电子烟及新型烟草制品广告，并禁止利用意见领袖进行推广。政策实施应明确违规举报机制，定期开展审计，并要求平台报告如何确保法律在其网站上得到有效执行。目前，主要由控烟相关方负责监控社交媒体平台上的TAPS内容，但这一责任应更多地转移至社交媒体公司[44]。社交媒体公司可以通过使用先进的人工智能系统来监督内容，从而大幅度自动识别烟草促销行为。

例如，在对4526名Instagram用户的分析中，这些用户创建了19951条与IQOS相关的帖子，其中42.1%的用户为Instagram授权的商业账户，其中59.0%来自个人物品及一般商品店铺，18.1%来自创作者和名人。网络中大多数活跃账户与IQOS直接相关（例如用户名中包含“IQOS”），或在账户简介中自称与烟草业务有关。这些结果清楚表明，当前社交媒体平台的自我监管远远不足以应对这一问题[45]。

6.2.4 烟草广告法的全球现状

WHO FCTC第13条明确指出，TAPS禁令在有效的烟草控制中起着至关重要的作用，并将禁止跨境TAPS作为综合控烟措施的一部分。WHO FCTC缔约方也意识到在监测和执行跨境TAPS禁令方面所面临的持续挑战，正准备制定第13条准则的补充文件，以应对自2008年准则通过以来媒体环境的显著变化。此外，缔约方呼吁建立更为有效的全球合作机制，以强化对跨境TAPS的管理[3]。各国在禁止本国境内的在线TAPS方面相对容易，但在缺乏国际合作的情况下，禁止那些从他国源发并通过数字边界传播的TAPS则较为困难。例如，欧盟要求其所有成员国禁止跨境烟草广告和赞助，并严格监督和执行这些规定[46]。

6.3 讨　　论

向WHO FCTC报告TAPS法规的国家中，约有一半（91/180，50.6%）表示其TAPS禁令涵盖本国互联网[47]。在线TAPS的跨境特性给监管机构带来了额外的挑战。它的跨境特性在于，一个国家创作、上传或广播的内容可能在另一个国家被消费或分享，从而跨越地理边界。服务提供商也可能位于与提供服务的国家不同的国家。内容还可能跨越“数字”边界，因为访问并不总是有效地限制在某个地理位置。跨境数字媒体消费为烟草行业及其利益相关者绕过TAPS控制提供了新的和新兴的渠道。

娱乐媒体可以通过支持互联网的设备（计算机、智能手机、平板电脑、智能电视）跨越国界，这些设备：

- 促进电影、电视连续剧或节目、视频游戏、音乐视频、体育、新闻、音乐、舞蹈和其他娱乐活动的在线流媒体；
- 允许访问国际和国内报纸和杂志的电子版本；
- 促进访问社交媒体帖子，包括商业和用户生成的内容和网站页面；
- 通过社交媒体为消费者和商业实体之间的互动提供机会；
- 可能包含烟草广告或提供嵌入式广告内容。

虽然在全面禁止TAPS的国家，烟草行业可能不会赞助电影中的烟草展示，但这些政策很少涵盖娱乐媒体中的未受赞助的烟草展示。为了规避这些限制，烟草公司可能会向制作公司的道具师提供免费产品样品，希望这些产品能够出现在演员手中。自2012年以来，印度要求凡是展示吸烟情节的电影都必须附带10秒的政府发布的反吸烟广告，并在烟草展示期间，在屏幕底部显示静态健康警示。任何出现在屏幕上的烟草制品品牌名称都必须被模糊处理。中国和泰国等国家也对电视和电影中允许的吸烟及烟草内容进行了监管。尽管多项研究已证实电影中的吸烟展示与年轻人吸烟风险增加之间的关联，但尚未对减少烟草描述的政策干预措施的影响进行研究或评估[48, 49]。

全球媒体内容制作商和流媒体服务商，如迪士尼[50]和Netflix[51]，已公开承诺减少新内容中烟草展示的频率，尤其是面向年轻观众的内容。这一举措是在发现Netflix受年轻人欢迎的节目中，烟草展示的数量随着时间推移而增加后才实施的。加州大学旧金山分校维护的媒体吸烟展示在线数据库记录了迪士尼、Netflix以及其他主要媒体公司[52]内容中持续展示烟草使用的情况。

6.3.1 跨境广告

致力于终止烟草制品促销的国家，不仅需要强化本国的TAPS禁令，还需与其他国家合作，协同减少跨境TAPS。这将需要更加有效的全球合作，并要求各国承诺定期更新TAPS法规，以应对新的媒体、通信平台和消费模式，以及行业不断演变的策略，这些策略融合了政治干预、广告和产品开发。为了达到全球最高标准，TAPS法规的改进和更新必须持续推进，并通过监管创新引领，例如全面停止烟草产品的零售销售，包括在线销售[53]。

法律专家提出了WHO FCTC缔约方可采取的共同行动，以减少跨境TAPS的可能途径[54]。它们包括：

- 建立机制，使缔约方能够报告来自其他缔约方领土的烟草广告（可直接通过指定联系人或中央公共卫生机构）；
- 同意在收到其他缔约方关于源自其领土的跨境烟草广告报告后采取适当行动，并向报告方通报所采取的措施；
- 同意协助其他缔约方调查、准备和起诉犯罪或潜在犯罪，例如协助获取相关证据和证人；
- 同意对居住在其领土内或拥有资产的个人或组织执行其他缔约方根据FCTC法律作出的判决，或说明拒绝执行的理由；
- 建立机制，使缔约方相互分享跨境广告的经验；
- 建立机制，缔约方可讨论为应对管辖和执法挑战而采取的合作措施的有效性，并根据需要作出新的安排；
- 建立机制，分享技术发展方面的经验和专门知识。

6.3.2 其他有害产品的网络营销监管

其他影响健康的商业因素，如酒精、食品、非酒精饮料和赌博，在有效监管在线营销方面面临与烟草类似的挑战。与烟草类似，社交媒体平台在赌博和酒精推广方面有着定义不清且执行不力的政策[55, 56]。对于食品和非酒精饮料的限制甚至更弱，社交媒体上充斥着不健康的加工食品、能量饮料和软饮料的促销活动[57]。

各国政府对多种形式不健康广告的应对措施差异很大。

6.3.3 烟草广告产品监管面临的挑战

评估全球TAPS禁令的一个限制是关于TAPS法律法规实施和执行的研究有限[58]。虽然WHO FCTC缔约方报告了其TAPS法律的范围，但报告中没有包含执法活动的细节，豁免条款和政策适用范围的限制也未得到充分说明。

对娱乐和在线媒体中的TAPS进行有效监管的其他挑战如下[4]：

- 内容共享平台的兴起（包括社交媒体）允许用户创建和分享内容。人们可以自由、便捷地观看和传播数字媒体。这种情况模糊了消费者与品牌所有者之间的界限，给跨境TAPS的监管带来了挑战。
- 媒体格局和TAPS形式的变化意味着法规可能需要更新，并使其具备应对新兴TAPS的前瞻性。
- 各国通常只禁止源自本国的跨境TAPS，而对从国外传入的TAPS则未作限制。
- 很难区分烟草使用和品牌展示的有偿与无偿呈现。
- 特别是在网络上，很难识别TAPS内容的来源地及其创建者或所有者。
- 系统地记录和追踪烟草行业的促销活动以及娱乐媒体中的烟草展示非常困难。
- 尚未批准世界卫生组织《烟草控制框架公约》的国家可能是跨境TAPS的来源。
- 年轻人和青少年是这类TAPS的理想目标人群；然而，针对保护这一群体免受此类内容影响的研究、资源和政策行动有限。

6.3.4 新产品和网络营销的相关挑战

法规和广告控制政策的执行是一项全球性挑战。例如，尽管泰国禁止所有ENDS产品的销售和进口[59]，但这些产品在网上非法销售[60]。巴西也存在类似情况，尽管不允许ENDS的营销、广告和进口，但它们仍在电子烟店、烟草店、互联网上和通过交付应用非法销售[61]。在南非，ENDS获批只能凭处方销售，但它们被大肆宣传为戒烟产品，并在无处方的情况下销售[62]。

在韩国，加热烟草设备被视为电子产品而非烟草制品[63]；因此，它们以生活方式为卖点进行宣传，包括英美烟草公司2019年在社交媒体上推广的活动，该活动以在年轻人中受欢迎的嘻哈音乐人为主角[64, 65]。该音乐视频由于只展示了加热设备而未涉及烟草荚，因此不受年龄限制，观看次数超过100万次[64]。

新型烟草、烟碱及其他雾化产品不断进入市场，并在全球范围内积极促销。在高收入国家，吸烟率下降，消费者能够负担昂贵的新产品；而在低收入和中等收入国家，广告禁令被规避[2]。

6.4 结　　论

各类媒体渠道，包括社交媒体，使用户暴露于TAPS。自2008年通过第13条实施指南以来，全球媒体格局发生了巨大变迁，并持续发展。娱乐媒体内容在区域和全球层面日益通过互联网提供，这可能引发跨境TAPS的暴露。技术转变的结果是，当前控制TAPS的方法已不足以应对。

快速发展的新媒体环境，加上对社交媒体平台监管的松懈与过度依赖自我监管，意味着全面禁烟要有效涵盖跨境TAPS，尽管具有挑战性，但仍是当务之急。除了强有力的国内监管行动外，还需要在区域和全球层面采取国际行动，包括各国间的合作，以减少TAPS[58]。媒体平台的迅速发展可能会造成监管漏洞，导致烟草品牌推广得以恢复。因此，监管限制必须具有前瞻性，预测可能的技术变革，并赋予监管机构快速应对变化的权力。

目前由社交媒体平台主导的自我监管是不够的。为了限制烟草制品促销内容的蔓延，有必要采取更精细、更严格的监管措施，尤其是限制官方账户、在线零售商和名人的营销行为。

6.5 研究空白和优先事项

已发表的研究主要集中在TAPS法律的必要性[66]、TAPS暴露的影响[67]、TAPS法律如何影响暴露程度[68, 69]以及其对吸烟态度、信仰和行为的潜在影响[70]；还探讨了烟草行业如何应对新通过的TAPS法律[71]并规避现有法律[72]。然而，关于TAPS禁令的实施及其后续监测和执行的数据较少。还需评估各国如何最有效地合作控制跨境TAPS。尽管在线TAPS的类型和模式正逐步得到系统监测，但仍应持续收集有关烟草行业的数据，尤其是考虑到新烟草和烟碱产品的开发及媒体环境的不断变化。评估TAPS禁令的全面性，特别是在涵盖在线TAPS形式和监测法规漏洞的利用方面，至关重要。

6.6 政 策 建 议

监管机构应制定全面战略，减少烟草和烟碱产品在社交媒体平台和在线数字娱乐媒体上的广告、促销和赞助力度。这一战略可以减少青少年和年轻人对烟草内容的接触，最终减少烟草使用。

- 确保TAPS法律全面，涵盖在线数字媒体平台，包括社交媒体，并具有足够的灵活性，以涵盖新媒体和平台。

- 在线数字TAPS的跨境特性需要国际合作，以实现有效的监督和执法。
- 要求烟草行业向政府当局披露全部TAPS活动，包括在线数字媒体平台上的所有活动，以加强监督和执法。
- 将新型和新兴烟碱和烟草产品纳入全面性法律，禁止烟草和非治疗性烟碱产品的广告、促销和赞助。
- 持续监测在线数字媒体平台以及新型和新兴烟碱和烟草产品的发展，确保TAPS法律的全面性，并禁止生活方式、时尚、创造力、身份、快乐和社交等广告主题。

参考文献

[1] Freeman B, Watts C, Astuti PAS. Global tobacco advertising, promotion and sponsorship regulation: What's old, what's new and where to next? Tob Control. 2022;31:216-21. doi:10.1136/tobaccocontrol-2021-056551.

[2] Ling PM, Kim M, Egbe CO, Patanavanich R, Pinho M, Hendlin Y. Moving targets: how the rapidly changing tobacco and nicotine landscape creates advertising and promotion policy challenges. Tob Control. 2022;31(2):222-8. doi:10.1136/tobaccocontrol-2021-056552.

[3] Tobacco advertising, promotion and sponsorship: depiction of tobacco in entertainment media. Report by the Convention Secretariat (FCTC/COP/7/38). Conference of the Parties to the WHO Framework Convention on Tobacco Control, seventh session. Geneva: World Health Organization; 2016 (FCTC_COP_7_38_EN.pdf).

[4] Conference of the Parties to the WHO Framework Convention on Tobacco Control. Decision FCTC/COP7(5). Tobacco advertising, promotion and sponsorship: depiction of tobacco in entertainment media. Geneva: World Health Organization; 2016 (https://fctc.who.int/docs/librariesprovider12/meetingreports/fctc_cop7_5_en.pdf?sfvrsn=f2653e3c_16&download=true).

[5] Blecher E. The impact of tobacco advertising bans on consumption in developing countries. J Health Econ. 2008;27(4):930-42. doi:10.1016/j.jhealeco.2008.02.010.

[6] Freeman B. New media and tobacco control. Tob Control. 2012;21(2):139-44. doi:10.1136/tobaccocontrol-2011-050193.

[7] WHO Framework Convention on Tobacco Control. Introduction. Article 1. Use of terms. Geneva: World Health Organization; 2003 (https://apps.who.int/iris/rest/bitstreams/50793/retrieve).

[8] WHO Study Group on Tobacco Product Regulation. Report on the scientific basis of tobacco product regulation: eighth report of a WHO study group. Geneva: World Health Organization; 2021 (https://www.who.int/publications/i/item/9789240022720).

[9] WHO Framework Convention on Tobacco Control Secretariat. Guidelines for implementation of Article 13 of the WHO Framework Convention on Tobacco Control (Tobacco advertising, promotion and sponsorship) 2008 (https://www.who.int/fctc/guidelines/article_13. pdf?ua=1).

[10] Zelefsky V. The differences between paid and organic content on social media. Jersey City (NJ): Forbes; 2022 (https://www.forbes.com/sites/forbescommunicationscouncil/2022/05/06/the-dif-

ferences-between-paid-and-organic-content-on-social-media/?sh=3ddfe0661526).

[11] Cavazos-Rehg P, Li X, Kasson E, Kaiser N, Borodovsky JT, Grucza R et al. Exploring how social media exposure and interactions are associated with ENDS and tobacco use in adolescents from the PATH study. Nicotine Tob Res. 2020;23(3):487-94. doi:10.1093/ntr/ntaa113.

[12] Donaldson SI, Dormanesh A, Perez C, Majmundar A, Allem JP. Association between exposure to tobacco content on social media and tobacco use: a systematic review and meta-analysis. JAMA Pediatr. 2022;176(9):878-85. doi:10.1001/jamapediatrics.2022.2223.

[13] The role of the media in promoting and reducing tobacco use (Tobacco Control Monograph No. 19. NIH Pub. No. 07-6242). Bethesda (MD): National Cancer Institute; 2008 (https://cancercontrol.cancer.gov/brp/tcrb/monographs/monograph-19).

[14] E-cigarette use among youth and young adults. A report of the Surgeon General. Atlanta (GA): Centers for Disease Control and Prevention, National Center for Chronic Disease Prevention and Health Promotion, Office on Smoking and Health; 2016 (https://www.cdc.gov/tobacco/sgr/e-cigarettes/index.htm).

[15] Yang Q, Clendennen SL, Loukas A. How does social media exposure and engagement influence college students' use of ENDS Pproducts? A cross-lagged longitudinal study. Health Commun. 2021:1-10. doi:10.1080/10410236.2021.1930671.

[16] Ad collections. Palo Alto (CA): Stanford Research into the Impact of Tobacco Advertising; 2023 (https://tobacco.stanford.edu/).

[17] Where there's smoke. Washington DC: Campaign for Tobacco-Free Kids; undated (https://www.takeapart.org/wheretheressmoke/).

[18] Klein EG, Czaplicki L, Berman M, Emery S, Schillo B. Visual attention to the use of #ad versus #sponsored on e-cigarette influencer posts on social media: a randomized experiment. J Health Commun. 2020;25(12):925-30. doi:10.1080/10810730.2020.1849464.

[19] Greenland S, Lužar K, Low D. Tobacco CSR, sustainability reporting, and the marketing paradox. In: Crowther D, Seifi S, editors. The Palgrave Handbook of Corporate Social Responsibility. London: Palgrave Macmillan; 2020:1-27 (https://link.springer.com/referenceworkentry/10.1007/978-3-030-22438-7_67-1).

[20] Legg T, Peeters S, Chamberlain P, Gilmore AB. The Philip Morris-funded Foundation for a Smoke-Free World: tax return sheds light on funding activities. Lancet. 2019;393(10190):2487-8. doi:10.1016/S0140-6736(19)31347-9.

[21] Meade A. Philip Morris-sponsored articles in the Australian could breach tobacco advertising laws. The Guardian, 18 November 2020 (https://www.theguardian.com/media/2020/nov/19/philip-morris-sponsored-articles-in-the-australian-could-breach-tobacco-advertising-laws).

[22] Watts C, Hefler M, Freeman B. "We have a rich heritage and, we believe, a bright future": how transnational tobacco companies are using Twitter to oppose policy and shape their public identity. Tob Control. 2019;28(2):227-32. doi:10.1136/tobaccocontrol-2017-054188.

[23] Driving addiction. Tobacco sponsorship in Formula One, 2021. New York (NY): STOP, Bloomberg Philanthropies; 2021 (https://exposetobacco.org/wp-content/uploads/TobaccoSponsorshipFormula-One-2021.pdf).

[24] Davies M, Stockton B, Chapman M, Cave T. The "unsmoke" screen: the truth behind PMI's cigarette-free future. London: Bureau of Investigative Journalism; 2020 (https://www.thebureau-

investigates.com/stories/2020-02-24/the-unsmoke-screen-the-truth-behind-pmiscigarette-free-future).

[25] Jackler RK. Propagands crusades by Philip Morris International & Altria: "Smoke-free future" and "Moving beyond smoke" campaigns. Exposing the hypocrisy of the claim: "A tobacco company that actually cares about health". Palo Alto (CA): Stanford University School of Medicine; 2022 (https://tobacco-img.stanford.edu/wp-content/uploads/2022/03/02103210/PMISFF-White-Paper-3-2-2022F-.pdf).

[26] Cranwell J, Opazo-Breton M, Britton J. Adult and adolescent exposure to tobacco and alcohol content in contemporary YouTube music videos in Great Britain: a population estimate. J Epidemiol Community Health. 2016;70:488-92. doi:10.1136/jech-2015-206402.

[27] Foundation for a Smoke-free World. Bath: University of Bath, Tobacco Tactics; 2022 (https://tobaccotactics.org/wiki/foundation-for-a-smoke-free-world).

[28] Freeman B, Hefler M, Hunt D. Philip Morris International's use of Facebook to undermine Australian tobacco control laws. Public Health Res Pract. 2019;29(3):e2931924. doi:10.17061/phrp2931924.

[29] Barker AB, Opazo Breton M, Cranwell J, Britton J, Murray RL. Population exposure to smoking and tobacco branding in the UK reality show 'Love Island'. Tob Control. 2018;27:709-11. doi:10.1136/tobaccocontrol-2017-054125.

[30] While you were streaming: smoking on demand. A surge in tobacco imagery is putting youth at risk. Washington DC: Truth Initiative; 2019 (https://truthinitiative.org/research-resources/tobacco-pop-culture/while-you-were-streaming-smoking-demand).

[31] Barker AB, Smith J, Hunter A, Britton J, Murray RL. Quantifying tobacco and alcohol imagery in Netflix and Amazon Prime instant video original programming accessed from the United Kingdom of Great Britain and Northern Ireland: a content analysis. BMJ Open. 2019;9:e025807. doi:10.1136/bmjopen-2018-025807.

[32] Some video games glamorize smoking so much that cigarettes can help players win. Washington DC: Truth Initiative; 2018 (https://truthinitiative.org/research-resources/tobacco-pop-culture/some-video-games-glamorize-smoking-so-much-cigarettes-can).

[33] BinDhim NF, Freeman B, Trevena L. Pro-smoking apps for smartphones: the latest vehicle for the tobacco industry? Tob Control. 2014;23:e4. doi:10.1136/tobaccocontrol-2012-050598.

[34] Advertising policies. Prohibited content. Menlo Park (CA): Facebook; 2021 (https://www.facebook.com/policies/ads/#).

[35] Ads on Instagram. Menlo Park (CA): Instagram; 2021 (https://help.instagram.com/1415228085373580).

[36] Twitter ads policies. San Francisco (CA): Twitter; 2021 (https://business.twitter.com/en/help/ads-policies.html).

[37] O'Brien EK, Hoffman L, Navarro MA, Ganz O. Social media use by leading US e-cigarette, cigarette, smokeless tobacco, cigar and hookah brands. Tob Control. 2020;29(e1):e87-97. doi:10.1136/tobaccocontrol-2019-055406.

[38] Freeman B, Hefler M, Hunt D. Philip Morris International's use of Facebook to undermine Australian tobacco control laws. Public Health Res Pract. 2019;29:e2931924 (https://www.phrp.com.au/issues/september-2019-volume-29-issue-3/philip-morris-internationals-use-facebook-un-

dermine-australian-tobacco-control-laws/).
[39] Advertising policies help. List of ad policies: dangerous products or services. Mountain View (CA): Google; 2021(https://support.google.com/adwordspolicy/answer/6014299?hl=en).
[40] Sun T, Lim CCW, Chung J, Cheng B, Davidson L, Tisdale C et al. Vaping on TikTok: a systematic thematic analysis. Tob Control. 2023;32(2):2514. doi:10.1136/tobaccocontrol-2021-056619.
[41] TikTok. Community guidelines. Illegal activities and regulated goods. Beijing: ByteDance; 2023 (https://www.tiktok.com/community-guidelines?lang=en#32).
[42] Kong G, Laestadius L, Vassey J, Majmundar A, Stroup AM, Meissner HI et al. Tobacco promotion restriction policies on social media. Tob Control. 2022 Nov 3;tobaccocontrol-2022-057348. doi:10.1136/tc-2022-057348.
[43] Gosh D. Are we entering a new era of social media regulation? Harvard Business Review, 14 January 2021 (https://hbr.org/2021/01/are-we-entering-a-new-era-of-social-media-regulation).
[44] About STOP. It's time to shine the light on the tobacco industry. New York (NY): Bloomberg Philanthropies; 2021 (https://exposetobacco.org/about/).
[45] Gu J, Abroms LC, Broniatowski DA, Evans WD. An investigation of influential users in the promotion and marketing of heated tobacco products on Instagram: a social network analysis. Int J Environ Res Public Health. 2022;19(3):1686. doi:10.3390/ijerph19031686.
[46] Ban on cross-border tobacco advertising and sponsorship. Brussels: European Commission; 2021 (https://ec.europa.eu/health/tobacco/advertising_en).
[47] WHO Framework Convention on Tobacco Control Secretariat. Article 13. Demand reduction measures. C2722-Ban covering the domestic internet. Report charts. Geneva: World Health Organization; 2020 (https://untobaccocontrol.org/impldb/indicator-report/?wpdtvar=3.2.7.2.b).
[48] Smoke-free movies: from evidence to action, third edition. Geneva: World Health Organization; 2015 (https://apps.who.int/iris/rest/bitstreams/850394/retrieve).
[49] Leonardi-Bee J, Nderi M, Britton J. Smoking in movies and smoking initiation in adolescents: systematic review and meta-analysis. Addiction. 2016;111(10):1750-63. doi:10.1111/add.13418.
[50] Barnes B. There's no smoking in Disney films. What about when it owns Fox? New York Times, 25 April 2018 (https://www.nytimes.com/2018/04/25/business/media/smoking-movies-disney-fox.html).
[51] Romo V. Netflix promises to quit smoking on (most) original programming. National Public Radio, 4 July 2019 (https://www.npr.org/2019/07/04/738719658/netflix-promises-to-quit-smoking-on-most-original-programming).
[52] Onscreen tobacco database. San Francisco (CA): Smokefree Media; 2021 (https://smokefreemedia.ucsf.edu/sfm-media).
[53] Smith EA, Malone RE. An argument for phasing out sales of cigarettes. Tob Control. 2020;29(6):703-8. doi:10.1136/tobaccocontrol-2019-055079.
[54] Kenyon AT, Liberman J. Controlling cross-border tobacco advertising, promotion and sponsorship-implementing the FCTC. Melbourne: Centre for Media and Communications Law; 2006 (https://www.researchgate.net/publication/228192102_Controlling_Cross-Border_Tobacco_Advertising_Promotion_and_Sponsorship_-_Implementing_the_FCTC).
[55] Carah N, Brodmerkel S. Alcohol marketing in the era of digital media platforms. J Stud Alcohol Drugs. 2021;82(1):18-27. PMID:33573719.

[56] Torrance J, John B, Greville J, O'Hanrahan M, Davies N, Roderique-Davies G. Emergent gambling advertising; a rapid review of marketing content, delivery and structural features. BMC Public Health. 2021;21(1):718. doi:10.1186/s12889-021-10805-w.

[57] McCarthy CM, de Vries R, Mackenbach JD. The influence of unhealthy food and beverage marketing through social media and advergaming on diet-related outcomes in children-A systematic review. Obesity Rev. 2022;23(6):e13441. doi:10.1111/obr.13441.

[58] Kennedy RD, Grant A, Spires M, Cohen JE. Point-of-sale tobacco advertising and display bans: policy evaluation study in five Russian cities. JMIR Public Health Surveillance. 2017;3(3):e52. doi:10.2196/publichealth.6069.

[59] 59 Patanavanich R, Glantz S. Successful countering of tobacco industry efforts to overturn Thailand's ends ban. Tob Control. 2021;30(e1):e10-9. doi:10.1136/tobaccocontrol-2020-056058.

[60] Phetphum C, Prajongjeep A, Thawatchaijareonying K, Wongwuttiyan T, Wongjamnong M, Yossuwan S et al. Personal and perceptual factors associated with the use of electronic cigarettes among university students in northern Thailand. Tob Induc Dis. 2021;19:31. doi:10.18332/tid/133640.

[61] de Pinho MCM, Riva MPR, de Souza Cury L, Andreis M. A promoção de novos produtos de tabaco nas redes sociais à luz da pandemia [Promotion of new tobacco products on social media in the pandemic]. Rev Bras Cancerol. 2020;66. doi:10.32635/2176-9745.RBC.2020v66nTemaAtual.1108.

[62] Agaku I, Egbe CO, Ayo-Yusuf O. Associations between electronic cigarette use and quitting-behaviours among South African adult smokers. Tob Control. 2022;31(3):464-72. doi:10.1136/tobaccocontrol-2020-056102.

[63] Kong J, Chu S, Park K, Lee S. The tobacco industry and electronic cigarette manufacturers enjoy a loophole in the legal definition of tobacco in South Korean law. Toba Control. 2021;30(4): 471-2.

[64] Yi J, Kim J, Lee S. British American Tobacco's 'Glo Sens' promotion with K-pop. Tobacco Control. 2021;30(5):594-6. doi:10.1136/tobaccocontrol-2020-055662.

[65] Comprehensive ban on cross-border tobacco advertising, promotions and sponsorship in ASEAN region. Bangkok: Southeast Asia Tobacco Control Alliance; undated (https://seatca.org/dmdocuments/Cross-border_final.pdf).

[66] Chido-Amajuoyi OG, Mantey DS, Clendennen SL, Pérez A. Association of tobacco advertising, promotion and sponsorship (TAPS) exposure and cigarette use among Nigerian adolescents: implications for current practices, products and policies. BMJ Glob Health. 2017;2(3). doi:10.1136/bmjgh-2017-000357.

[67] Tan AS, Hanby EP, Sanders-Jackson A, Lee S, Viswanath K, Potter J. Inequities in tobacco advertising exposure among young adult sexual, racial and ethnic minorities: examining intersectionality of sexual orientation with race and ethnicity. Tob Control. 2021;30(1):84-93. doi:10.1136/tobaccocontrol-2019-055313.

[68] Kahnert S, Demjén T, Tountas Y, Trofor AC, Przewoźniak K, Zatoński WA et al. Extent and correlates of self-reported exposure to tobacco advertising, promotion and sponsorship in smokers: Findings from the EUREST-PLUS ITC Europe surveys. Tob Induc Dis. 2018;16. doi:10.18332/tid/94828.

[69] Li L, Borland R, Yong HH, Sirirassamee B, Hamann S, Omar M et al. Impact of point-of-sale to-

bacco display bans in Thailand: findings from the international tobacco control (ITC) Southeast Asia survey. Int J Environ Res Public Health. 2015;12(8):9508-22. doi:10.3390/ijerph120809508.

[70] Nicksic NE, Bono RS, Rudy AK, Cobb CO, Barnes AJ. Smoking status and racial/ethnic disparities in youth exposure to tobacco advertising. J Ethn Subst Abuse. 2020;21(3):959-74. doi:10.1080/15332640.2020.1815113.

[71] da Silva ALO, Grilo G, Branco PAC, Fernandes AMMS, Albertassi PGD, Moreira JC. Tobacco industry strategies to prevent a ban on the display of tobacco products and changes to health warning labels on the packaging in Brazil. Tob Prev Cessat. 2020;6:66. doi:10.18332/tpc/128321.

[72] Astuti PAS, Assunta M, Freeman B. Raising generation 'A': a case study of millennial tobacco company marketing in Indonesia. Tob Control. 2018;27(e1):e41-9. doi:10.1136/tobaccocontrol-2017-054131.

7. 综 合 建 议

世界卫生组织烟草制品管制研究小组发布了一系列报告，为烟草制品管制提供科学依据。它们是世界卫生组织的技术系列（以前称为全球公共卫生产品）；如上所述，这些是世界卫生组织开发的产品（或资源），对全球或许多区域的许多国家都有益处[1]。TobReg报告根据世界卫生组织《烟草控制框架公约》第9、10条[2]，为烟碱和烟草制品的含量、释放物和设计特征的监管提供了基于证据的方法。研究组第十次会议的上一份报告[3]提供了关于新烟草和烟碱产品的基于可靠科学的建议。研究组的建议在各种情况下都具有相关性，具体取决于国家的监管环境、烟草和非治疗性烟碱产品的使用流行率，以及其他对这些产品监管有影响的相关因素，如政策目标和监管能力（包括禁令）。

本报告中提到的TobReg第九次会议的审议、成果和建议，涉及向包括儿童和青少年在内的不同年龄人群推广和分发非治疗性烟碱（特别是烟碱产品）的新方式。世界卫生组织《烟草控制框架公约》[2]是第一个旨在打击烟草流行的国际公共卫生条约，在过去20年中挽救了许多生命。各国越来越多地根据世界卫生组织《烟草控制框架公约》采取强有力的烟草控制政策措施，包括供需措施，以保护其公民并减少烟草制品的使用率。这导致全球卷烟和其他烟草制品的销售减少。因此，烟草行业利用技术和创新手段，通过引入营销和推广烟草和烟碱产品的新方式[3, 4]来提高利润，特别是通过针对儿童和青少年的战略和战术，使烟草和非治疗性烟碱产品更具吸引力[5, 6]，以维持产品的使用。鉴于其任务，TobReg在主题专家的协助下，综合已发表文献的证据，制定基于证据的产品监管建议。这些建议通过世界卫生组织总干事提供给各国，以协助应对烟草控制的监管挑战，而烟草控制仍然是全球优先事项。

监管机构需要谨记，烟草每年造成800多万人死亡[7, 8]。其中700多万人直接死于烟草使用，约130万人为死于接触二手烟的非吸烟者[9, 10]。烟草导致多达一半的使用者死亡，仍然是全球卫生紧急情况。合成烟碱的使用为烟草控制带来了新的复杂问题，并对监管机构构成了挑战，而一些烟草控制法律未必涵盖合成烟碱。烟草和烟碱产品的在线营销，以及尼古丁袋的引入推广，尤其是面向儿童和青少年的推广，进一步增加了问题的复杂性。因此，本报告的建议不应孤立地看待，而应放在更广泛的烟草控制背景中，以补充研究小组在其他报告[11-18]中关于卷烟、无烟烟草、水烟、设计特征、口味以及新型产品的建议。

烟草控制界非常清楚烟草行业蓄意破坏烟草控制并减缓世界卫生组织《烟草控制框架公约》的实施。烟草行业积极营销和推广对烟草控制构成严重威胁的新型烟草和烟碱产品，并通过隐蔽策略在网上宣传推广其产品。研究组审查了成员国提出的关于监管机构感兴趣的技术援助请求，并提出了若干建议。然而，TobReg重申其在上一份报告中的结论，即监管机构应继续关注更广泛的烟草控制，不应被烟草及相关行业的策略或其为维持烟草和非治疗性烟碱产品使用而采取的积极促销策略分散注意力。

本报告强调了以下方面的重要性：

- 继续关注烟草控制，以降低烟草使用率；
- 制定全面的烟草控制法律，适用于所有烟草制品及所有形式的烟草和非治疗性烟碱产品；
- 开展国际合作，解决烟草和烟碱产品的跨境营销问题；
- 制定全面法律，规范烟草广告、促销和赞助活动，包括推广烟草和非治疗性烟碱产品的新方法；
- 加强对烟草和烟碱产品法规的监测和执行，包括烟草和相关行业的活动；
- 弥补可能被烟草和相关行业利用的监管空白；
- 落实研究组的建议。

本报告第2~6章提供了科学信息、关于在线营销的证据以及填补烟草控制监管空白的政策建议和指导。报告还确定了进一步工作的重点领域，关注各国的监管需求，并考虑区域差异，为所有国家，特别是世界卫生组织成员国，提供持续、有针对性的技术支持战略。研究组的主要建议概述如下。

7.1 主要建议

向决策者和所有其他有关各方提出的主要建议如下：

- 研究组注意到全球积极推广烟草和烟碱产品，敦促成员国确保继续关注世界卫生组织《烟草控制框架公约》所概述的循证减少烟草使用的措施，不要被烟草行业或其他既得利益分散注意力；
- 确保烟草制品的法规涵盖并适用于所有形式的烟草和烟碱产品，而不仅限于传统卷烟；
- 要求制造商披露有关这些产品的相关信息：选定有害化学品的释放水平和上市前评估中使用的生物标志物水平；
- 确保烟草广告、促销和赞助的法律全面，至少符合世界卫生组织《烟草控制框架公约》的标准，并涵盖在线数字媒体平台，包括社交媒体及其他形式的直接或间接营销；

- 加强监测和执法，并开展国际合作，以应对烟草及相关行业的跨境做法，包括在线数字烟草广告、促销和赞助；
- 要求烟草及相关行业向政府当局披露所有广告、促销和赞助活动，包括在线数字媒体平台上的活动；
- 在制定或更新烟草制品法规时，根据世界卫生组织《烟草控制框架公约》第9、10条，处理烟草制品的成分和释放物问题，并支持产品评估、监测和信息披露；
- 禁止在非治疗性烟碱产品和所有烟草制品中添加薄荷醇及其他有助于吸入的成分，包括化学结构或生理和感官效果与薄荷醇相似的合成冷却剂；
- 修订国家烟草控制法律，以填补合成烟碱产品的监管空白，确保所有合成烟碱产品都纳入其范围，包括当前上市的药理类似物及未来可能出现的任何产品；
- 要求制造商实施统一的标签规则，对含有合成烟碱或各种天然或合成来源烟碱混合物的产品，分开申报不同烟碱形式或类似物的含量；
- 建立或扩大对产品及其使用者的监测，包括人口统计、其他烟草及相关产品的使用情况、尼古丁袋的品牌、类型和口味，以获取知识并评估使用的普及性和用户特征；
- 规范尼古丁袋，以防止所有形式的营销，并采取一切必要措施，以尽量减少青少年获取这些产品的机会以及这些产品对青少年的吸引力及其起始使用；
- 以与外观、成分及用途类似的产品相同的方式规范管制非治疗性烟碱产品；
- 确保尼古丁袋不被归类为药品，除非通过严格的制药途径被证明为烟碱替代疗法，并获得国家监管机构的许可；
- 在制定有关电子烟碱传输系统、加热型烟草制品和其他新烟草和烟碱产品的政策决策时，使用与行业无关的生物标志物数据和国家经验；
- 落实研究组就规范管制非治疗性烟碱及所有形式烟草制品所构成的具体挑战提出的建议。

敦促各国执行上述建议，因为各国已经掌握了足够的信息，可以采取行动来保护其公民，尤其是年轻一代的健康。虽然该报告承认，对于一些主题（包括合成烟碱及评估ENDS、ENNDS和HTP的生物标志物）仍有许多需要进一步研究的内容，但需要继续进行独立研究，以获取更多相关信息。所需数据包括尼古丁袋的使用率、这些产品的特性、烟碱产品中合成烟碱的使用情况及供应情况、合成烟碱的科学基础以及烟草及相关行业的促销策略。鉴于全球有13亿人使用烟草，烟草控制界应继续加快采用循证政策和建议，例如世界卫生组织《烟草控制框架公约》、WHO MPOWER措施以及世界卫生组织《烟草控制框架公约》缔约方会议相关报告中概述的措施。因此，各国应实施行之有效的政策措施，并考虑实施本报告中的建议。关于所审议的每个专题的具体建议见第2.10、3.4、4.11、5.9和6.6节。

7.2 对公共卫生政策的意义

研究组的报告为理解关于烟碱和烟草制品（包括卷烟、无烟烟草和水烟）监管的科学基础、研究和证据提供了指导。研究组第八份报告[18]涉及新烟草和烟碱产品，特别是电子烟碱传输系统、电子非烟碱传输系统和加热型烟草制品。第九份报告重点介绍了促进吸入的添加剂的作用，社交媒体和数字营销对公共卫生的影响，与尼古丁袋和合成烟碱营销相关的挑战及其监管影响，以及评估电子烟碱传输系统、电子非烟碱传输系统和加热型烟草制品的暴露、效应和易感性生物标志物的现有证据。报告还考虑了这些产品的引入对烟草控制的潜在影响，确定了研究差距并提出了建议。这些建议直接应对了成员国面临的一些独特的监管挑战，包括产品市场的直接和间接广告，以及尼古丁袋和合成烟碱等产品在全球市场的渗透。此外，报告还为成员国更新了相关知识，并为制定烟草和烟碱产品的有效监管战略提供了指导。

研究组虽然由监管、技术和科学专家组成，但其独特之处在于能够梳理和提炼复杂的数据与研究，并将其综合为政策建议，为国家、区域和全球各层级的政策制定提供支持。本报告由具有烟草制品监管相关学科专业知识的科学家撰写，阐述了各国政府在有效监管烟草和非治疗性烟碱产品方面所面临的挑战。监管机构、政府和其他相关方可以依据所提供的科学依据和证据，制定政策以加强烟草控制，并酌情弥补监管漏洞。在本次审议中，针对尼古丁袋、合成烟碱、烟草和烟碱产品的在线数字营销等主题，研究组查明了政策与研究方面的差距，指出了信息不足的领域。各国在制定研究议程时，可以将重点放在与其政策目标、国情和监管环境相关的关键领域上。这是研究组的重要作用，尤其对于那些在烟草制品监管技术信息方面资源或能力不足的政府而言。

研究组的建议旨在促进国际监管协调，并推广烟草及非治疗性烟碱产品监管的最佳做法，加强世界卫生组织六大地区的产品监管能力，为成员国提供基于可靠科学的现成资源，支持缔约方实施世界卫生组织《烟草控制框架公约》。鉴于全球范围内烟草和烟碱产品的强力推广，研究组敦促成员国继续集中精力在世界卫生组织《烟草控制框架公约》中提出的循证措施，减少烟草使用，而不受烟草及相关行业的干扰。

烟草制品的监管措施是对世界卫生组织《烟草控制框架公约》其他减少烟草需求规定的补充。研究小组的建议如果得到有效实施，将有助于减少烟草使用的流行并改善公众健康。

7.3 对 WHO 规划的影响

该报告履行了世界卫生组织烟草制品管制研究小组的任务，即向总干事提供科学合理、基于证据的建议[1]，帮助成员国应对烟草制品管制这一技术含量极高、充满复杂挑战的领域。研究组的审议结果和主要建议将帮助成员国更好地理解传统烟草制品及新型产品，以及制造商采用的促销策略。

该报告对烟草和非治疗性烟碱产品监管知识的贡献，在为秘书处提供信息方面发挥了关键作用，尤其是在向成员国提供技术支持时。同时，报告还通过全球烟草监管机构论坛为各国监管机构提供最新信息，并通过世界卫生组织向2023年11月召开的缔约方大会第十次会议提交的报告，更新世界卫生组织《烟草控制框架公约》缔约方的相关技术信息，尤其是关于该公约第9、10条的技术事项。报告将涵盖研究组第九份报告的主要信息和建议。所有这些行动将为实现可持续发展目标中的3.a（加强在所有国家实施世界卫生组织《烟草控制框架公约》）以及世界卫生组织第十三个全球工作计划中的“三十亿”目标作出贡献。

作为世界卫生组织的一项技术产品（全球公共卫生产品[1]），本报告面向所有国家，以推动在全球和国家层面减少烟草使用，改善整体公共卫生。

参 考 文 献

[1] Thirteenth General Programme of Work 2019–2023. Geneva: World Health Organization; 2019(https://apps.who.int/iris/bitstream/handle/10665/324775/WHO-PRP-18.1-eng.pdf,accessed 29 December 2020).

[2] WHO Framework Convention on Tobacco Control. Geneva: World Health Organization; 2003(http://www.who.int/fctc/en/, accessed 10 January 2021).

[3] Tobacco industry tactics: advertising, promotion and sponsorship. Geneva: World Health Organization;2019(https://applications.emro.who.int/docs/FS-TFI-202-2019-EN.pdf?ua=1).

[4] Driving addiction: F1, Netflix and cigarette company advertising. STOP. A global tobacco industry watchdog. Paris: International Union Against Tuberculosis and Lung Disease; https://exposetobacco.org/wp-content/uploads/F1-Netflix-Driving-Addiction.pdf.

[5] Modern Addiction: Myths and Facts about how the Tobacco Industry Hooks Young Users.

1 2003 年 11 月，理事长正式确定了前科学咨询委员会的地位从一个科学咨询委员会到一个研究组

Brief,August 2021. STOP. A Global Tobacco Industry Watchdog. https://exposetobacco.org/wp-content/uploads/Modern-Addiction-Mythbuster.pdf.

[6] Flavours (including Menthol) in Tobacco Products. Brief, May 2022. STOP. A Global Tobacco Industry Watchdog. https://exposetobacco.org/wp-content/uploads/Flavors-Including-Menthol-In-Tobacco-Products-FINAL.pdf.

[7] WHO report on the global tobacco epidemic, 2019. Geneva: World Health Organization; 2019(http://www.who.int/tobacco/global_report/en, accessed 19 December 2020).

[8] GBD 2019 Risk Factors Collaborators. Global burden of 87 risk factors in 204 countries and territories, 1990-2019: a systematic analysis for the Global Burden of Disease Study 2019.Lancet. 2020;396:1223-49.

[9] Tobacco. Fact sheet. Geneva: World Health Organization; 2020 (https://www.who.int/newsroom/fact-sheets/detail/tobacco, accessed 23 December 2020).

[10] Findings from the Global Burden of Disease Study 2017: GBD Compare. Seattle (WA): Institute for Health Metrics and Evaluation; 2018 (http://vizhub.healthdata.org/gbd-compare, accessed 19 December 2020).

[11] The scientific basis of tobacco product regulation. Report of a WHO study group (WHO Technical Report Series, No. 945). Geneva: World Health Organization; 2007 (https://www.who.int/tobacco/global_interaction/tobreg/who_tsr.pdf, accessed 10 January 2021).

[12] The scientific basis of tobacco product regulation. Second report of a WHO study group (WHO Technical Report Series, No. 951). Geneva: World Health Organization; 2008 (https://apps.who.int/iris/bitstream/handle/10665/43997/TRS951_eng.pdf?sequence=1, accessed 10 January 2021).

[13] WHO Study Group on Tobacco Product Regulation. Report on the scientific basis of tobacco product regulation: Third report of a WHO study group (WHO Technical Report Series,No. 955). Geneva: World Health Organization; 2009 (https://apps.who.int/iris/bitstream/handle/10665/44213/9789241209557_eng.pdf?sequence=1, accessed 10 January 2021).

[14] WHO Study Group on Tobacco Product Regulation. Report on the scientific basis of tobacco product regulation: Fourth report of a WHO study group (WHO Technical Report Series, No.967).Geneva:World Health Organization; 2012 (https://apps.who.int/iris/bitstream/handle/10665/44800/9789241209670_eng.pdf?sequence=1, accessed 10 January 2021).

[15] WHO Study Group on Tobacco Product Regulation. Report on the scientific basis of tobacco product regulation: Fifth report of a WHO study group (WHO Technical Report Series,No. 989). Geneva: World Health Organization; 2015 (https://apps.who.int/iris/bitstream/handle/10665/161512/9789241209892.pdf?sequence=1, accessed 10 January 2021).

[16] WHO Study Group on Tobacco Product Regulation. Report on the scientific basis of tobacco product regulation: Sixth report of a WHO study group (WHO Technical Report Series,-No.1001).Geneva: World Health Organization; 2017 (https://apps.who.int/iris/bitstream/handle/10665/260245/9789241210010eng.pdf?sequence=1, accessed 10 January 2021).

[17] WHO Study Group on Tobacco Product Regulation. Report on the scientific basis of tobacco product regulation: Seventh report of a WHO study group (WHO Technical Report Series,No. 1015). Geneva: World Health Organization; 2019 (https://apps.who.int/iris/bitstream/handle/10665/329445/9789241210249-eng.pdf, accessed 10 January 2021).

[18] WHO study group on tobacco product regulation. Report on the scientific basis of tobacco product regulation: eighth report of a WHO study group. Geneva: World Health Organization;2021 (WHO Technical Report Series, No. 1029; (https://www.who.int/publications/i/item/9789240022720).

1. Introduction

Tobacco is a global public health threat and kills more than 8 million people a year globally *(1)*, about 1.2 million of those deaths resulting from exposure of non-smokers to second-hand smoke *(2)*. Comprehensive tobacco control is therefore essential to tackle the global tobacco epidemic and prevent needless deaths. Product regulation can play a role in reducing the demand for tobacco, and effective tobacco product regulation is an essential component of a comprehensive tobacco control programme *(3)*. It includes regulation of contents and emissions by mandated testing, disclosure of test results, setting limits as appropriate, disclosure of information on products and imposing standards for product packaging and labelling. Tobacco product regulation is covered under Articles 9, 10 and 11 of the WHO Framework Convention on Tobacco Control (WHO FCTC) *(4)* and in the partial guidelines for implementation of Articles 9 and 10 of the WHO FCTC *(5)*. Other WHO resources, including the basic handbook on tobacco product regulation *(3)*, the handbook on building laboratory testing capacity *(6)* and the online modular courses based on the handbooks, are available on the WHO website *(7)*, support Member States in this respect. Additionally, the WHO Study Group on Tobacco Product Regulation (TobReg) has published a number of reports and advisory notes that provide guidance on several aspects of regulating tobacco products and non-therapeutic nicotine products.

The Study Group was formally constituted by the WHO Director-General in 2003 to address gaps in the regulation of tobacco products. Its mandate is to provide evidence-based recommendations on policy for tobacco product regulation to the Director-General. TobReg is composed of national and international scientific experts on product regulation, treatment of tobacco dependence, toxicology and laboratory analyses of tobacco product ingredients and emissions. The experts are from countries in all six regions of WHO *(8)*. As a formal entity of WHO, TobReg submits technical reports that provide the scientific basis for tobacco product regulation to the WHO Executive Board, through the Director-General, to draw the attention of Member States to WHO's work in this field. The reports, which are part of the WHO Technical Report Series, include previously unpublished background papers that synthesize published scientific literature and have been discussed, evaluated and reviewed by TobReg. In accordance with Articles 9 and 10 of the WHO FCTC, relevant decisions of the Conference of the Parties (COP) to the WHO FCTC and relevant WHO reports submitted to the COP, the TobReg reports identify evidence-based approaches to regulating all forms of tobacco products and non-therapeutic nicotine products, including new and emerging products such as electronic nicotine delivery systems (ENDS), electronic non-nicotine delivery systems (ENNDS), heated tobacco products (HTPs) and nicotine pouches. These reports, now considered to be WHO technical products (formally known as WHO global public health goods), respond to World Health Assembly resolutions WHA53.8 (2000), WHA53.17 (2000) and WHA54.18 (2001). WHO technical products or global public health goods are initiatives developed or undertaken by WHO that are of benefit either globally or to many countries in several regions *(9)*. This designation presents a unique opportunity for TobReg to engage directly with Member States and contribute to national, regional and global policy.

The ninth meeting of TobReg took place from 13–15 December 2022 in Tbilisi, Georgia, hosted by the Georgian National Centre for Disease Control and Public Health and organized by the WHO Tobacco Free Initiative Unit of the Health Promotion Department. About 40 participants, including TobReg members, WHO staff, the Secretariat of the WHO FCTC and invited experts, discussed the scientific literature on pertinent topics in product regulation, including emerging issues, according to Member States' requests to WHO and requests of the COP. Topics previously considered by the Global Tobacco Regulators Forum, such as synthetic nicotine and nicotine pouches, which have recently presented regulatory challenges because of the way in which they are marketed and used to exploit regulatory loopholes, are also addressed in the report. In response to repeated requests by Member States to the Secretariat to provide technical assistance and authoritative guidance on emerging issues in tobacco product regulation, the report focuses on newer ways in which non-therapeutic nicotine in nicotine and tobacco products is delivered and promoted to people of different ages, including children and adolescents. The meeting thus provided a platform for discussing six background papers:

- Additives that facilitate inhalation, including cooling agents, nicotine salts and flavourings;
- Synthetic nicotine: science, global legal landscape and regulatory considerations;
- Nicotine pouches: characteristics, use, harmfulness and regulation;
- Biomarkers of exposure, effect and susceptibility for assessing electronic nicotine delivery devices and heated tobacco products, and their possible prioritization;
- Internet, influencer and social media marketing of tobacco and non-therapeutic nicotine products and associated regulatory considerations; and
- The WHO Study Group on Tobacco Product Regulation: two decades of recommendations – translating evidence into policy action.

The sixth background paper was included to inform the future work of the Study Group in translating science into policy and will be considered separately by the Study Group. The report therefore includes the first five background papers. The requests of Members States in all WHO regions, the knowledge of the Secretariat and the Study Group in these areas and relevant literature formed the basis of the content of the five background papers. The information in these background papers updates current knowledge and will advance nicotine and tobacco product regulation and inform national and global policy.

The background papers were prepared by experts according to the terms of reference or an outline drawn up by the WHO secretariat for each paper and were reviewed and revised by TobReg members and by expert reviewers identified by WHO. The period of the literature search is indicated in each paper; for most, this was the second quarter of 2022 or the first quarter of 2023. The papers were subject to several rounds of review before and after the meeting by independent technical experts, the WHO secretariat, people in other relevant WHO departments, colleagues at regional offices and members of the Study Group before compilation into the technical report.

The secretariat, in consultation with the Study Group, invited experts who contributed to discussions and provided the most recent empirical scientific evidence and regulations on the topics under consideration. This ninth report of TobReg on the scientific basis of tobacco product regulation is designed to guide Member States in achieving the most effective evidence-based means to bridge regulatory gaps in tobacco control and to develop coordinated regulatory frameworks for tobacco products. Additionally, it identifies future areas of work, focusing on the regulatory needs of countries, thus providing a strategy for continued technical support to Member States. All experts and other participants in the meeting, including members of the Study Group, were required to complete declarations of interests, which were evaluated by WHO.

The report comprises this introduction to the context of the report, five papers on topics pertinent to tobacco control regulations and civil society organizations, and concludes with a summary of the recommendations in each section. The recommendations, which represent syntheses of complex research and evidence, promote international coordination of regulation and adoption of best practices in product regulation, and capacity-building for product regulation in all WHO regions, represent a ready resource for Member States, based on sound science, for implementation of the WHO FCTC by its Parties. Given the aggressive promotion of nicotine and tobacco products globally, the Study Group urges Member States to continue their focus on evidence-based measures to reduce tobacco use, as outlined in the WHO FCTC, and to avoid distraction by the tobacco and related industries.

This ninth report of the Study Group addresses additives that facilitate inhalation, synthetic nicotine, nicotine pouches, biomarkers for assessing ENDS, ENNDS and HTPs and social media marketing of tobacco and non-therapeutic nicotine products. It does not cover all the emerging issues in nicotine and tobacco product regulation, including flavours and design features such as filters and flavour accessories. The Study Group will continue to cover other aspects of product regulation, including other products of interest (such as waterpipes, cigarettes and smokeless tobacco) and other emerging issues that directly impact tobacco control in subsequent reports, guided by countries' regulatory requirements and pertinent issues in tobacco product regulation. The Group will thus ensure continued, timely technical support to all countries and address non-therapeutic nicotine and tobacco products broadly and factors with regulatory implications for product regulation, especially those that affect the attractiveness, addictiveness and toxicity of these products.

In summary, the outcomes of TobReg's deliberations and its recommendations will improve Member States' understanding of the evidence on the topics considered in the report, including synthetic nicotine, online marketing of tobacco products and nicotine pouches, contribute to the body of knowledge on product regulation, inform WHO's work, especially in providing technical support to Member States, and keep Member States, regulators, civil society organizations, research institutions and other interested parties up to date on product regulation through various platforms. Parties to the WHO FCTC will be updated by a comprehensive report to be submitted to COP10, via the Convention Secretariat, on technical matters related to implementation of Articles 9 and 10 of the WHO FCTC, which will include the messages and recommendations in this report. Thus, the Study Group's activities will contribute to meeting target 3.a of the Sustainable Development Goals: strengthening implementation of the WHO FCTC *(9)*.

References

1. WHO report on the global tobacco epidemic 2021: addressing new and emerging products. Geneva: World Health Organization; 2021 (https://www.who.int/publications/i/item/9789240032095).
2. Tobacco control to improve child health and development: thematic brief. Geneva: World Health Organization; 2021 (who.int/publications/i/item/9789240022218).
3. Tobacco product regulation: basic handbook. Geneva: World Health Organization; 2018 (https://www.who.int/tobacco/publications/prod_regulation/basic-handbook/en/, accessed 10 January 2021).
4. WHO Framework Convention on Tobacco Control. Geneva: World Health Organization; 2003 (http://www.who.int/fctc/en/, accessed 10 January 2021).
5. Partial guidelines on implementation of Articles 9 and 10. Geneva: World Health Organization; 2012 (https://www.who.int/fctc/guidelines/Guideliness_Articles_9_10_rev_240613.pdf, accessed January 2019).

6. Tobacco product regulation: building laboratory testing capacity. Geneva: World Health Organization; 2018 (https://www.who.int/tobacco/publications/prod_regulation/building-laboratory-testing-capacity/en/, accessed 14 January 2019).
7. Tobacco product regulation courses. Geneva: World Health Organization (https://openwho.org/courses/TPRS-building-laboratory-testing-capacity/items/3S11LKUFGyoTksZ5RSZbID; https://openwho.org/courses/TPRS-tobacco-product-regulation-handbook/items/7zq-7S1jxAtpbUWiZdfH98I).
8. TobReg members. In: World Health Organization [website]. Geneva: World Health Organization (https://www.who.int/groups/who-study-group-on-tobacco-product-regulation/about, accessed 10 January 2021).
9. Feacham RGA, Sachs JD. Global public goods for health. The report of working group 2 of the Commission on Macroeconomics and Health. Geneva: World Health Organization; 2002 (https://apps.who.int/iris/bitstream/handle/10665/42518/9241590106.pdf?sequence=1, accessed 10 January 2021).
10. Decision FCTC/COP8(22). Novel and emerging tobacco products. Decision of the Conference of the Parties to the WHO Framework Convention on Tobacco Control, eighth session. Geneva: World Health Organization; 2018 (https://www.who.int/fctc/cop/sessions/cop8/FCTC__COP8(22).pdf, accessed 7 November 2020).
11. Sustainable development goals. Geneva: World Health Organization (https://www.who.int/health-topics/sustainable-development-goals#tab=tab_2, accessed 10 January 2021).

2. Additives that facilitate inhalation, including cooling agents, nicotine salts and flavourings

Reinskje Talhout[1], Centre for Health Protection, National Institute for Public Health and the Environment, Bilthoven, Netherlands (Kingdom of the)

Adam M. Leventhal, Institute for Addiction Science, University of Southern California, Los Angeles (CA), USA

Contents

Abstract

Objective: Some additives counteract the harshness and bitterness of the aerosols of tobacco and nicotine products (TNPs), making them easier to inhale. This is a problem for public health, as it may stimulate the uptake and continued use of TNPs, especially by young people. This paper provides a conceptual framework of the processes, mechanisms and methods for assessing inhalation facilitation (IF). Specific additives in TNPs that may promote IF are reviewed and their potential health impact discussed.

Methods: A targeted (non-systematic) search of PubMed and other bibliographic sources with no restrictions on time period, up to September 2022, included terms related to IF processes (e.g. "harshness", "puff duration"), candidate additives (e.g. "menthol") or candidate mechanisms (e.g. "TRPM8 [transient receptor potential cation channel, family 8] receptor"). Inclusion of studies in the review was agreed by consensus by the two authors.

Results: We defined IF as a modification to a TNP that improves the user's sensory experience of inhaling the product's aerosol (reduced bitterness and harshness) and may alter inhalation behaviour, particularly more intense inhalation (e.g. deeper puffs, faster inhalation, larger puff volume) and also restoration of breathing patterns that are disturbed by inhaled irritants. The review showed that: (a) menthol and synthetic coolants decrease the irritation caused by nicotine and other TNP aerosol constituents by activating TRPM8 and other receptors and may promote dependence in inexperienced users; (b) acid additives and sugars, which yield acids upon combustion, lower the "pH" of TNP aerosol, resulting in higher levels of protonated nicotine, which is perceived as less harsh than free-base nicotine and may increase blood nicotine levels; (c) sweet flavourings in e-cigarettes reduce perceptions of bitterness and may escalate use, although their effects on perceived harshness are inconclusive; (d) sugars in tobacco impart sweet flavour sensations, but limited industry-independent data preclude a strong conclusion for IF; (e) some effects of additives on IF are amplified in non-smokers and younger populations; and (f) studies should be conducted on inhalation behaviour.

Conclusions: Several additives may facilitate inhalation of tobacco smoke and/or e-cigarette aerosol by improving the sensory experience. IF additives may increase nicotine blood levels, dependence and, in some cases, inhalation behaviour, especially in young people and non-smokers. Further research on the effects of TNP additives on sensory attributes and inhalation behaviour may provide useful evidence for regulatory policy.

Keywords: tobacco and nicotine products (TNPs), product attractiveness/appeal, inhalation facilitation, additives, cooling effects, pH lowering, masking bitter taste

2.1 Introduction

Research on internal tobacco industry documents has shown that cigarette manufacturers have manipulated product design, including appearance, flavour and smoke characteristics, to enhance their appeal and consumer acceptance *(1,2)*. The mechanisms included increasing nicotine delivery and facilitating smoke inhalation *(3,4)*. More than 100 cigarette additives have been found that camouflage the odour of environmental tobacco smoke emitted from cigarettes, enhance or maintain nicotine delivery, could increase the addictiveness of cigarettes, and mask undesirable sensory effects associated with smoking, such as irritation *(2)*. The products also include additives that may facilitate inhalation from tobacco and nicotine products (TNPs) such as e-cigarettes, cigars and hookah tobacco water pipes *(5,6)*.

Additives that facilitate inhalation may affect individual and population health in several ways. Especially for people starting use, nicotine and tobacco are aversive because of unpleasant sensory sensations, which can deter regular tobacco use *(7,8)*. Thus, for young people who try inhalable TNPs, additives may increase a product's attractiveness and thereby their odds of becoming a regular user, contributing to a higher prevalence of use. Furthermore, inhalation facilitation (IF) may promote nicotine addiction and the risk of long-term, heavy use *(7)*. Together, they constitute higher abuse liability. Increased inhalability of e-cigarettes may also, however, make them more satisfying nicotine substitutes for some adult smokers and encourage them to quit smoking and switch to vaping. Policy-makers have identified additives that facilitate inhalation of TNPs as key determinants of use and therefore potential targets for regulation *(8,9)*. A science-based framework to guide research and regulatory policy on IF from TNPs is, however, lacking.

This paper describes operationalization of IF in terms of effects, underlying mechanisms and studies addressing IF; reviews and weighs the evidence for a targeted set of additives that plausibly promote IF; and discusses the findings in terms of their potential health impact. First, we describe how nicotine and other sensory irritants decrease the inhalability of smoke, especially for novice users. Next, we propose a definition and conceptual model of IF, including factors other than additives. We also describe study designs in which IF of additives can be assessed. The subsequent section provides an overview of the categories of additives that facilitate inhalation and their effects in TNPs, with a focus on tobacco cigarettes and e-cigarettes (considered to be an inhalable nicotine product). Other inhalable products, such as heated tobacco products, were not included in our search, nor did we include other factors such as physical design (e.g. filter ventilation) that may also affect IF. Finally, we discuss our findings, identify gaps in the evidence and describe the potential impact of banning or setting upper limits on such additives and existing legislation on additives that facilitate inhalation.

2.2 Methods

A search was conducted in the bibliographic database PubMed and other sources (e.g. conference proceedings, general web search), with no restrictions on time period, up to September 2022 (with one exception, a paper in January 2023 with additional evidence on organic acids in e-liquids). A targeted (non-systematic) strategy was used that included search terms related to IF (e.g. "harshness," "puff duration"), candidate additives (e.g. "menthol") and candidate mechanisms (e.g. "TRPM8 receptor"). Papers were also obtained in exploratory "snowball sampling", in which the reference sections of articles were examined and potentially relevant articles were obtained and reviewed. As the review was not exhaustive, studies were included in the review by consensus between the co-authors. Priority was given to studies with stronger designs and greater relevance.

2.3 Nicotine and other sensory irritants that affect inhalability

Often, the first encounter with a tobacco product is unpleasant, as cigarette smoke contains numerous irritants that stimulate chemosensory nerves, leading to unpleasant burning and tingling sensations and reflex responses such as coughing, sneezing and avoidance *(7,10)*. As nicotine is the main irritant in tobacco smoke, many of these effects also occur with use of oral (such as nicotine chewing gum) and inhalable (such as e-cigarettes) nicotine products *(11)*. Nicotine also has rewarding effects in both humans and animals, even at low concentrations *(7,10)*. Nicotine activates brain systems that control reward by binding to nicotinic acetylcholine receptors (nAChRs) located within the mesolimbic dopaminergic pathway and the antinociception (pain reduction) system. Nicotine also elicits aversive sensory effects in the oral cavity and throat, including irritation, pain, a bitter taste, nausea and dizziness *(11,12)*. In response, smokers titrate their nicotine intake in order to experience the rewarding effects while avoiding aversion *(13)* by mixing smoke with air to allow inhalation without too much irritation *(11)*. Although the initial harshness of nicotine and tobacco is aversive, especially to novice users, and therefore can deter the uptake of regular tobacco use *(7,14)*, with repeated use, sensory stimuli that are paired repeatedly with the central effects of nicotine (unconditioned stimuli) can acquire motivational significance and promote smoking-related behaviour due to the association with a pending nicotine reward *(1,15)*. Sensory cues arise from various neural responses, including smell (via the olfactory nerve), irritation (trigeminal nerve) and taste (facial, glossopharyngeal and vagal nerves), and the cues may develop incentive value through a learnt association with the centrally mediated drug reward *(1)*. Other components of cigarette smoke, discussed below, can reinforce this effect.

The irritating properties and aversive bitter taste of nicotine are mediated mainly by activation of nAChRs located in nociceptive nerve endings, such as in the oral or nasal mucosa and lungs *(1,11,13,16)*. The nociceptors excite neurons in the trigeminal subnucleus caudalis and other brainstem regions *(11,12)*. Upon subsequent exposure, these neurons decrease firing, with desensitization of peripheral sensory neurons and progressively decreasing oral irritation *(12)*. Nicotine also elicits an nAChR-mediated bitter taste by excitation of gustatory afferents. In studies in rodents, the animals avoided nicotine solutions, even when sweeteners were added *(12)*.

Transient receptor potential (TRP) cation channels are involved in the local irritation and pain induced by nicotine, in particular the subfamilies TRPV1, TRPA1 and TRPM8, which are widely expressed in the human oropharynx and larynx *(1,11)*. Several compounds that target these TRP channels, such as menthol, can modify the oral irritation and pain elicited by nicotine *(7,11)*. TRPM5, a signal mediator in chemosensory cells and a key component of taste transduction, has been implicated in the bitter taste of nicotine *(17)*.

Other compounds in tobacco smoke are also involved in smoke-induced pain and irritation, such as reactive aldehydes (e.g. formaldehyde and acrolein), acids (acetic acid) and volatile organic hydrocarbons (cyclohexanone) *(7)*. For example, acrolein activates chemosensory nerves via the TRPA1 irritant receptor, and acetic acid and cyclohexanone probably act through acid-sensing ion channels, TRPV1 receptors and other classes of sensory receptors *(7,18)*.

2.4 Additives that facilitate inhalation

2.4.1 Definition and conceptual framework

We define IF as a modification to a TNP that improves the user's sensory experience of inhaling the product's aerosol (reduced bitterness and harshness) and may alter their inhalation behaviour (in particular, more intense inhalation [e.g. deeper puffs, faster inhalation, larger puff volume], but also restores breathing patterns that are disturbed by inhalant irritants). A conceptual model of IF is shown in Fig. 1 and described below, with supporting evidence reviewed in section 4.2. It should be noted that evidence is not available for all the factors in the conceptual model for all the additives reviewed. The illustration rather depicts the authors' proposal of the concepts involved in IF.

Fig. 1. Conceptual model of the effects of additives in facilitating inhalation and of the corresponding effects on health

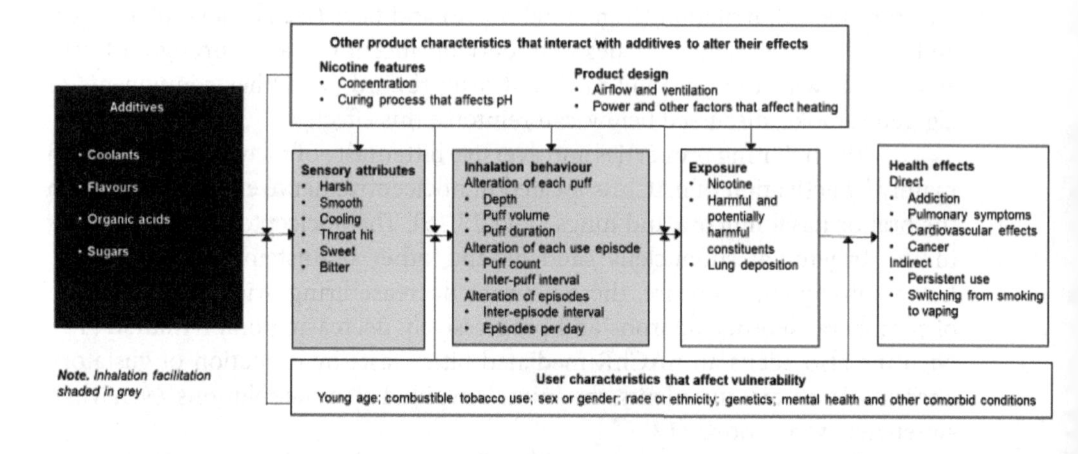

Additives to TNPs (e.g. flavourings, cooling agents, organic acids, sugars) are the focus of this paper. Other factors can affect IF, including extraction of compounds from tobacco, the nicotine concentration and design manipulations (e.g. filter ventilation, airflow, heating element, curing process); however, these factors are not directly addressed.

IF processes

Additives can improve the sensory experience of inhaling TNP aerosol by affecting airway sensations (increased smoothness or coolness, reduced harshness or irritation, and a pleasant "throat hit"), potentially making the aerosol easier to inhale. Olfactory and oro-sensory features (increased sweetness and decreased bitterness) elicited by additives may peripherally promote IF by increasing the appeal of a product, so that more aerosol is inhaled. IF-related increases in inhalation include instantaneous effects, such as greater inhalation depth, volume, velocity and duration per puff. Such effects may be strongest for nicotine-naive users who are not accustomed to inhaling harsh and bitter TNP aerosols. For inexperienced and younger users, TNPs with desirable sensory features may also shorten the inter-puff interval and increase the number of puffs per use episode (e.g. lighting and then putting out a cigarette), because they may need less time to "recover" from sensory irritation between puffs. This could escalate use and dependence. For established daily TNP users with severe nicotine dependence who are used to maintaining nicotine blood levels and avoiding withdrawal symptoms, additives that promote IF may result in inhalation of more nicotine per puff, which could promote faster satisfaction per puff and reduce the number of puffs necessary to achieve nicotine satiation.

Consequences of IF

Altered inhalation behaviour may increase exposure in two ways. First, IF-related changes in inhalation behaviour can directly increase the total quantity of aerosol consumed per puff, per use episode and per day and deeper inhalation. Secondly, IF-related increases in inhalation may alter pulmonary deposition, allowing more nicotine absorption and rendering the exposed parts of the lung more vulnerable. Two reviews of the weight of evidence for a causal relation between filter ventilation and lung adenocarcinoma showed that deeper inhalation of cigarette smoke may increase the rate of adenocarcinoma *(19)*. IF-related increases in exposure can have numerous direct health effects, including on the cardiovascular and pulmonary systems and increased risks of various cancers.

Additionally, higher puff volume, shorter inter-puff intervals and deeper inhalation affect the rate and volume of nicotine delivered to the blood, which corresponds directly to the product's reinforcing effects *(20,21)*. The pleasant sensory attributes of a tobacco product also contribute to its reinforcing effects and increase reinforcement synergistically with nicotine *(22)*. A product's reinforcing effects are directly related to its addiction potential and the likelihood of persistent use. IF is harmful in any inhalable TNP for youth and adults who are not TNP users, as IF may stimulate uptake and continuation of TNP use. IF in e-cigarettes could, however, be useful for adult smokers who wish to switch to e-cigarettes. Additives that promote IF in e-cigarettes may increase their nicotine yield and reinforcement and thus increase adoption and switching to e-cigarette products and cessation of tobacco smoking. In adults who switch completely from tobacco cigarettes to e-cigarettes, however, additives that promote IF in e-cigarettes might promote sustained vaping and potentially greater exposure to harmful constituents. Additional data on the net effect at population level are necessary.

Interaction with other products and user characteristics

The quality of a product's sensory attributes that promote IF may depend on user characteristics. On the one hand, youth and never-smokers may be deterred by harsh and bitter tastes, while additives that promote IF would suppress the deterrence. On the other hand, long-term adult smokers who wish to switch to e-cigarettes may seek products to replace the sensory attributes of cigarettes and provide a suitable throat hit and robust tastes. Hence, additives that suppress the bitterness and harshness of e-cigarettes may have less effect in promoting IF among smokers who are already accustomed to inhaling harsh, bitter tobacco smoke. Additional user characteristics (e.g. genetics, mental health, other comorbid conditions, race or ethnicity, sex or gender) may also affect their sensitivity to the sensory attributes of TNPs and their vulnerability to the effects of exposure to nicotine or harmful or potentially harmful constituents. Other product characteristics can interact with additives that promote IF by amplifying their effect on inhalation behaviour and on exposure and outcomes. For instance, additives that suppress the harsh, bitter taste of nicotine may have a particularly strong effect in e-cigarettes with a very high nicotine concentration.

2.4.2 Evidence review and integration

The literature on several classes of additives and their role in the IF processes depicted in the model is summarized below. We considered primary evidence of IF as that which demonstrated effects of additives on sensory experience and/or inhalation behaviour. Studies of IF-related mechanisms of action and the consequences of IF (biomarkers of exposure and health outcomes) were reviewed to provide supporting evidence for the biological plausibility and health significance of the IF scientific framework.

Additives and their effects and putative mechanisms are summarized in Table 1, which is based on the following types of evidence: (1) basic mechanistic studies of the effects of additives on the sensory and pain pathways that putatively underlie IF; (2) animal models of exposure to tobacco-product aerosol on IF-related sensory processes, exposure and inhalation behaviour; (3) human clinical laboratory experiments on the effect of self-administration of tobacco product aerosol with various additives on IF-related sensory processes, product appeal, exposure and inhalation behaviour; and (4) observational studies on whether use of products that contain additives is associated with altered inhalation behaviour.

Table 1. Classes of additives implicated in IF: mechanisms and effects

Mechanism	Additive	Reported sensory effects	Comments
TRPM-8 activation	Menthol	Increase cooling, reduce harshness of nicotine, minty flavour	Evidence available for both tobacco products and e-cigarettes Increased inhalation behaviour in rodents but inconclusive effects on inhalation behaviour in humans
	Wilkinson Sword (WS) compounds such as WS-3, WS-5, WS-14 and WS-23	Increase cooling, reduce harshness, reduce bitterness	Evidence predominantly for e-cigarettes Often combined with other flavours in "ice" hybrid flavours in e-cigarettes
pH lowering	Organic acids and nicotine salts in e-cigarettes	Increase mildness, reduce irritation	Higher blood nicotine levels
	Sugars in tobacco	Combust to acids, increase mildness	Mainly industry data
	Organic acids in tobacco	Increase mildness, decrease irritation	Mainly industry data
Olfactory and oro-sensory mechanisms	Flavourings with sweet properties	Increase sweetness, reduce bitterness, partial evidence of increased smoothness and reduced harshness	Predominantly in e-cigarettes, hookahs and cigars
	Sugars	Impart a sweet flavour	Predominantly in cigarettes; mainly industry data

2.5 Additives with cooling effects

2.5.1 Menthol

Menthol is a naturally occurring compound in the mint plant (*Mentha* spp.). It is used as an additive in various food, medicinal and cosmetic products and in TNPs. Menthol has been detected not only in "menthol flavoured" TNPs but also in TNPs that are not explicitly marketed as "menthol-containing" *(23)*. Menthol affects the central nervous system by activating nAChRs in the brain; however, its role in IF is mediated by its anti-irritant, cooling, analgesic properties *(24)*. The sensory effects of menthol are mediated mainly by its interactions with the TRP cation channel melastatin 8 (TRPM8) in cold-sensitive sensory neurons lining the airways and the TRP ankyrin 1 (TRPA1), a sensory irritant receptor *(12)*. Menthol may also have analgesic and cross-desensitizing properties, in which pre-treatment with menthol may reduce the irritating effects of nicotine, even after its acute cooling effect dissipates *(11)*. Evidence from studies in rodent models indicates that the effects of respiratory irritants in TNP aerosol can be suppressed by menthol, resulting in more frequent breathing, shorter inter-breath intervals and faster respiratory flow rate *(7,25)*. Rodent inhalation behaviour is an analogue of increased puff count, shorter inter-puff intervals and faster puff velocity associated with IF.

The US Food and Drug Administration (USFDA) conducted a comprehensive review of the literature on the effects of menthol in tobacco cigarettes, including human clinical experiments and observational studies *(24)*.

The conclusion was that menthol increases the palatability of cigarettes by masking the harsh taste of tobacco smoke and reducing aversive sensory responses associated with initial smoking experiences (e.g. irritation, coughing) and thus promotes continuation of smoking. The conclusion was strongest for the role of menthol in uptake and dependence in youth, difficulty in quitting smoking and a disproportionate impact on Black smokers *(24)*. For example, in one observational cross-sectional study, young adult smokers of menthol and non-menthol cigarettes, particularly African Americans, reported on their positive and negative subjective responses to smoking; greater positive subjective responses were associated with more frequent smoking *(26)*. The USFDA found mixed evidence for an association between menthol and dependence in adults and with measures of smoking topography *(24)*. As adults with an established smoking habit have strong preferences for certain brands of cigarettes and there is a natural selection bias for menthol flavours, it is difficult to draw strong conclusions from human clinical and observational studies of menthol-flavoured cigarettes in this population.

Several clinical laboratory experiments with e-cigarettes have shown that menthol increases perceptions of coolness and a pleasant taste *(27–31)*. Four studies showed that menthol interacts with nicotine to alter some of the sensory attributes or appeal of e-cigarette aerosol *(28–31)*. For example, a study of young adult e-cigarette users showed that menthol flavour interacted with nicotine at a concentration of 6 mg/mL to counteract the aversive sensory features of nicotine *(31)*. The study also showed that the direct, interactive effects of menthol with nicotine in increasing the appeal of e-cigarettes were more pronounced in vapers who had never smoked than in dual users or vapers who had previously smoked *(29)*. In a study of adolescents, however, no evidence was found that menthol e-cigarettes altered the effects of nicotine level on sensory attributes or appeal; menthol increased perceived coolness at two nicotine concentrations *(27)*. None of the studies indicated that menthol affected puffing behaviour or short-term exposure to nicotine *(27,30)*.

2.5.2 Synthetic cooling agents

Synthetic coolants, including compounds such as WS-3, WS-5, WS-14 and WS-23, have been detected in various types of TNPs. Tobacco industry documents show that, in the 1970s and 1980s, major tobacco manufacturers, including RJ Reynolds and Phillip Morris, tested but initially did not widely market tobacco cigarettes containing WS synthetic coolants *(32–34)*. Synthetic coolants have, however, been identified recently in cigarette products in Germany *(35,36)*. Synthetic coolants may be present in tobacco cigarette products with cooling features that are marketed as "non-menthol" in certain US markets in which menthol tobacco cigarette sales have been banned (e.g. California) *(37)*. They have been detected in e-cigarette products in the past few years *(38,39)*, including products marketed as "ice" hybrid flavours that combine constituents with a fruit, mint or other characterizing flavour with the synthetic coolant (e.g. "raspberry ice"). Recent studies indicate that ice-hybrid-flavoured e-cigarettes that may contain synthetic coolants are commonly used by young people and young adults in the USA *(40–42)*, where sales have recently increased *(43)*.

Several synthetic coolants are based on the *p*-menthane structure of menthol. Like menthol, WS-3 and WS-23 are pharmacologically active at the TRPM8 cold receptors lining the airways and oral cavity *(7,16,25,44)*. Some evidence indicates that WS-3 is more active at TRPM8 receptors than menthol, generates stronger cooling sensations *(45, 46)* and may activate the sensory irritant receptor, TRPA1 *(47–49)*.

In view of the pharmacological properties of these synthetic cooling agents, in TNPs containing these compounds, they may generate cooling sensations that mask the harshness of nicotine without providing a strong minty flavour, unlike products that contain menthol *(41)*. Anecdotal reports by users on social media and online discussions indicate a substantial cooling effect of WS-23 or WS-3 in e-cigarettes, without the strong minty taste of menthol *(41)*. In a human laboratory study, administration of e-cigarettes flavoured with nicotine salt and with WS-23 (vs no cooling agent) to adult users of TNPs increased the e-cigarettes' appeal, smoothness and coolness and reduced their bitterness and harshness *(50)*. Additionally, e-cigarettes flavoured with WS-23 were perceived as smoother, cooler and less harsh than those with menthol. The effects of cooling agent additives did not significantly differ between fruit, tobacco or mint; 2% vs 4% nicotine concentration; or smoking status. The possible IF effects might explain why young adult users of ice-hybrid flavoured e-cigarettes reported more symptoms of nicotine dependence than with other flavours in an observational study *(40)*.

2.6 Additives that lower pH

The tobacco industry has conducted research on the effects of manipulating pH levels on tolerance to cigarette smoking *(3)*. In TNPs, the extent of nicotine absorption across membranes and nicotine-mediated harshness depend on the extent of nicotine protonation *(51)*. The fraction of protonated vs unprotonated (free-base) nicotine depends on the pH of the product and thus can be influenced by adding acidic or basic additives. In its free-base state, nicotine permeates membranes and is then converted to the protonated state, which is the ligand of nAChRs *(52–54)*. At pH > 7–12, above the physiological level, nicotine is present in a free-base form, which is more readily absorbed across membranes and also provides a stronger throat hit and is experienced as harsher. Free-base nicotine can be aversive especially at high concentrations, because it is absorbed preferentially in the upper respiratory tract, causing irritation, whereas protonated nicotine is less irritating and hence can be inhaled more deeply, resulting in deposition deeper in the respiratory tract *(55)*. This results in greater net absorption of nicotine into the systemic circulation. Cigarette smoke is usually slightly acidic, with a pH of about 6, which makes the smoke less harsh and easier to inhale than smoking products with higher pH, such as cigars *(3,51)*. Once cigarette smoke reaches the pulmonary alveoli, nicotine leaves the smoke and, at the physiological pH of the lungs, is readily absorbed through the pulmonary capillaries into the systemic circulation *(51)* due to the larger absorptive surface of the lung at pH 7.4 and the high local buffering capacity of the lung *(56,57)*. This effect is expected only for inhalable products and not for products in which nicotine is absorbed in the oral cavity, such as nicotine pouches.

Protonated nicotine is thus less harsh and bitter on inhalation than free-base nicotine, so that high amounts of nicotine are more palatable *(5)*. Thus, at lower pH, overall nicotine delivery may be higher *(3)*. As the irritation due to the free base is largely attenuated, protonated nicotine is less aversive at high concentrations, increasing the attractiveness of the product. Furthermore, greater nicotine absorption, with faster, higher peak blood nicotine levels, probably predicts greater abuse liability *(51)*.

This mechanism and its consequences for sensory appeal and smoking behaviour are discussed below for acid additives that lead to nicotine salts (with protonated nicotine) in e-liquids. Other examples are also touched upon, such as laevulinic acid as an example of a tobacco additive and sugars in tobacco that result in increased acid levels in smoke upon combustion.

2.6.1 Organic acids in e-liquids

In the USA, marketing of Juul and similar e-cigarettes led to a rapid increase in e-cigarette use by young non-smokers *(58,59)*. These products contain high levels of aerosol nicotine, and the e-liquids contain protonated nicotine instead of free-base nicotine due to the addition of organic acids *(60)*. Several organic acids have been used in salt-based e-cigarettes, including lactic, salicylic, benzoic, laevulinic, ditartaric and maleic *(61)*. The effects of nicotine protonation on nicotine blood levels have been studied by several groups *(62–65)*. Some studies have shown that e-cigarettes with nicotine salt solutions, unlike e-cigarettes filled with free-base liquid, result in nicotine blood profiles similar to those of smokers of tobacco cigarettes *(62–65)*. Secondary data analysis of a randomized clinical trial of e-cigarettes also showed that smokers who switched to nicotine salt pod-style system e-cigarettes (similar to Juul) maintained their nicotine levels and transferred their dependence, suggesting that these products have a reinforcement potential similar to that of cigarettes and facilitate switching *(66)*. Observational data show that adolescents who use Juul and other pod-style e-cigarettes that contain nicotine salts experience similar levels of nicotine dependence as adolescent smokers *(67)* but greater dependence than young users of other e-cigarette products that may not contain nicotine salts *(68)*, emphasizing the need to regulate access and marketing to this age group.

Thus, use of nicotine salts increases the addiction potential of TNPs *(64)*, and the effect increases with nicotine dose. A study funded by Juul Labs of liquids containing protonated nicotine showed that higher levels of protonated nicotine give rise to significantly higher plasma nicotine levels and relief from craving than lower levels *(69)*. A version of Juul produced in the European Union, with nicotine at 18 mg/mL, delivered less nicotine and reduced the urge to smoke or vape less strongly than tobacco cigarettes *(70)*. A comparison with the US Juul product, containing 59 mg/mL, gave similar results *(71)*.

Three studies were conducted to compare the sensory effects of nicotine salts with those of free-base nicotine. A randomized clinical trial in the USA showed that formulations containing salt nicotine at 24 mg/mL had significantly higher ratings than free-base nicotine for appeal, sweetness and smoothness and lower ratings for bitterness and harshness. The effects of nicotine salt on enhancing smoothness and reducing harshness were stronger in people who had never smoked cigarettes than in those who had ever smoked cigarettes *(72)*. Nicotine salts improved the sensory experience and thereby the attractiveness of vaping, particularly among never smokers unaccustomed to inhaling free-base nicotine. These findings are in accordance with observational data from England, which indicate that Juul, which contains nicotine salts, is more commonly used by never smokers than by current smokers, whereas tank devices, which typically include free-base nicotine, are more commonly used by current or former smokers, although other confounding factors (e.g. age) could explain the association *(73)*. A clinical laboratory study in the USA found that nicotine lactate and benzoate (protonated) e-liquids had greater appeal, smoothness and sweetness and less harshness and bitterness than free-base nicotine. There was some evidence that e-liquids that are highly protonated had stronger effects than e-liquids that were moderately protonated. The effects of nicotine formulation did not differ by tobacco use status or flavours *(74)*. In Netherlands (Kingdom of the), a study of home use showed no significant difference in scores for appeal, harshness and topography of nicotine salts versus free-base nicotine at a concentration of 12 mg/mL *(75)*. This is the only study of the effects of nicotine protonation state on topography. Apart from the lower nicotine levels, users in the Dutch study could vape freely, with monitoring of puffing parameters, whereas a set puffing protocol was used in the study in the USA.

2.6.2 Laevulinic and other organic acids in cigarettes

Many different acid additives have been used in the production of conventional cigarettes to increase their smoothness and decrease the throat hit *(56,76)*. For example, lactic acid has been used to decrease harshness and bitterness and produce a sweeter flavour. Citric additives have been used not only to reduce harshness and modify flavours but also to modify the pH of smoke and to neutralize the throat hit. Tartaric and lactic acids have also been used to modify the pH of smoke.

A review of internal tobacco industry documents indicated that laevulinic acid was used to increase nicotine yields while enhancing perceptions of smoothness and mildness *(3)*. Laevulinic acid reduces the pH of cigarette smoke and desensitizes the upper respiratory tract, promoting inhalation of cigarette smoke deeper into the lungs. Industry studies also found significantly increased peak plasma nicotine levels in smokers of ultralight cigarettes with added laevulinic acid.

2.6.3 Sugars

The pH of smoke can also be affected by sugars in tobacco. An industry document mentioned that harshness can be reduced by adding a suitable organic acid or by increasing the sugar level in tobacco *(76)*. In cigarettes, 0.5% of the sugars in the tobacco are transferred into mainstream smoke, where most is combusted, pyrolysed or pyrosynthesized *(77–79)*. Addition of sugars to cigarette tobacco has been reported to increase the acidity of smoke *(77,78,80)*; however, combustion of sugar during smoking results in acids that reduce the pH *(81)*, thus decreasing the harshness and irritation of the smoke *(82,83)*, increasing the palatability of the product and facilitating inhalation. Sugars have been referred to by the tobacco industry as "ameliorants", to "... smooth out harshness and bitterness and/or eliminate pungent aromas from tobaccos" *(84)*.

2.7 Additives with flavouring properties that may mask bitter taste

2.7.1 Flavourings with sweet features

Hundreds of flavouring constituents have been identified in various types of inhalable TNPs *(85,86)*, many classified in categories that could be considered as having sweet features (e.g. fruit, mint, dessert) *(85,86)*. Given the wide variety of such constituents, it is difficult to identify one or several biological pathways for the effects of sweet flavourings. From a psychosensory perspective, there is some evidence that TNPs with sweet features may exert their effects via olfaction and not by their oro-sensory impact alone *(87,88)*.

Studies of the effects of flavours with sweet properties on the processes of IF provide some evidence of possible effects, but the results for specific outcomes are not consistent. A wide variety of additives with sweet elements identified in tobacco products (e.g. carob bean extract, liquorice) may facilitate inhalation indirectly by pyrolysis of sugars *(9)*, as reviewed in more detail in section 4.5.2. A systematic review of qualitative studies indicates that sweet flavours in cigars, hookahs, e-cigarettes and cigarettes reduce perceptions of harshness and make the products more tolerable *(89)*. Flavours with sweet properties in e-cigarettes (e.g. fruit flavours) have been shown to reduce perceptions of bitterness in clinical experiments *(17)*. The results for an effect of sweet flavours in reducing harshness and increasing the smoothness of e-cigarettes are inconsistent *(31,90–92)*. There is also evidence that sweet flavours reduce the bitter-enhancing effects of nicotine in e-cigarettes *(31,90)*. In a laboratory clinical study of adolescent e-cigarette users, green apple e-cigarette flavour increased the acute puff count and puff duration to a greater extent than menthol or no flavour *(27)*. A longitudinal observational cohort study of adolescent e-cigarette users showed that sweet or fruit rather than menthol, tobacco or mint flavours were associated with more self-reported puffs per vaping episode 6 months later but not in the number of vaping episodes per day *(93)*; cross-sectional associations were not reported. In a cross-sectional observational population-based study of US residents, vaping of sweet-flavoured e-cigarettes was more common among adolescents and young adults than among older adults *(94)*.

Use of strawberry rather than tobacco flavoured e-cigarettes (19–20 mg/mL nicotine) was assessed in a clinical laboratory study of 14 adult e-cigarette users *(95,96)*. The effects were similar for the amount of nicotine inhaled and systematically retained, but, in the standardized 15-puff protocol, the plasma nicotine level was significantly higher. In an ad-libitum protocol, the average puff duration was significantly longer with the strawberry e-liquid than with the tobacco e-liquid. There were no differences in subjective measures of abuse liability between the two flavours. Although inferences are limited by the small sample size, this study provides some evidence that sweet-flavoured e-liquids may be associated with increased nicotine exposure and inhalation behaviour.

In a population-based observational study of biomarkers of exposure in the USA in 2015–2016 of 211 exclusive e-cigarette users who reported having used their product within the past 24 h, the biomarker for acrylonitrile was higher in users of fruit-only flavoured e-cigarettes than in users of any other non-tobacco flavour (mint, clove, chocolate or other); however, the concentration of acrylonitrile did not differ. Concentrations of biomarkers of exposure to nicotine (cotinine), benzene and acrolein did not differ significantly by flavour group *(97)*. Because this was an observational study, which did not account for differences in user behaviour (e.g. frequency of vaping), device or e-liquid (e.g. nicotine concentration), it is difficult to determine whether the biomarkers of exposure of users of different flavours were influenced by these external factors.

2.7.2 Sugars and sweeteners

In addition to their effect on pH (see 4.4.3), sugars also contribute to the flavour of TNPs *(78,83,98–100)* and e-cigarettes *(101,102)*. In smoked products, the sweet taste of caramel flavours generated by the combustion of sugars improves the taste and smell of the tobacco smoke for both users and bystanders *(78,82,103–105)*. Furthermore, during curing and smoking of tobacco, sugars can participate in Maillard reactions to produce flavouring that gives TNPs a characteristic woody, caramel and baking flavour *(15,106)*. One class of compounds resulting from sugars via Maillard reactions is pyrazines, which are also used as tobacco additives, especially in low-tar cigarettes with cocoa, nutty or popcorn-type flavours *(15,106)*. It has been hypothesized that they may reduce noxious sensations such as irritation in the upper airways or have chemosensory effects that reinforce the learnt behaviour of smoking *(15)*. Sweet flavours probably also lower a smoker's cough threshold. Rinsing the mouth with sucrose solution modulates sensitivity to the cough reflex, and it has been suggested that this is due to release of endogenous opioids in response to a sweet taste *(107)*.

Industry documents indicate that the acceptance of tobacco smoke by smokers is proportional to the sugar level in the tobacco, which could be due to their flavours and their effect on pH (see section 4.3.3) *(78,99)*. When the ratio between sugars and tobacco alkaloids such as nicotine is increased, the impact is decreased and "liking" increased to a certain optimum *(76)*. Addition of sugars to cigarettes to enhance the sensory attributes of cigarette smoke and encourage smoking initiation and maintenance have been discussed by industry as part of their marketing strategy *(108,109)*. In e-cigarettes, addition of sucralose, an artificial sweetener, increased overall flavour and sweetness but had no significant effect on harshness or irritation *(87)*. High-intensity sweeteners like saccharine and glycyrrhizin are also added to the mouthpiece and wrapper of tobacco products such as cigarillos *(110)*.

2.8 Discussion

2.8.1 Main findings

Taken together, the literature reviewed in this paper partially validates the proposed IF framework. We found evidence that several additives facilitate inhalation of tobacco smoke and/or e-cigarette aerosol, for example by providing more desirable sensory attributes. Evidence was also found that some additives that improve the sensory attributes of TNPs result in higher nicotine blood levels or maintenance of nicotine dependence. Few studies were found on the effects of additives on objective measures of puffing topography and inhalation behaviour. We found evidence for the biological plausibility of the framework in studies that showed that several additives impact pathways implicated in sensation and respiration.

These findings indicate that research on the effects of additives in TNPs on sensory attributes and inhalation behaviour may provide useful evidence for regulatory policy. This was particularly the case of studies of harshness and smoothness in humans and the biological pathways of airway irritation in animal models.

Menthol and synthetic cooling agents have been found to reduce aversive sensory responses to both tobacco cigarettes and e-cigarettes. The biological plausibility of their effect on IF is based on studies showing that the sensory effects of cooling agents are mediated mainly by their interactions with TRPM8 in the cold-sensitive sensory neurons lining the airways and by TRPA1. In a study of inhalation by rodents, menthol resulted in deeper inhalation of cigarette smoke and higher blood cotinine levels. The evidence for more intense or otherwise altered puffing behaviour in humans is, however, mixed. Direct experimental evidence of the effect of synthetic coolants was observed in one study of e-cigarettes. The similarity of the effects of synthetic coolants to those of menthol is biologically plausible, as they share an underlying mechanism, although synthetic coolants and menthol differ in potency, with potentially stronger effects of synthetic coolants on coolness and pleasant respiratory sensations. Some evidence indicates that the effect of menthol on IF is more robust in younger populations who are not regular tobacco cigarette smokers and who differ in other population characteristics (e.g. sex, race).

With regard to additives that lower pH, many studies suggest that acid additives in e-cigarettes facilitate inhalation of e-cigarette aerosol. While more research should be conducted, with lower nicotine levels, the available studies show that, at higher nicotine concentrations (> 20 mg/mL), protonation of nicotine (with organic acid additives) in e-liquids increases several IF processes over that with free-base nicotine at the same concentration, including more desirable sensory attributes, which, in turn, result in higher nicotine blood levels and maintenance of nicotine dependence. More information is needed on whether acid additives result in more intense or otherwise altered inhalation behaviour. It has been reported that acid additives and sugars, which yield acids upon combustion, lower the pH of cigarette smoke, and other studies indicate that such compounds decrease the harshness and increase the smoothness of tobacco smoke. Most of the evidence on the effects on human perception has been found in older internal industry documents. Thus, even though it is likely that similar effects as presented in section 4.3.1 for e-cigarettes will also be found in cigarettes, as they share the same mechanism, i.e. lowering the pH, additional and independent research is necessary. Some evidence indicates that the effect of pH on IF may be stronger for non-smokers and younger populations.

Additive flavours with sweet properties in e-cigarettes (e.g. fruit flavours) consistently reduce perceptions of bitterness, although evidence that they reduce perceptions of harshness and increase smoothness is inconclusive, as is evidence on the relative contribution of olfactory and gustatory effects. Two studies with different designs found that adolescent use of fruit-flavoured e-cigarettes was associated with increased levels of most inhalation behaviour, including puff duration and count. In e-cigarettes, addition of sucralose, an artificial sweetener, increased the overall flavour intensity and sweetness but had no effect on harshness or irritation. Sugars have been reported to impart a sweet or caramel taste to cigarette smoke, but most of the evidence on effects on human perception was in older internal industry documents. Thus, even though it is likely that effects similar to those of sweet flavourings will be found, independent research is necessary. While some effect on desirable sensory attributes has been found for all additives with flavouring properties, data are lacking on effects on nicotine blood levels, maintenance of nicotine dependence and intensity of puffing.

2.8.2 Regulatory mechanisms in the European Union and North America for additives that facilitate inhalation

The European Tobacco Products Directive (TPD), Article 7.6.d, stipulates that European Union Member States shall prohibit the placing on the market of tobacco products for smoking and e-cigarettes containing additives that facilitate inhalation or nicotine uptake *(111)*. The TPD does not, however, provide a definition of IF or nicotine uptake facilitation. Belgium *(112)* and Germany *(113)* already prohibit use of menthol for its IF properties at any level, which is further supported by advice from the European Union Joint Action Tobacco Control *(114)*, which concluded that all menthol analogues, including geraniol, have a TRPM8-dependent cooling effect and may act cumulatively. As this effect is an intrinsic property of the compounds, products containing menthol and its analogues at any level do not comply with Article 7.6.d of the TPD, even if their level of application in tobacco does not induce measurable effects. Belgium has banned all activators of the TRPM8 thermoreceptor, and Germany has also banned other specific TRPM8 activators.

Canada does not permit use of additives with any flavouring properties, sweeteners, colouring agents or several other compounds that increase the attractiveness of tobacco products, although there are a few exceptions (guar gum, alcohol flavours) *(115)*. The USA has planned Federal product standards that would prohibit all characterizing flavours in cigarettes and cigars, including menthol *(8,116)*. Whether these regulations will extend to non-menthol synthetic coolants is unclear. The USA has no other specific product standards that ban other types of additives that may facilitate inhalation. The USFDA decides case by case on legal marketing of e-cigarettes and other novel products for each brand and product line. The US Tobacco Control Act stipulates that any regulatory decision take into consideration the impact on the population as a whole. Thus, regulatory restrictions should be designed to minimize TNP use by young people and non-users and should, if possible, not deter adult smokers from quitting use of conventional tobacco products. To date, the USA has denied applications for numerous e-cigarette products marketed with characterizing flavours and sweet features on the basis of evidence that they attract young people *(117)*. Decisions on marketing of menthol-flavoured e-cigarettes are pending. The USFDA has authorized the marketing of several e-cigarette products that contain organic acids and protonated nicotine *(118,119)*.

2.9 Recommended research

To provide actionable evidence for regulatory decisions, a wider evidence base on most additives is necessary. In view of gaps in the evidence, we recommend the following:

- clinical studies of the effects of additives on inhalation behaviour, such as those measured by topography devices attached to TNPs;
- prospective longitudinal studies of users to determine whether use of TNPs with additives is associated with more pleasant sensory perceptions and/or increases in measures of inhalation behaviour;
- preclinical research with animal models to address specific questions that cannot be investigated in humans, such as the effect of introducing TNP-naive research subjects to TNPs with or without additives;
- comparisons of the effects of additives in e-cigarettes that promote IF in adult smokers, adult non-smokers and young people; and
- research on the IF of a wider range of products, other than tobacco cigarettes and e-cigarettes, including hookahs, cigars and heated tobacco.

Studies should also be conducted of products that contain possible inhalation facilitating additives, including specific brands and flavours, which could be triangulated with measures of additives in those brands. Survey instruments could be used to ask participants which flavours they use, the sensory attributes of their preferred product (e.g. how harsh it is), and, for e-cigarettes, the device type or nicotine formulation (salt or free base).

Research on potentially less harmful TNPs, in particular e-cigarettes, to determine the effects of additives on IF processes should include comparison of the effects on young non-smoking populations and older adult smokers. For example, additives in e-cigarettes that promote IF and adverse exposure in young non-smokers but do not encourage switching to e-cigarettes by adult smokers would be priorities for regulatory restrictions. For example, additives in e-cigarettes that promote IF and adverse exposure in young non-smokers but do not encourage switching to e-cigarettes by adult smokers would be priorities for regulatory restrictions.

Several fundamental aspects of what constitutes IF merit further research. It is unclear whether additives that promote sweetness or reduce bitterness directly increase inhalation behaviour or simply make products more attractive. Experimental studies in which the sweetness-enhancing or bitterness-reducing properties of additives (e.g. blocking olfaction, bitterness receptor knockout rodent models) on inhalation behaviour might be useful. It is unclear which study design is optimal for assessing whether additives increase inhalation behaviour. Studies of inhalation behaviour in established users are at risk of selection bias, because participants have pre-existing preferences. Animal models of inhalation behaviour in which exposure to TNP is controlled may be especially useful, although not in accordance with the ambition to limit studies in animals. Research should be conducted on whether increased inhalation is necessary or sufficient to increase exposure to nicotine and other harmful constituents and to increase the risk of adverse health outcomes, including addiction. Comparison of the effects of altered puff duration, count, velocity, volume, inter-puff interval and inter-episode interval on exposure and outcomes would be useful. Such research will indicate which sensory and inhalation behaviour outcomes are critical for inclusion in studies of the impact of new additives.

Testing might be conducted by asking research participants to use the product as intended and to report on their sensory experience during use. Studies with unblinded and blinded testing, in which the participant does not know the name of the product or see the marketing materials, to elicit subjective harshness, sweetness, coolness or other sensory attributes during self-administration of the product, might be valuable. Additional outcomes related to inhalation (e.g. puff duration, velocity, volume; inter-puff interval) would also be useful. Such data (with the scientific literature) could be triangulated with lists of ingredients and marketing materials to determine whether a product is in violation of a ban or product standard that restricts additives that promote IF.

2.10 Policy recommendations

We make the following recommendations to policy-makers on all inhalable TNPs.

- Ban ingredients that facilitate inhalation, as they facilitate use of inhaled tobacco products (cigarette, cigars, hookah, heated tobacco products or any other inhaled product containing tobacco). There is no justification for permitting the use of ingredients, such as flavouring agents, which make tobacco products more attractive.

The partial guidelines for implementation of Articles 9 and 10 of the WHO Framework Convention on Tobacco Control *(120)* state that, from the perspective of public health, there is no justification for permitting the use of ingredients such as flavouring agents that make TNPs more attractive. The partial guidelines therefore recommend that "Parties should regulate all tobacco product design features that increase the attractiveness of tobacco products, in order to decrease the attractiveness of tobacco products". Consequently, given the WHO definition of attractiveness (factors such as taste, smell and other sensory attributes, ease of use, flexibility of the dosing system, cost, reputation or image, assumed risks and benefits, and other characteristics of a product designed to stimulate use), policy-makers should ban ingredients that facilitate inhalation, which facilitates use of a product. Such a ban is included in the European TPD for smoked tobacco products and e-cigarettes, in Article 7.6.d *(95)*.

The evidence reviewed here provides support for the definition of IF that we propose for use by policy-makers to regulate additives permitted in TNPs: IF as a modification to a TNP that improves the user's sensory experience of inhaling the product's aerosol (reduced bitterness and harshness) and may alter inhalation behaviour (in particular, more intense [e.g. deeper puffs, faster inhalation, larger puff volume]) and also restoration of breathing patterns that are normally disturbed by inhalant irritants.

In addition to a general ban on ingredients that facilitate inhalation, it is recommended that policy-makers include a non-limited list of such compounds, particularly inhalable tobacco products. It is recommended that this list be included in legislation such that it can easily be adapted when new scientific insights necessitate addition of compounds to the list. A list of specific compounds would facilitate surveillance and enforcement. For example, Belgium *(112)* and Germany *(113)* already prohibit use of menthol for its inhalation facilitating properties at any level. Belgium then banned all activators of the TRPM8 thermoreceptor *(112)*. Another straightforward approach is to provide a list of additives that are permitted in TNPs that do not include any compound with IF effects, which is similar to the policy in Canada *(115)*.

It is recommended that ingredients that facilitate inhalation from conventional cigarettes be banned, as use of conventional cigarettes is not beneficial for any type of user, smoker or non-smoker. Legislation of e-cigarettes and other inhalable products that are potentially less harmful than conventional cigarettes may depend on each country's circumstances and policy aims. Policy-makers may consider effects at population level and weigh the evidence for whether additives that promote IF could make these products more satisfying nicotine substitutes for some adult smokers on the one hand and whether they increase appeal, risk of dependence and other adverse outcomes in young people and non-smokers on the other hand. The aim of the recommendations below is to prevent young people and never smokers from taking up any type of inhaled TNP, including e-cigarettes. Thus, we propose that all additives in all inhaled TNPs be banned when they facilitate inhalation.

- Ban the addition to TNPs of menthol at any level and also of both synthetic (e.g. WS) and natural (e.g. geraniol) coolant chemicals with similar chemical structure or physiological and sensory effects to avoid substitution.

A previous exhaustive review *(24)* provides sufficient evidence that menthol additives facilitate IF in tobacco cigarettes, and the conclusion is reinforced by the evidence in the current paper. We recommend that any regulatory agency that aligns its policies with the TPD and similar legislative frameworks impose a ban on the addition of menthol at any level to all inhalable TNPs. Chemicals with a similar chemical structure or similar physiological and sensory effects should be included in the ban to avoid substitution. These would include synthetic analogues such as WS compounds with cooling properties similar to those of menthol and natural compounds such as geraniol, which have similar properties and may also facilitate inhalation. In view of the evidence that coolants in e-cigarettes facilitate inhalation more strongly in younger populations and non-users of smoked tobacco products, regulatory restrictions on cooling agent additives for inhalable TNPs merit consideration.

- Ban nicotine salts in e-liquids at levels that exceed 20 mg/mL to protect children, adolescents and non-smokers. Setting minimal levels of pH in e-liquids and tobacco products would reduce the bioavailability of nicotine and reduce the addictiveness of products.

By triangulating evidence on additives that lower pH in e-cigarettes and to amplify their impact on IF in younger populations and non-smokers, regulators should consider banning acid additives and nicotine salts in e-liquids at nicotine levels > 20 mg/mL. While sufficient evidence is not available for nicotine levels < 20 mg/mL, regulators might consider banning such additives at any nicotine level as a precaution. Furthermore, setting minimal levels of pH in e-liquids could be a pragmatic application of regulation of these products. Measures to ban acid additives or acid-generating additives in cigarettes should also be considered. Some TNPs may have other additives or modifications that lower pH, such as certain tobacco leaf curing processes or sugar additives. Regulatory restrictions on inhalable tobacco products according to a pH threshold rather than the presence of a particular additive merit consideration.

- Ban all flavourings that impart a sweet taste, including sugars, in all TNPs.

Our findings on additives with flavouring properties that mask the bitter taste indicate that all flavourings that impart a sweet taste be banned, including sugars. In e-liquids, this refers to flavourings that facilitate IF at any level of addition, including all constituents used in any e-cigarettes with non-tobacco characterizing flavours. In tobacco products, regulators may consider banning such flavourings at levels that impart a characterizing flavour other than tobacco, although flavourings may also have effects at concentrations below the threshold for a clearly noticeable flavour other than tobacco. Assessment of characterizing flavours requires sensory panels for surveillance and enforcement; a less time-consuming approach would be to ban addition of such flavourings at any level. Although the current paper focuses on additives that facilitate IF, natural tobacco leaves may also contain sugars and flavourings. For example, sugar is naturally present in many tobacco types. Regulators could also consider banning sugars and flavourings that are naturally present in tobacco, as they also impart a flavour. For the consumer, it is immaterial whether sugars or flavourings are added or naturally present.

References

1. Megerdichian CL, Rees VW, Wayne GF, Connolly GN. Internal tobacco industry research on olfactory and trigeminal nerve response to nicotine and other smoke components. Nicotine Tob Res. 2007;9(11):1119–29. doi:10.1080/14622200701648458.
2. Rabinoff M, Caskey N, Rissling A, Park C. Pharmacological and chemical effects of cigarette additives. Am J Public Health. 2007;97(11):1981–91. PMID: 17666709.
3. Keithly L, Ferris Wayne G, Cullen DM, Connolly GN. Industry research on the use and effects of levulinic acid: a case study in cigarette additives. Nicotine Tob Res. 2005;7(5):761–71. doi:10.1080/14622200500259820.
4. Ferris Wayne G, Connolly GN. Application, function, and effects of menthol in cigarettes: a survey of tobacco industry documents. Nicotine Tob Res. 2004;6 Suppl 1:S43–54. doi:10.1080/14622203310001649513.
5. Duell AK, Pankow JF, Peyton DH. Nicotine in tobacco product aerosols: It's deja vu all over again". Tob Control. 2020;29(6):656–62. doi:1136/tobaccocontrol-2019-055275.
6. Stanton CA, Villanti AC, Watson C, Delnevo CD. Flavoured tobacco products in the USA: synthesis of recent multidiscipline studies with implications for advancing tobacco regulatory science. Tob Control. 2016;25(Suppl 2):ii1–3. doi:10.1136/tobaccocontrol-2016-053486.
7. Willis DN, Liu B, Ha MA, Jordt SE, Morris JB. Menthol attenuates respiratory irritation responses to multiple cigarette smoke irritants. FASEB J. 2011;25(12):4434–44. doi:10.1096/fj.11-188383.
8. Proposed rule on tobacco product standard for menthol in cigarettes. Silver Spring (MD): Food and Drug Administration; 2022.
9. Nair JN. Additives in tobacco products: contribution of carob bean extract, cellulose fibre, guar gum, liquorice, menthol, prune juice concentrate and vanillin to attractiveness, addictiveness and toxicity of tobacco smoking. Heidelberg: German Cancer Research Center; 2012 (https://www.researchgate.net/publication/272180111_Additives_in_Tobacco_Products_Contribution_of_Carob_Bean_Extract_Cellulose_Fibre_Guar_Gum_Liquorice_Menthol_Prune_Juice_Concentrate_and_Vanillin_to_Attractiveness_Addictiveness_and_Toxicity_of_Tobacco).
10. Alarie Y. Sensory irritation by airborne chemicals. CRC Crit Rev Toxicol. 1973;2(3):299–363. doi:10.3109/10408447309082020.
11. Arendt-Nielsen L, Carstens E, Proctor G, Boucher Y, Clavé P, Nielsen KA et al. The role of TRP channels in nicotinic provoked pain and irritation from the oral cavity and throat: translating animal data to humans. Nicotine Tob Res. 2022;24(12):1849–60. doi:10.1093/ntr/ntac054.

12. Carstens E, Carstens MI. Sensory effects of nicotine and tobacco. Nicotine Tob Res. 2022;24(3):306–15. PMID: 33955474.
13. De Biasi M, Dani JA. Reward, addiction, withdrawal to nicotine. Annu Rev Neurosci. 2011;34:105–30. doi:10.1146/annurev-neuro-061010-113734.
14. Mead EL, Duffy V, Oncken C, Litt MD. E-cigarette palatability in smokers as a function of flavorings, nicotine content and propylthiouracil (PROP) taster phenotype. Addict Behav. 2019;91:37–44. doi:10.1016/j.addbeh.2018.11.014.
15. Alpert HR, Agaku IT, Connolly GN. A study of pyrazines in cigarettes and how additives might be used to enhance tobacco addiction. Tob Control. 2016;25(4):444–50. doi:10.1136/tobaccocontrol-2014-051943.
16. Fan L, Balakrishna S, Jabba SV, Bonner PE, Taylor SR, Picciotto MR et al. Menthol decreases oral nicotine aversion in C57BL/6 mice through a TRPM8-dependent mechanism. Tob Control. 2016;25(Suppl 2):ii50–4. doi:10.1136/tobaccocontrol-2016-053209.
17. Johnson NL, Patten T, Ma M, De Biasi M, Wesson DW. Chemosensory contributions of e-cigarette additives on nicotine use. Front Neurosci. 2022;16:893587. doi:10.3389/fnins.2022.893587.
18. Kichko TI, Kobal G, Reeh PW. Cigarette smoke has sensory effects through nicotinic and TRPA1 but not TRPV1 receptors on the isolated mouse trachea and larynx. Am J Physiol Lung Cell Mol Physiol. 2015;309(8):L812–20. doi:10.1152/ajplung.00164.2015.
19. Song MA, Benowitz NL, Berman M, Brasky TM, Cummings KM, Hatsukami DK et al. Cigarette filter ventilation and its relationship to increasing rates of lung adenocarcinoma. J Natl Cancer Inst. 2017;109(12). doi:10.1093/jnci/djx075.
20. Benowitz NL. Pharmacology of nicotine: addiction, smoking-induced disease, and therapeutics. Annu Rev Pharmacol Toxicol. 2009;49:57–71. doi:10.1146/annurev.pharmtox.48.113006.094742.
21. Benowitz NL. Pharmacokinetic considerations in understanding nicotine dependence. Ciba Found Symp. 1990;152:186–200;200–9. doi:10.1002/9780470513965.ch11.
22. Chaudhri N, Caggiula AR, Donny EC, Palmatier MI, Liu X, Sved AF. Complex interactions between nicotine and nonpharmacological stimuli reveal multiple roles for nicotine in reinforcement. Psychopharmacology (Berl). 2006;184(3-4):353-66. doi:10.1007/s00213-005-0178-1.
23. Schneller LM, Bansal-Travers M, Mahoney MC, McCann SE, O'Connor RJ. Menthol, nicotine, and flavoring content of capsule cigarettes in the US. Tob Regul Sci. 2020;6(3):196–204. doi:10.18001/trs.6.3.4.
24. Scientific review of the effects of menthol in cigarettes on tobacco addiction: 1980–2021. Rockville (MD): Department of Health and Health Services; 2021 (https://www.fda.gov/media/157642/download).
25. Ha MA, Smith GJ, Cichocki JA, Fan L, Liu YS, Caceres AI et al. Menthol attenuates respiratory irritation and elevates blood cotinine in cigarette smoke exposed mice. PLoS One. 2015;10(2):e0117128. doi:10.1371/journal.pone.0117128.
26. Cohn AM, Alexander AC, Ehlke SJ. Affirming the abuse liability and addiction potential of menthol: differences in subjective appeal to smoking menthol versus non-menthol cigarettes across African American and white young adult smokers. Nicotine Tob Res. 2022;24(1):20–7. doi:10.1093/ntr/ntab137.
27. Jackson A, Green B, Erythropel HC, Kong G, Cavallo DA, Eid T et al. Influence of menthol and green apple e-liquids containing different nicotine concentrations among youth e-cigarette users. Exp Clin Psychopharmacol. 2021;29(4):355–65. doi:10.1037/pha0000368.
28. Rosbrook K, Green BG. Sensory effects of menthol and nicotine in an e-cigarette. Nicotine Tob Res. 2016;18(7):1588–95. doi:10.1093/ntr/ntw019.
29. Leventhal AM, Goldenson NI, Barrington-Trimis JL, Pang RD, Kirkpatrick MG. Effects of non-tobacco flavors and nicotine on e-cigarette product appeal among young adult never, former,

and current smokers. Drug Alcohol Depend. 2019;203:99-106.
30. Krishnan-Sarin S, Green BG, Kong G, Cavallo DA, Jatlow P, Gueorguieva R, et al. Studying the interactive effects of menthol and nicotine among youth: An examination using e-cigarettes. Drug Alcohol Depend. 2017;180:193-9.
31. Leventhal AM, Goldenson NI, Barrington-Trimis JL, Pang RD, Kirkpatrick MG. Effects of non-tobacco flavors and nicotine on e-cigarette product appeal among young adult never, former, and current smokers. Drug Alcohol Depend. 2019;203:99–106. doi:10.1016/j.drugalcdep.2019.05.020.
32. Leffingwell JC. Wilkinson Sword "CWM", 1975. Los Angeles (CA): University of California at Los Angeles, Industry Documents Library (https://www.industrydocuments.ucsf.edu/docs/tkjn0089).
33. Newman FS. Memorandum from Philip Morris in-house counsel (Newman) to Philip Morris in-house counsel (Holtzman) providing legal advice and analysis regarding content of draft press release for new product introduction. Northwind comments on 3rd tier inquires, 1981. Los Angeles (CA): University of California at Los Angeles, Industry Documents Library; 2021 (https://www.industrydocuments.ucsf.edu/docs/kgxy0101).
34. Daylor FL. Accomplishments of 81000, 1982. Los Angeles (CA): University of California at Los Angeles, Industry Documents Library; 2022 (https://www.industrydocuments.ucsf.edu/docs/xyjx0112).
35. Reger L, Moß J, Hahn H, Hahn J. Analysis of menthol, menthol-like, and other tobacco flavoring compounds in cigarettes and in electrically heated tobacco products. Beitr Tabakforsch Int. 2018;28(2):93–102. doi:10.2478/cttr-2018-0010.
36. Jabba SV, Jordt SE. Turbocharged Juul device challenges European tobacco regulators. Eur Respir J. 2020;56(2).
37. Jewett C, Baumgaertner E. R.J. Reynolds pivots to new cigarette pitches as flavor ban takes effect. New York Times, 11 January 2023 (https://www.nytimes.com/2023/01/11/health/cigarettes-flavor-ban-california.html).
38. Erythropel HC, Anastas PT, Krishnan-Sarin S, O'Malley SS, Jordt SE, Zimmerman JB. Differences in flavourant levels and synthetic coolant use between USA, EU and Canadian Juul products. Tob Control. 2020;30(4):453–5. doi:10.1136/tobaccocontrol-2019-055500.
39. Jabba SV, Erythropel HC, Torres DG, Delgado LA, Anastas PT, Zimmerman JB et al. Synthetic cooling agents in US-marketed e-cigarette refill liquids and disposable e-cigarettes: chemical analysis and risk assessment. Nicotine Tob Res. 2022;24(7):1037–46. doi:10.1093/ntr/ntac046.
40. Leventhal A, Dai H, Barrington-Trimis J, Sussman S. "Ice" flavoured e-cigarette use among young adults. Tob Control. 2023;32(1):114–7. doi:10.1136/tobaccocontrol-2020-056416.
41. Leventhal AM, Tackett AP, Whitted L, Jordt SE, Jabba SV. Ice flavours and non-menthol synthetic cooling agents in e-cigarette products: a review. Tob Control. 2022. doi:10.1136/tobaccocontrol-2021-057073.
42. Chaffee BW, Halpern-Felsher B, Croker JA, Werts M, Couch ET, Cheng J. Preferences, use, and perceived access to flavored e-cigarettes among United States adolescents and young adults. Drug Alcohol Depend Rep. 2022;3:100068. doi:10.1016/j.dadr.2022.100068.
43. Ali FRM, Seaman EL, Diaz MC, Ajose J, King BA. Trends in unit sales of cooling flavoured e-cigarettes, USA, 2017–2021. Tob Control. 2022. doi:10.1136/tc-2022-057395.
44. Lemon CH, Norris JE, Heldmann BA. The TRPA1 ion channel contributes to sensory-guided avoidance of menthol in mice. eNeuro. 2019;6(6). doi:10.1523/ENEURO.0304-19.2019.
45. Leffingwell J, Rowsell D. Wilkinson Sword cooling compounds: from the beginnning to now. Perfumer Flavorist. 2014;39:34–43 (https://img.perfumerflavorist.com/files/base/allured/all/document/2014/02/pf.PF_39_03_034_10.pdf).
46. Symcool® cooling agents. Infinite sensations. Brunswick (GA): Symrise, Inc.' undated (https://

www.symrise.com/fileadmin/symrise/Marketing/Scent_and_care/Aroma_molecules/symrise-symcool-A4-pages-eng.pdf).

47. Johnson S, Tian M, Sheldon G, Dowd E. Trigeminal receptor study of high-intensity cooling agents. J Agric Food Chem. 2018;66(10):2319–23. doi:10.1021/acs.jafc.6b04838.
48. Bandell M, Story GM, Hwang SW, Viswanath V, Eid SR, Petrus MJ et al. Noxious cold ion channel TRPA1 is activated by pungent compounds and bradykinin. Neuron. 2004;41(6):849–57. doi:10.1016/s0896-6273(04)00150-3.
49. Klein AH, Iodi Carstens M, McCluskey TS, Blancher G, Simons CT, Slack JP et al. Novel menthol-derived cooling compounds activate primary and second-order trigeminal sensory neurons and modulate lingual thermosensitivity. Chem Senses. 2011;36(7):649–58. doi: doi:10.1093/chemse/bjr029.
50. Tackett AP, Han DH, Peraza N, Whaley R, Leventhal AM. Effects of "Ice" flavored e-cigarettes with synthetic cooling agent WS-23 or menthol on user-reported appeal and sensory attributes. In: Annual meeting of the Society for Research on Nicotine and Tobacco, San Antonio (TX); 2023.
51. Benowitz NL. The central role of pH in the clinical pharmacology of nicotine: implications for abuse liability, cigarette harm reduction and FDA regulation. Clin Pharmacol Ther. 2022;111(5):1004–6. doi:10.1002/cpt.2555.
52. Wittenberg RE, Wolfman SL, De Biasi M, Dani JA. Nicotinic acetylcholine receptors and nicotine addiction: A brief introduction. Neuropharmacology. 2020;177:108256. doi:10.1016/j.neuropharm.2020.108256.
53. Gholap VV, Kosmider L, Golshahi L, Halquist MS. Nicotine forms: Why and how do they matter in nicotine delivery from electronic cigarettes? Expert Opin Drug Deliv. 2020;17(12):1727–36. doi:10.1080/17425247.2020.1814736.
54. Xiu X, Puskar NL, Shanata JA, Lester HA, Dougherty DA. Nicotine binding to brain receptors requires a strong cation-pi interaction. Nature. 2009;458(7237):534–7. doi:10.1038/nature07768.
55. Pankow JF. A consideration of the role of gas/particle partitioning in the deposition of nicotine and other tobacco smoke compounds in the respiratory tract. Chem Res Toxicol. 2001;14(11):1465–81. doi:10.1021/tx0100901.
56. Addictiveness and attractiveness of tobacco additives. Brussels: Scientific Committee on Emerging and Newly Identified Health Risks; 2010 (https://ec.europa.eu/health/scientific_committees/emerging/docs/scenihr_o_029.pdf).
57. Willems EW, Rambali B, Vleeming W, Opperhuizen A, van Amsterdam JG. Significance of ammonium compounds on nicotine exposure to cigarette smokers. Food Chem Toxicol. 2006;44(5):678–88. doi:10.1016/j.fct.2005.09.007.
58. Ramamurthi D, Chau C, Jackler RK. JUUL and other stealth vaporisers: hiding the habit from parents and teachers. Tob Control. 2018. doi:10.1136/tobaccocontrol-2018-054455.
59. Kavuluru R, Han S, Hahn EJ. On the popularity of the USB flash drive-shaped electronic cigarette Juul. Tob Control. 2019;28(1):110–2. doi:10.1136/tobaccocontrol-2018-054259.
60. Jackler RK, Ramamurthi D. Nicotine arms race: JUUL and the high-nicotine product market. Tob Control. 2019;28(6):623–8. doi:10.1136/tobaccocontrol-2018-054796.
61. Pennings JLA, Havermans A, Pauwels C, Krüsemann EJZ, Visser WF, Talhout R. Comprehensive Dutch market data analysis shows that e-liquids with nicotine salts have both higher nicotine and flavour concentrations than those with free-base nicotine. Tob Control. 2022. doi:10.1136/tobaccocontrol-2021-056952.
62. Hajek P, Pittaccio K, Pesola F, Myers Smith K, Phillips-Waller A, Przulj D. Nicotine delivery and users' reactions to Juul compared with cigarettes and other e-cigarette products. Addiction. 2020;115(6):1141–8. doi:10.1111/add.14936.

63. O'Connell G, Pritchard JD, Prue C, Thompson J, Verron T, Graff D et al. A randomised, open-label, cross-over clinical study to evaluate the pharmacokinetic profiles of cigarettes and e-cigarettes with nicotine salt formulations in US adult smokers. Intern Emerg Med. 2019;14(6):853–61. doi:10.1007/s11739-019-02025-3.
64. Prochaska JJ, Vogel EA, Benowitz N. Nicotine delivery and cigarette equivalents from vaping a JUULpod. Tob Control. 2021. doi:10.1136/tobaccocontrol-2020-056367.
65. Reilly SM, Bitzer ZT, Goel R, Trushin N, Richie JP. Free radical, carbonyl, and nicotine levels produced by Juul electronic cigarettes. Nicotine Tob Res. 2019;21(9):1274–8. doi:10.1093/ntr/nty221.
66. Leavens ELS, Nollen NL, Ahluwalia JS, Mayo MS, Rice M, Brett EI, et al. Changes in dependence, withdrawal, and craving among adult smokers who switch to nicotine salt pod-based e-cigarettes. Addiction. 2022;117(1):207–15. doi:10.1111/add.15597.
67. Kechter A, Cho J, Miech RA, Barrington-Trimis JL, Leventhal AM. Nicotine dependence symptoms in US youth who use JUUL e-cigarettes. Drug Alcohol Depend. 2021;227:108941. doi:10.1016/j.drugalcdep.2021.108941.
68. Tackett AP, Hebert ET, Smith CE, Wallace SW, Barrington-Trimis JL, Norris JE et al. Youth use of e-cigarettes: Does dependence vary by device type? Addict Behav. 2021;119:106918. doi:10.1016/j.addbeh.2021.106918.
69. Goldenson NI, Fearon IM, Buchhalter AR, Henningfield JE. An open-label, randomized, controlled, crossover study to assess nicotine pharmacokinetics and subjective effects of the JUUL system with three nicotine concentrations relative to combustible cigarettes in adult smokers. Nicotine Tob Res. 2021;23(6):947–55. doi:10.1093/ntr/ntab001.
70. Mallock N, Rabenstein A, Gernun S, Laux P, Hutzler C, Karch S et al. Nicotine delivery and relief of craving after consumption of European JUUL e-cigarettes prior and after pod modification. Sci Rep. 2021;11(1):12078. doi:10.1038/s41598-021-91593-6.
71. Phillips-Waller A, Przulj D, Smith KM, Pesola F, Hajek P. Nicotine delivery and user reactions to Juul EU (20 mg/ml) compared with Juul US (59 mg/ml), cigarettes and other e-cigarette products. Psychopharmacology (Berl). 2021;238(3):825–31. doi:10.1007/s00213-020-05734-2.
72. Leventhal AM, Madden DR, Peraza N, Schiff SJ, Lebovitz L, Whitted L et al. Effect of exposure to e-cigarettes with salt vs free-base nicotine on the appeal and sensory experience of vaping: a randomized clinical trial. JAMA Netw Open. 2021;4(1):e2032757. doi:10.1001/jamanetworkopen.2020.32757.
73. Tattan-Birch H, Brown J, Shahab L, Jackson SE. Trends in use of e-cigarette device types and heated tobacco products from 2016 to 2020 in England. Sci Rep. 2021;11(1):13203. doi:10.1038/s41598-021-92617-x.
74. Han DH, Wong M, Peraza N, Vogel EA, Cahn R, Mason TB et al. Dose–response effects of two nicotine salt formulations on electronic cigarette appeal and sensory attributes. Tob Control. 2023. doi:10.1136/tc-2022-057553.
75. Pauwels CGGM, Pennings JLA, Boer K, Baloe EP, Hartendorp APT, van Tiel L et al. Sensory appeal and puffing intensity of e-cigarette use: influence of nicotine salts versus free-base nicotine in e-liquids. submitted. 2022.
76. Hale R, Christatis K, Lin SS,Wynn R. Basic flavor investigation: low tar/high flavor literature review, 1990. Los Angeles (CA): University of California at Los Angeles, Industry Documents Library; 2009 (https://www.industrydocuments.ucsf.edu/docs/myjj0045).
77. Klus H, Scherer G, Muller L. Influence of additives on cigarette related health risks. Beitr. Tabakforsch Int. 2012;25(3):412–93. doi:10.2478/cttr-2013-0921.
78. Talhout R, Opperhuizen A, van Amsterdam JG. Sugars as tobacco ingredient: Effects on mainstream smoke composition. Food Chem Toxicol. 2006;44(11):1789–98. doi:10.1016/j.fct.2006.06.016.

79. Gager FL JR, Nedlock JW, Martin WJ. Tobacco additives and cigarette smoke. Part I. Transfer of d-glucose, sucrose, and their degradation products to the smoke. Carbohydrate Res. 1971;17(2):327–33. doi:10.1016/s0008-6215(00)82540-9.
80. Seeman JI, Dixon M, Haussmann HJ. Acetaldehyde in mainstream tobacco smoke: formation and occurrence in smoke and bioavailability in the smoker. Chem Res Toxicol. 2002;15(11):1331–50. doi:10.1021/tx020069f.
81. Elson LA, Betts TE. Sugar content of the tobacco and pH of the smoke in relation to lung cancer risks of cigarette smoking. J Natl Cancer Inst. 1972;48(6):1885–90. PMID: 5056275.
82. Rodgman A. Some studies of the effects of additives on cigarette mainstream smoke properties. II. Casing materials and humectants. Beitr Tabakforsch Int. 2002;20(4):279–99. doi.org/10.2478/cttr-2013-0742,
83. Leffingwell JC. Leaf chemistry: basic chemical constituents of tobacco leaf and differences among tobacco types. In: Davis DL, Nielsen MT, editors. Tobacco: production, chemistry and technology. Oxford: Blackwell Science; 1999:265–84 (http://www.leffingwell.com/download/Leffingwell%20-%20Tobacco%20production%20chemistry%20and%20technology.pdf).
84. Jenkins CR, Boham N, Burden AJ, Dixon M, Dowle M, Fiebelkorn RT et al. British American Tobacco Company Limited product seminar. Tobacco documents 760076123-760076408. 1997:760076123-408.
85. Krusemann EJZ, Pennings JLA, Cremers J, Bakker F, Boesveldt S, Talhout R. GC-MS analysis of e-cigarette refill solutions: A comparison of flavoring composition between flavor categories. J Pharm Biomed Anal. 2020;188:113364. doi:10.1016/j.jpba.2020.113364.
86. Bakker't Hart IME, Bakker F, Pennings JLA, Weibolt N, Eising S, Talhout R. Flavours and flavourings in waterpipe products: a comparison between tobacco, herbal molasses and steam stones. Tob Control. 2022. doi:10.1136/tobaccocontrol-2021-056955.
87. Rosbrook K, Erythropel HC, DeWinter TM, Falinski M, O'Malley S, Krishnan-Sarin S et al. The effect of sucralose on flavor sweetness in electronic cigarettes varies between delivery devices. PLoS One. 2017;12(10):e0185334. doi:10.1371/journal.pone.0185334.
88. Krusemann EJZ, Wenng FM, Pennings JLA, de Graaf K, Talhout R, Boesveldt S. Sensory evaluation of e-liquid flavors by smelling and Vvaping yields similar results. Nicotine Tob Res. 2020;22(5):798–805. doi:10.1093/ntr/ntz155.
89. Kowitt SD, Meernik C, Baker HM, Osman A, Huang LL, Goldstein AO. Perceptions and experiences with flavored non-menthol tobacco products: a systematic review of qualitative studies. Int J Environ Res Public Health. 2017;14(4). doi:10.3390/ijerph14040338.
90. Pullicin AJ, Kim H, Brinkman MC, Buehler SS, Clark PI, Lim J. Impacts of nicotine and flavoring on the sensory perception of e-cigarette aerosol. Nicotine Tob Res. 2020;22(5):806–13. doi:10.1093/ntr/ntz058.
91. Hayes JE, Baker AN. Flavor science in the context of research on electronic cigarettes. Front Neurosci. 2022;16:918082. doi:10.3389/fnins.2022.918082.
92. Baker AN, Bakke AJ, Branstetter SA, Hayes JE. Harsh and sweet sensations predict acute liking of electronic cigarettes, but flavor does not affect acute nicotine intake: a pilot laboratory study in men. Nicotine Tob Res. 2021;23(4):687–93. doi:10.1093/ntr/ntaa209.
93. Leventhal AM, Goldenson NI, Cho J, Kirkpatrick MG, McConnell RS, Stone MD et al. Flavored e-cigarette use and progression of vaping in adolescents. Pediatrics. 2019;144(5). doi:10.1542/peds.2019-0789.
94. Soneji SS, Knutzen KE, Villanti AC. Use of flavored e-cigarettes among adolescents, young adults, and older adults: findings from the Population Assessment for Tobacco and Health Study. Public Health Rep. 2019;134(3):282–92. doi:10.1177/0033354919830967.
95. St Helen G, Dempsey DA, Havel CM, Jacob P 3rd, Benowitz NL. Impact of e-liquid flavors on nicotine intake and pharmacology of e-cigarettes. Drug Alcohol Depend. 2017;178:391–8.

doi:10.1016/j.drugalcdep.2017.05.042.

96. St Helen G, Shahid M, Chu S, Benowitz NL. Impact of e-liquid flavors on e-cigarette vaping behavior. Drug Alcohol Depend. 2018;189:42–8. doi:10.1016/j.drugalcdep.2018.04.032.
97. Smith DM, Schneller LM, O'Connor RJ, Goniewicz ML. Are e-cigarette flavors associated with exposure to nicotine and toxicants? Findings from wave 2 of the Population Assessment of Tobacco and Health (PATH) study. Int J Environ Res Public Health. 2019;16(24). doi:10.3390/ijerph16245055
98. Seeman JI, Laffoon SW, Kassman AJ. Evaluation of relationships between mainstream smoke acetaldehyde and "tar" and carbon monoxide yields in tobacco smoke and reducing sugars in tobacco blends of US commercial cigarettes. Inhal Toxicol. 2003;15(4):373–95. doi:10.1080/08958370304461.
99. Bernasek PF, Furin OP, Shelar GR. Sugar/nicotine study. Industry documents library. 1992;ATP 92-210:22. (sljb0079–Truth Tobacco Industry Documents) (ucsf.edu).
100. Weeks WW. Relationship between leaf chemistry and organoleptic properties of tobacco smoke. In: Davis DL, Nielson MT, editors. Tobacco: production, chemistry and technology. Oxord: Blackwell Science; 1999:304–12 (https://www.wiley.com/en-us/Tobacco%3A+Production%2C+Chemistry+and+Technology-p-9780632047918).
101. Fagan P, Pokhrel P, Herzog TA, Moolchan ET, Cassel KD, Franke AA et al. Sugar and aldehyde content in flavored electronic cigarette liquids. Nicotine Tob Res. 2018;20(8):985–92. doi:10.1093/ntr/ntx234.
102. Patten T, De Biasi M. History repeats itself: Role of characterizing flavors on nicotine use and abuse. Neuropharmacology. 2020;177:108162. doi:10.1016/j.neuropharm.2020.108162.
103. Final opinion on additives used in tobacco products. Brussels: Scientific Committee on Emerging and Newly Identified Health Risks; 2016 (https://health.ec.europa.eu/latest-updates/scenihr-final-opinion-additives-used-tobacco-products-2016-01-29-1_en).
104. Bates C, Jarvis M, Connolly G. Tobacco additives. Cigarette engineering and nicotine addiction. London: Action on Smoking and Health, Imperial Cancer Research Fund; 1999 (https://www.researchgate.net/publication/242598454_Tobacco_Additives_Cigarette_Engineering_and_Nicotine_Addiction).
105. Fowles J. Chemical factors influencing the addictiveness and attractiveness of cigarettes in New Zealand. Auckland: Ministry of Health Libaray; 2001 (http://www.moh.govt.nz/notebook/nbbooks.nsf/0/D0B68B2D9CB811ABCC257B81000D9959?OpenDocument).
106. Banožić M, Jokić S, Ačkar Đ, Blažić M, Šubarić D. Carbohydrates – key players in tobacco aroma formation and quality determination. Molecules. 2020;25(7):1734. doi:10.3390/molecules25071734.
107. Wise PM, Breslin PA, Dalton P. Sweet taste and menthol increase cough reflex thresholds. Pulm Pharmacol Ther. 2012;25(3):236–41. doi:10.1016/j.pupt.2012.03.005.
108. Truth tobacco industy documents. Los Angeles (CA): University of California at Los Angeles, Industry Documents Library (https://www.industrydocumentslibrary.ucsf.edu/tobacco/).
109. Ferreira CG, Silveira D, Hatsukami DK, Paumgartten FJ, Fong GT, Gloria MB et al. The effect of tobacco additives on smoking initiation and maintenance. Cad Saude Publica. 2015;31(2):223-5. doi:10.1590/0102-311XPE010215.
110. Erythropel HC, Kong G, deWinter TM, O'Malley SS, Jordt SE, Anastas PT et al. Presence of high-intensity sweeteners in popular cigarillos of varying flavor profiles. JAMA. 2018;320(13):1380–3. doi:10.1001/jama.2018.11187.
111. Directive 2014//40/Eu of the European Parliament and of the Council of 3 April 2014 on the approximation of the laws, regulations and administrative provisions of the Member States concerning the manufacture, presentation and sale of tobacco and related products and repealing Directive 2001/37/EC. Off J Eur Unuon. 2014;L127/1 (https://eur-lex.europa.eu/le-

gal-content/EN/TXT/PDF/?uri=CELEX:32014L0040).

112. No more menthol in cigarettes and smoking tobacco. Berlin: Federal Institute for Risk Assessment; 2020 (https://www.bfr.bund.de/en/press_information/2020/19/no_more_menthol_in_cigarettes_and_smoking_tobacco-246948.html)
113. Verduidelijking van artikel 5 van het Koninkilijk besluit van 5/02/2016 [Clarification of Article 5 of the Royal Decree of 5/02/2016]. The Hague: Federale Overheidsdienst Volksgezondheid; 2022 (https://www.health.belgium.be/sites/default/files/uploads/fields/fpshealth_theme_file/nl_art5_inhalation_facilitation.pdf).
114. Agreement No. 761297-JATC-JA-03-2016. WP9: D9.3 Report on the peer review of the enhanced reporting information on priority additives. Brussels: European Union, Joint Action on Tobacco Control; 2020) (https://jaotc.eu/wp-content/uploads/2021/04/D9.3-Report-on-the-peer-review-of-the-enhanced-reporting-information-on-priority-additives.pdf).
115. Regulating tobacco and vaping products: Tobacco regulations. Ottawa: Government of Canada; 2020 (https://www.canada.ca/en/health-canada/services/smoking-tobacco/regulating-tobacco-vaping/tobacco.html).
116. Tobacco product standard for characterizing flavors in cigars (FDA-2021-N-1309). Silver Spring (MD): Food and Drug Administration; 2020 (https://fda.report/media/158013/Tobacco+Product+Standard+for+Characterizing+Flavors+in+Cigars.pdf).
117. Decision summary for marketing denial of flavored e-cigarettes. Silver Spring (MD): Food and Drug Administration; 2021.
118. Marketing order and post-authorization marketing restrictions and requirements LogicTobacco Products. Silver Spring (MD): Food and Drug Administration; 2022.
119. Marketing order and post-authorization marketing restrictions and requirements Vuse Tobacco Products. Silver Spring (MD): Food and Drug Administration; 2021.
120. WHO Framework Convention on Tobacco Control. Partial guidelines for implementation of Articles 9 and 10 (FCTC/16.3.2017). Geneva: World Health Organization; 2017 (https://fctc.who.int/publications/m/item/regulation-of-the-contents-of-tobacco-products-and-regulation-of-tobacco-product-disclosures#:~:text=Whereas%20Article%209%20deals%20with,governmental%20authorities%20and%20the%20public).

3. Synthetic nicotine: science, global legal landscape and regulatory considerations

Micah L. Berman, College of Public Health, Moritz College of Law and Cancer Control Program, James Comprehensive Cancer Center, The Ohio State University, Columbus (OH), USA

Patricia J. Zettler, Moritz College of Law and Cancer Control Program, James Comprehensive Cancer Center, The Ohio State University, Columbus (OH), USA

Sven-Eric Jordt, Departments of Anesthesiology, Pharmacology and Cancer Biology, Duke University School of Medicine, Durham (NC) and Tobacco Center of Regulatory Science, Department of Psychiatry, Yale School of Medicine, New Haven (CT), USA

Contents

Preface: key findings

- Synthetic nicotine products – including nicotine pouches, e-liquids, disposable e-cigarettes, gums, toothpicks and infused combustible products – are marketed and sold throughout the world.
- Synthetic nicotine products are sold with marketing claims (e.g. "tobacco-free") that may suggest they are safer than products containing tobacco-derived nicotine, and some products are sold with flavour concepts (e.g. "chocolate dream," "pink lemonade") that are likely to appeal to young people.
- Synthetic nicotine is added to marketed products in two forms, S- and R-nicotine. S-Nicotine is the primary form of nicotine in tobacco plants. The pharmacological, metabolic and toxicological effects of R-nicotine and of mixtures and R- and S-nicotine, however, are poorly understood.

- No standard methods for the chemical analysis of synthetic nicotine are available, and adulteration of products with tobacco-derived nicotine is a concern.
- Whether synthetic nicotine products are regulated under current regulations for tobacco control depends on how the laws define the products covered by the regulations. Laws that apply only to "tobacco products" or "tobacco-derived" products may not be broad enough to cover synthetic nicotine products, because synthetic nicotine is not derived from tobacco plants.
- Tobacco companies are aware that some tobacco control laws do not cover synthetic nicotine products and have sought to take advantage of such regulatory gaps.
- Some countries have amended their tobacco control laws so that they apply to products containing nicotine that is not made or derived from tobacco, such as synthetic nicotine. The tobacco control laws of many countries do not, however, clearly apply to such products or do not apply to the full range of currently marketed products.

3.1 Introduction

Companies are increasingly marketing a wide range of synthetic nicotine products, which contain or are promoted as containing nicotine that is chemically synthesized rather than derived from tobacco plants. These products have not been shown to pose fewer risks than products containing tobacco-derived nicotine, although their marketing sometimes claims or implies that they do. In many countries, synthetic nicotine products are not clearly subject to current tobacco control regulations, although they may be subject to other laws, such as for consumer protection, in some countries. In other countries, however, tobacco control laws have been updated to cover these products in various ways. Countries should consider legal adjustments that they might make to close regulatory gaps for synthetic nicotine products, covering both the broad range of products currently on the market and those products that might emerge in the future.

3.1.1 Background

The rise of novel and emerging tobacco products, such as electronic nicotine and non-nicotine delivery systems, imitation tobacco products and nicotine pouches, has led to new forms of nicotine use, including by young people. The success of tobacco control measures and the social stigma associated with consuming conventional tobacco products (including cigarettes, cigars, waterpipe tobacco and smokeless tobacco products) contributed to motivating the industry to develop e-cigarettes and other novel products distinct from conventional products. Lately, companies have begun to sell versions of these novel or unconventional products with claims that they contain synthetic rather than tobacco-derived nicotine *(1)*. These products are sometimes sold with flavours that appeal to young people. Additionally, although there is currently no evidence that products containing synthetic nicotine have different health effects from or are less addictive than products containing tobacco-derived nicotine, synthetic nicotine products are being sold with marketing claims that may suggest that they are safer than tobacco-derived nicotine products *(2)*.

The recent appearance of products promoted as containing synthetic nicotine or "tobacco-free" nicotine on many markets, including products sold worldwide over the Internet, has caused many WHO Member States to consider sharing regulatory information on this topic. Many Member States have requested technical assistance from WHO to address this issue and to provide a synthesis of the available evidence and authoritative advice on addressing products that are claimed to contain synthetic nicotine. This report, commissioned by WHO, was prepared to clarify those issues.

The report covers synthetic nicotine products marketed for recreational use, rather than for medical use, such as smoking cessation. It provides an overview of the types of synthetic nicotine products being sold, the claims with which they are marketed, and the science of synthetic nicotine production, toxicology, pharmacology and detection. It also provides information about the global legal landscape for synthetic nicotine products, focused on tobacco control laws. Specifically, we reviewed and coded the laws of 210 countries and the European Union (EU) Tobacco Products Directive on the Tobacco Control Laws website (www.tobaccocontrollaws.org). Of the 211 jurisdictions, 21 did not have a law or had no English translation. In 52 of the remaining 190 jurisdictions, the laws provided definitions broad enough to cover at least certain synthetic nicotine products (e.g. e-cigarettes but not other synthetic nicotine products), 29 provided definitions that covered a broader range of synthetic nicotine products, 92 had definitions that did not apply to any type of synthetic nicotine product, and in 17 jurisdictions, it was unclear whether the laws cover synthetic nicotine.

3.1.2 Types of synthetic nicotine products

News reports *(3)* suggest that the United States of America (USA) is currently the largest market for synthetic nicotine products, although this may change as a result of an amendment to US law in March 2022 that brings synthetic nicotine products within the purview of tobacco products authorities of the US Food and Drug Administration (USFDA). The second largest market is that of the Republic of Korea *(3)*. Currently, most products marketed as containing synthetic nicotine are either e-cigarettes, e-liquids or nicotine pouches. These are not, however, the only kinds of synthetic nicotine products being sold *(2)*. For example, several companies sell chewing-gum products described as containing synthetic or "tobacco-free" nicotine *(4,5)*; at least two companies are marketing synthetic nicotine toothpicks *(6,7)*; and a Canadian company, PODA, announced plans in 2021 to launch a "heat-not-burn product" containing "pelletized tea leaves infused with synthetic nicotine" *(8)*. This company has since been purchased by Philip Morris, and it is not clear whether its products will be marketed. At least one company, Outlaw Dip, is offering "100% tobacco free" moist snuff "that does NOT come from tobacco" *(9)*, and at least one other company, Ronin, is selling a combustible cannabidiol cigarette infused with "non-tobacco nicotine" *(10)*. There is, therefore, a wide variety of products sold as containing synthetic rather than tobacco-derived nicotine, and new types of products may continue to emerge.

Additionally, many of these synthetic or "tobacco-free" nicotine products contain flavours that are likely to appeal to young people. For instance, some toothpicks are sold with flavours such as "butterscotch cake" and "strawberry cheesecake" *(7)*, and certain disposable e-cigarettes are sold with flavour concepts such as "banana ice" and "blue razz" *(11)*.

3.1.3 Marketing and promotion of synthetic nicotine products

Many companies that market synthetic nicotine products make claims that may suggest, implicitly or explicitly, that their products are "safer" than products containing tobacco-derived nicotine. These include claims that synthetic nicotine contains fewer impurities than tobacco-derived nicotine and that synthetic nicotine is equivalent to pharmaceutical-grade nicotine. Companies also claim that synthetic nicotine products have other advantages over products with tobacco-derived nicotine, such as that they provide more satisfaction and a better taste experience and that they are more environmentally friendly. Some synthetic nicotine products are marketed as effective aids for smoking cessation or as equivalent to approved nicotine replacement therapy, sometimes with a disclaimer that the product is not a smoking cessation product. Table 1 provides a few examples.

Table 1. Examples of promotional claims about synthetic nicotine products

Product	Owner or manufacturer	Claim
Juice Head pouches	Juice Head (USA)	"...may offer higher nicotine satisfaction with potentially less risk than tobacco-derived nicotine. In addition, while tobacco nicotine often features a strong pungent odor and taste, synthetic nicotine is virtually tasteless and odorless." "...it is important to note that tobacco cultivation (which is commonly very subsidized) can be very damaging to the environment and is often a process that is highly labor-intensive, cumbersome, and wasteful." "It should be noted that tobacco-derived nicotine may come along with more risks of side effects than pouches made without tobacco." (12)

Pacha Mama vape pen	Charlie's Holdings, Inc. (USA)	"increased purity and consistency over traditionally harvested nicotine" (13)
Outlaw Dip	Outlaw Dip Company (USA)	"pharmaceutical grade" (9)
Bidi Pouch	Kaival Brands Innovations Group, Inc. (USA)	"aims to help adult smokers take their first steps in going smokeless" (14)
ZIA gum	Next Generation Labs LLC (USA)	"the Only Nicotine Gum Developed with Synthetic Nicotine" (4) "offers the same nicotine satisfaction as any tobacco-derived product containing nicotine" (4) and "ZIA™ gum is not intended to assist in quitting efforts" (15)
VaporX e-juice and disposable e-cigarettes	Vaporex Co., Ltd (Republic of Korea)	"We are committed to protecting the health of smokers by providing them with a valuable and appropriate vaping experience" (16)

3.2 The science of synthetic nicotine

The rapid, poorly regulated introduction of synthetic nicotine products (electronic cigarettes, oral pouches and other product categories) in the USA and other countries raises questions about their safety and potential differences in the addictive and reinforcing properties of synthetic nicotine. In this section, we review the strategies for chemical synthesis, the different forms of synthetic nicotine in products, the manufacturers and patent landscape, and the toxicological, pharmacological and metabolic properties of synthetic nicotine.

3.2.1 Methods

Research databases, including PubMed and Web of Science, were searched with terms such as "synthetic nicotine", "*R*-nicotine", "(+)-nicotine", "L-nicotine", "D-nicotine", "racemic nicotine" and "nicotine synthesis" for journal articles on synthetic nicotine and studies of the effects of nicotine enantiomers (defined below). Patents were sought on patents.google.com with combinations of terms such as "nicotine" "synthesis" and/or "stereoselective", "enantioselective". The tobacco legacy database at www.industrydocuments.ucsf.edu/tobacco/ was searched with terms such as "synthetic nicotine", "nicotine synthesis", "synthesis of nicotine" and "*R*-nicotine".

3.2.2 Results

Synthetic nicotine: what is it and how it differs from tobacco-derived nicotine

Nicotine exists in two chemical forms that are structural mirror images. The two forms, termed enantiomers, are *S*- and *R*-nicotine (Fig. 1A). Nicotine in tobacco plants consists of > 99% *S*-nicotine and only minimal amounts of *R*-nicotine *(17)*. Chemists first synthesized nicotine in 1904 *(18)*, resulting in a mixture containing both *S*- and *R*-nicotine in a 50:50 ratio *(18,19)*, known as a racemic mixture. This mixture differs from tobacco-derived nicotine in that it has a much higher *R*-nicotine content and a lower *S*-nicotine content.

Fig. 1. Structure and chemistry of synthetic nicotine

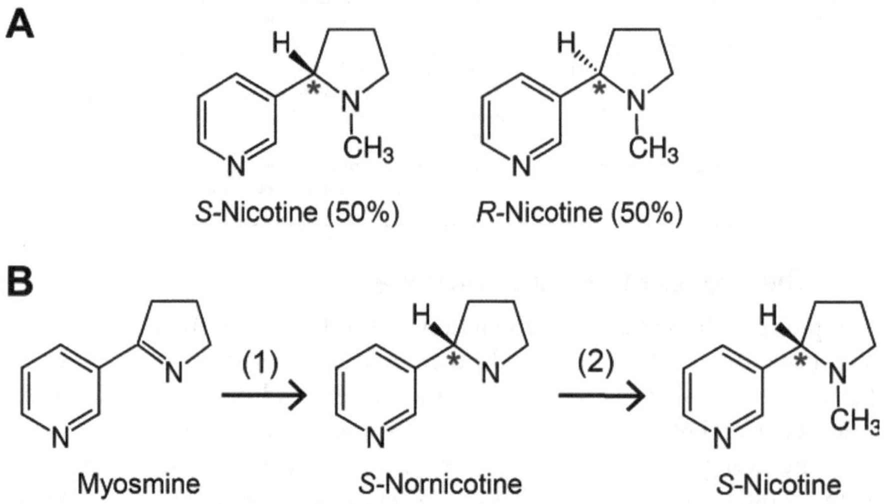

A

S-Nicotine (50%) *R*-Nicotine (50%)

B

Myosmine (1) → *S*-Nornicotine (2) → *S*-Nicotine

(A) Structures of *S*- and *R*-nicotine. The compounds differ in their configuration at the carbon atom labelled with a red asterisk, a chiral centre. In tobacco leaf, > 99% of nicotine is present as *S*-nicotine. Synthetic "tobacco-free nicotine" marketed by Next Generation Labs is racemic, containing 50% *S*-nicotine and 50% *R*-nicotine. Pure synthetic *S*-nicotine is chemically indistinguishable from *S*-nicotine purified from tobacco.
(B) Synthesis of *S*-nicotine as described in a patent assigned to Zanoprima involving a biotechnological step. The starting material is myosmine, which is first converted to *S*-nornicotine with a recombinant enzyme *(1)*, a NADH/NADPH-dependent imine reductase by a stereoselective reaction. *S*-Nornicotine is then converted to *S*-nicotine by methylation *(2)*.

A search in the Truth Tobacco Industry Documents (the database of tobacco industry internal corporate documents compiled during litigation in the USA) with the term "synthetic nicotine" revealed that the industry had already considered use of synthetic nicotine in the 1960s. Employees of British American Tobacco proposed addition of synthetic nicotine to increase the nicotine:tar ratio in combustible cigarettes *(20)*; however, the proposal was not pursued further because of concern that synthetic nicotine was available only as a racemic mixture, with unknown health effects. Furthermore, the price of synthetic nicotine was much higher than that of tobacco-derived nicotine *(20)*. Employees of RJ Reynolds and Liggett & Myers also considered use of synthetic nicotine to adjust nicotine levels in cigarettes; however, the idea was abandoned for the same reasons *(21,22)*. The Truth Tobacco Industry Documents database provides no further evidence after 1978 of consideration of the use of synthetic nicotine use by the major US tobacco companies. Subsequently, chemists developed new strategies for synthesizing nicotine, including methods to produce pure *S*-nicotine, the form of nicotine prevalent in tobacco leaf *(19,23)*.

The synthetic nicotine marketplace: manufacturers, patents and pricing

In 2015, the company Next Generation Labs (NGL) began marketing synthetic nicotine in the USA under the trademarks TFN® (Tobacco Free Nicotine) for consumer products and PHARMANIC® for pharmaceutical products. In the same year, NGL filed an application for US and world-wide patents with the title "Process for the preparation of (RS)-nicotine" *(24)*. The US patent, assigned to NGL in 2017, describes a synthetic pathway with ethyl nicotinate as the starting material. Ethyl nicotinate is derived from nicotinic acid (niacin), a synthetic chemical produced from petrochemical sources. It is reacted with *N*-vinyl-2-pyrrolidinone to form myosmine, a tobacco alkaloid. Myosmine is then converted to nornicotine. Subsequent methylation of nornicotine results in a racemic (50:50) nicotine mixture of *S*- and *R*-nicotine (Fig. 1A) *(24)*. NGL also filed a patent for use of their synthetic nicotine in smoking cessation products *(25)*. In 2016, Hellinghausen et al. analysed the nicotine content of electronic cigarette liquids marketed in the USA and containing TFN-branded synthetic nicotine manufactured by NGL and confirmed that the product is racemic nicotine *(26)*. While vaping products containing synthetic nicotine have been marketed in the USA since 2015, they attracted public attention only in 2021, when the popular vaping company Puff Bar announced a switch to synthetic nicotine in their products *(27)*. Analysis of these products showed that they contained racemic nicotine *(28)*. The source of the synthetic racemic nicotine in Puff Bar products has not been revealed.

At the same time, advances in chemistry resulted in optimization of strategies for manufacturing pure *S*-nicotine. Several companies have filed patent applications for the synthesis of *S*-nicotine. Contraf-Nicotex-Tobacco (Germany), the world's largest supplier of pharmaceutical-grade nicotine, developed a process for synthesizing racemic nicotine from ethyl nicotinate and *n*-vinylpyrrolidone nicotinic acid; subsequent selective purification enriches the compound to produce pure *S*-nicotine *(29,30)*. Vaping products containing Contraf-Nicotex-Tobacco's synthetic *S*-nicotine have been marketed in the USA since 2020 *(31,32)*. Zanoprima Life Sciences Ltd (London, United Kingdom) also manufactures synthetic *S*-nicotine *(33)* and was granted a US patent for a process involving a biotechnological step for the synthesis of *S*-nicotine in 2021 *(34)*. The starting material is myosmine, which is first converted to *S*-nornicotine with a commercially available recombinant enzyme, an NADH/NADPH-dependent imine reductase. *S*-Nornicotine is then converted to *S*-nicotine by methylation (Fig. 1B). This product is currently marketed under the brand name SyNic *(33)*. Hangsen International Group, a major manufacturer of vaping devices and e-liquids, applied for a Chinese and a world patent for a similar process and markets synthetic *S*-nicotine under the brand name "Motivo" *(35,36)*. NJOY, a major Ecigarette manufacturer (soon to be owned by the cigarette-maker Altria *(37)*), was also awarded a patent for nicotine synthesis and purification *(38)*. Some patents describe the resulting nicotine as "> 99.9% pure", with a chiral purity of > 99.6% *S*-nicotine or more. Wholesale products are listed as having a purity of 99.9% *S*-nicotine *(39)*.

In 2019, a representative of NGL stated that the company's synthetic racemic nicotine product, the racemic mix of *R*- and *S*-nicotine, "is only three to four times the current cost of tobacco-derived nicotine" *(40)*. As of 21 March 2023, the wholesale price of 1 L of NGL TFN racemic synthetic nicotine was quoted as US$ 1800, while the same wholesaler offered 1 L of tobacco-derived nicotine for US$ 229.99–429.99, depending on the brand. Thus, the price of the synthetic version is four to eight times higher than that for tobacco-derived nicotine *(41,42,43)*. Zanoprima's SyNic synthetic *S*-nicotine was marketed at a price of US$ 999.99 per litre, while the same seller quoted a price of US$ 229.99 for tobacco-derived nicotine, a difference of about four times *(39, 44)*. Thus, although the price of synthetic nicotine remains substantially higher than that of tobacco-derived nicotine, synthetic nicotine products continue to be marketed, including electronic cigarette products and oral nicotine pouches, also known as "white snus". These products are often advertised with claims that they are purer and healthier than products containing tobacco-derived nicotine.

Manufacturers of synthetic nicotine (Table 2) have begun to enforce their intellectual property, leading to legal conflicts and market consolidation. NGL's intellectual property was recently confirmed by Chinese authorities, enabling the company to enforce its patents in the country, where the large majority of e-cigarette products are manufactured *(45)*. Zanoprima sued a major ecigarette and liquid manufacturer, Hangsen, for infringement of its patent in a US district court *(46)*. Nicotine manufactured by Hangsen was added to "Geekbar" products marketed in the USA in 2021; however, Hangsen ceased marketing its "Motivo"-brand synthetic *S*-nicotine in the USA after the lawsuit was filed, while continuing sales outside the USA *(35,47)*.

Table 2. Major manufacturers of synthetic nicotine and their synthesis routes

Manufacturer	Starting material	Product	Stereoselective step
Next Generation Labs LLC (NGL)	Ethyl nicotinate	Racemic (50:50) *R*-:*S*-nicotine	Not applicable
Contraf-Nicotex-Tobacco	Ethyl nicotinate	*S*-Nicotine	Stereoselective recrystallization
Zanoprima Lifesciences Ltd	Myosmine	*S*-Nicotine	Enzymatic stereoselective step
Hangsen International Group	Myosmine	*S*-Nicotine	Enzymatic stereoselective step
NJOY LLC	Racemic (50:50) R-:*S*-nicotine	*S*-Nicotine	Stereoselective recrystallization

Health claims by manufacturers of synthetic nicotine

Like the sellers of ENDS, the companies that manufacture synthetic nicotine promote their products with health-related statements. NGL claims that "TFN is devoid of many of the residual impurities that tobacco derived nicotine contains

... TFN is virtually tasteless and odorless ... there is no need to mask the off-flavor and aroma of tobacco-based nicotine" *(48)*. NGL also claims that "specific ratios of the 'R' to the 'S' isomers could potentially offer nicotine use at satisfying but non-addictive or less addictive levels". Contraf-Nicotex-Tobacco opposes this notion, claiming that its synthetic *S*-nicotine is superior to the racemic version, stating "If you look at the European and the US pharmacopoeias, the percentage of *S*-isomers in nicotine must be higher than 99 percent" *(40,49)*. Zanoprima claims that its synthetic *S*-nicotine "is free of related tobacco alkaloids, TSNAs [tobacco-specific nitrosamines], odour, and harsh taste" *(33)*. These statements may represent claims that their synthetic nicotine has superior, drug-like properties. Companies also claim that they use a sustainable "green chemistry" approach for production that is environmentally more friendly than agricultural tobacco production, which requires pesticides, fertilizers, extensive land use and hazardous production methods.

Toxicological, pharmacological and metabolic properties of synthetic nicotine

As described above, there are currently two forms of synthetic nicotine in marketed products, *S*-nicotine and racemic nicotine, the latter consisting of 50% *S*-nicotine and 50% *R*-nicotine. As synthetic *S*-nicotine is chemically identical to tobacco-derived *S*-nicotine, its toxicological, metabolic and pharmacological properties should also be identical, especially if they are added at the purity claimed by the major manufacturers of synthetic products (> 99.9%). Nevertheless, even at this high degree of purity, trace amounts of other chemicals remaining from the chemical process might be present, which deserve further attention.

If a consumer uses a product containing synthetic racemic nicotine, 50% of their nicotine intake is *R*-nicotine. Less is known about *R*-nicotine's toxicological, metabolic and pharmacological effects than about those of *S*-nicotine. A study in mice established that the dose necessary to have a lethal effect in 50% of the animals (LD_{50}) 60 min after intravenous injection was 0.33 mg/kg for *S*-nicotine and 6.15 mg/kg for *R*-nicotine, which is > 18 times higher, suggesting that *R*-nicotine is less acutely toxic than *S*-nicotine under those conditions *(50)*. The study also established that a higher dose of *R*-nicotine than of *S*-nicotine is necessary to induce convulsions.

Pharmacological studies have shown that *R*-nicotine is about 10 times less potent as an agonist of nicotine receptors than *S*-nicotine *(51)*. A study of nicotine binding in the brain showed that *S*-nicotine is 10 or more times more potent than *R*-nicotine *(52)*. Long-term administration of either form of nicotine was shown to increase the number of nicotine binding sites in rat brain *(53)*.

In an operant behavioural study of the capability of rats to discriminate injected *R*- or *S*-nicotine from saline, *S*-nicotine was nine times more potent than *R*-nicotine *(54)*. A study to characterize the locomotor stimulant action of nicotine in rats showed that *S*-nicotine was at least 10 times more potent in stimulating motor activity *(55)*. *S*-Nicotine was four to five times more potent than *R*-nicotine in conditioned taste aversion assays in rats *(56)*. In contrast to *S*-nicotine, *R*-nicotine did not induce weight loss in rats and did not trigger epinephrine release *(51,53)*. Pharmacological studies of the enantiomers in standard experimental paradigms for nicotinic pharmacology showed that *S*-nicotine, the prevalent nicotine enantiomer in tobacco (> 99%), is 4–28 times more potent than *R*-nicotine, which is present at high levels (50%) in synthetic racemic nicotine *(51,52,54,55,57,58,59,60)*.

S- and *R*-nicotine also differ in their metabolism. Studies in guinea pigs showed that *S*-nicotine formed only oxidative metabolites, whereas *R*-nicotine formed both oxidative and *N*-methylated metabolites *(61)*. The degradation kinetics of the resulting *S*- and *R*-cotinine also differed. Studies of metabolism in various laboratory animal species showed strong differences between *S*-nicotine and *R*-nicotine in degradation and excretion and also sex differences in *R*-nicotine metabolism *(60,61,62)*. Species differences were also observed in *N*-methylation of *S*- and *R*-nicotine in human, rat and guinea pig liver cytosol extracts *(63)*. While the human extract catalysed *N*-methylation of both forms of nicotine, rat extract did not form any *N*-methylation products, and guinea pig extract transformed only *R*-nicotine and not *S*-nicotine *(63)*. It is not known whether these *N*-methylation products are bioactive and whether *S*- and *R*-nicotine methylation products act differently. These findings indicate that human metabolism of *R*-nicotine and its behavioural effects should be investigated further, and that predictions of the toxicological outcomes of *R*-nicotine consumption should not be based on animal models alone. The absence of such key data and the observed species differences preclude assessment of the toxicological risk of *R*-nicotine to humans.

In addition to the differences in nicotinic receptor-mediated pharmacological effects, *R*- and *S*-nicotine have differential effects on other pharmacological targets. For example, a tobacco industry-sponsored study on acetylcholinesterase, the enzyme that degrades the neurotransmitter acetylcholine in the synaptic cleft to terminate neurotransmission, revealed that *R*-nicotine is a more potent inhibitor of the enzyme than *S*-nicotine, binding to a different site on the enzyme protein *(64)*. The experiments were performed with acetylcholine esterase isolated from electric eels and at nicotine concentrations much higher than those received by smokers. Whether such effects occur in humans and how they affect acetylcholine levels and neurotransmission should be studied further. Both forms of nicotine interfere with the production of certain lipid mediators involved in regulation of inflammation, with similar potency, showing that some biological processes are equally affected by the two forms of nicotine *(65)*.

Psychophysical studies

Psychophysical studies were conducted to determine whether *R*- and *S*-nicotine elicit different odour or irritant sensations. People perceive nicotine vapour as aversive when they are exposed through the nose. At higher concentrations, nicotine vapour causes nasal irritation, including stinging and burning sensations, mediated by the trigeminal nerve, which transmits pain signals to the brain. Test subjects reported lower thresholds for detection of *S*- than for *R*-nicotine and greater burning and stinging intensity, while olfactory perceptions were elicited at similar levels. In electrical recordings of mucosal potential, *S*-nicotine elicited stronger responses than *R*-nicotine. Smokers perceived *S*-nicotine as more hedonic than non-smokers, probably because of previous experience *(66)*. This appears, so far, to be the only systematic study of human responses to *S*- and *R*-nicotine. The experiments were very short, as individual vapour stimuli were applied for only 250 ms.

Analytical detection of synthetic nicotine

Hellinghausen et al. *(26)* developed a method to validate the presence of synthetic racemic nicotine in vaping products labelled as containing the compound. They used a chiral stationary phase for separation of *R*- and *S*-nicotine by high-pressure liquid chromatography, followed by circular dichroism detection and electrospray ionization mass spectrometry. They reported that one product contained twice more total nicotine (sum of *R*- and *S*-nicotine) than the content stated on the product label, effectively listing only the strength of *S*-nicotine, while the nicotine content listed on other labels was equivalent to that measured, half of which was *S*-nicotine. These observations suggest that uniform product labelling practices should be imposed by regulators, to prevent unknowing users from exposure to higher levels of *R*-nicotine or to lower levels of *S*-nicotine than they are used to. Inappropriate labelling of nicotine content could motivate consumers to purchase products with a higher total nicotine content, potentially resulting in significantly higher *S*-nicotine intake. The authors also detected impurities that require further characterization *(26)*. Analysis of Puff Bar vaping products for the presence of *S*- and *R*-nicotine by ^{1}H-nuclear magnetic resonance spectroscopy, polarimetry and gas chromatography–mass spectrometry (GC/MS) confirmed the presence of both nicotine forms, but a slightly higher content of *S*-nicotine than *R*-nicotine. The authors speculated that the manufacturer might have added tobacco-derived nicotine, although further analysis would be necessary *(28)*.

Several methods have been proposed to differentiate nicotine derived from tobacco from synthetic nicotine. As synthetic *S*-nicotine is now available at high purity, it is difficult to differentiate the two; as the compounds are chemically identical, they cannot be differentiated by standard analytical techniques. Carbon isotope analysis has been proposed as a solution. Carbon has three isotopes,

^{12}C, ^{13}C and ^{14}C. ^{14}C has a half-life of 5700 years, a property that is used in radiocarbon dating of biological materials. ^{14}C is constantly replenished in the atmosphere by the sun's radiation and is then integrated into living plant matter, including tobacco plants and their natural products, such as nicotine. Synthetic nicotine is produced from petrochemical precursors that were formed in the earth millions of years ago and have a much lower ^{14}C content. For example, a ^{14}C analytical method has been developed to differentiate between natural and fossil chemical-derived vanillin, a popular flavour chemical *(67)*. Depending on the metabolic pathways involved, natural products may also contain a higher ratio of ^{13}C:^{12}C. High-temperature liquid chromatography coupled with isotope ratio mass spectrometry ("HT RPLC/IRMS") has become the standard approach for identifying foods adulterated with synthetic additives and can be used to differentiate between natural and synthetic caffeine, ethanol, sugars and other chemicals *(67,68)*. Cheetham et al. *(69)* used a ^{14}C method to compare samples of tobacco-derived and synthetic nicotine and found that the tobacco-derived samples contained 100% "modern" biocarbon, such that their carbon isotope distribution is identical to the current distribution in the earth's atmosphere. The synthetic nicotine samples contained only about 35% biocarbon, indicating that some natural precursors were probably used in their synthesis. The commercial synthetic nicotine preparations tested were found to be of high purity (> 99.9% nicotine content), containing only minor amounts of nicotine derivatives and degradants, fulfilling the US Pharmacopeia criteria for pharmaceutical-grade nicotine *(70)*. The commercial purified tobacco-derived nicotine samples were of similarly high purity, also fulfilling the US Pharmacopeia criteria for pharmaceutical-grade nicotine. The authors also devised methods to identify products containing mixtures of synthetic and tobacco-derived nicotine and a method for purifying nicotine from electronic cigarette liquids, an essential first step in the analysis of marketed products to detect the presence of carbon-based solvents (propylene glycol, glycerol), flavour chemicals and other additives in marketed products.

Qualitative and quantitative methods for analysis of hydrogen isotopes (hydrogen and deuterium) and nitrogen isotopes (^{15}N) in nicotine also revealed substantial differences between tobacco-derived nicotine obtained from various locations and from synthetic nicotine *(71,72)*. Thus, while significant advances have been made in analytical methods to discriminate synthetic from tobacco-derived nicotine, no standard method is yet available. The instrumentation and skills necessary to apply such methods are costly, and few countries have such capability.

3.2.3 Summary and discussion

Manufacturers have developed several methods for more efficient, more economical production of synthetic nicotine. Currently, two forms of synthetic nicotine are added to marketed products – racemic nicotine, consisting of 50% *S*-nicotine and 50% *R*-nicotine, and pure *S*-nicotine. The price of synthetic nicotine remains significantly higher than that of tobacco-derived nicotine. Consumers who use products containing racemic nicotine inhale much higher amounts of *R*-nicotine than users of tobacco-derived nicotine or pure *S*-nicotine, which raises questions about the long-term safety of such products. While *R*-nicotine is significantly less potent than *S*-nicotine in standard pharmacological assays and behavioural tests, the only toxicological studies of the effects of *R*-nicotine are studies of acute effects. There is evidence that *R*-nicotine differentially affects other pharmacological and toxicological targets, raising concern about unexpected toxicological effects. In none of the published pharmacological studies were animals exposed for longer than 1–2 weeks, and in none were subsequent pathological effects examined. None of the published studies addressed the effects of racemic nicotine, in which both *R*- and *S*-nicotine are present, and in none were their effects compared after inhalation and after ingestion, the routes through which consumer products dispense nicotine. Most of the studies of the effects of *R*- and *S*-nicotine were published between the 1970s and the 1990s. Toxicological methods have advanced significantly since then and should be used to examine the long-term effects of *R*-nicotine intake. Chemical analytical methods allow differentiation between synthetic and tobacco-derived nicotine; however, the methods have not been standardized and require substantial investment in advanced equipment and training. Analytical studies raise concern about inaccurate labelling of marketed products and undisclosed addition of tobacco-derived nicotine, probably added to increase addictiveness and increase profits, while health claims for synthetic nicotine are maintained. Tested commercial preparations of both purified tobacco-derived nicotine and synthetic nicotine fulfil US Pharmacopeia criteria for the purity of pharmaceutical-grade nicotine; however, not all currently marketed preparations have been compared. Because of the closely similar purity of synthetic and tobacco-derived preparations, claims of health attributed to synthetic nicotine and to purified tobacco-derived nicotine should be based on strong scientific evidence.

If regulators restrict the use of synthetic nicotine in marketed products, the chemical synthesis methods developed by manufacturers could be modified rapidly to generate nicotine analogues *(19)*. The tobacco industry has a long history of studying the addictive and reinforcing effects of nicotine-related tobacco alkaloids, including anabasine, nornicotine, anatabine, cotinine and myosmine *(19,59,73–76)*. Regulators should be aware that these analogues might be used to replace nicotine in marketed products.

3.3 The legal landscape

An unregulated market of synthetic nicotine products risks undermining public health progress in mitigating the harm of tobacco use *(2,77,78)*. For instance, lawmakers in the USA wrote a letter to the USFDA in November 2021, expressing concern that unregulated sale of synthetic nicotine products was "undermin[ing] efforts to reduce the continued popularity of youth vaping" *(79)*. Additionally, current marketing claims for certain synthetic products may mislead people who use those products by suggesting, for example, that they are safer than tobacco-derived nicotine products, even though such a claim is not supported by evidence.

If synthetic nicotine products remain unregulated, companies are likely to make a business choice to sell products containing synthetic rather than tobacco-derived nicotine (or at least claim to be doing so), undermining comprehensive regulation of novel tobacco and nicotine products *(77)*. Companies are aware that synthetic nicotine products are not covered by tobacco control laws in some countries. Two of the major global suppliers of synthetic nicotine, Hangsen and NGL, have both touted "[f]ewer restrictions for new market introductions" as one of the key benefits of synthetic nicotine *(48)*. Before recent changes to US law, an investment analyst in the USA referred to synthetic nicotine as a potential "golden ticket", as use of synthetic nicotine instead of tobacco-derived nicotine might mean "no FDA regulation, no tobacco taxes, no flavor restrictions, and no restrictions on direct to consumer e-commerce".[1] Puff Bar, which produces disposable e-cigarettes that are popular among young people, took advantage of the former regulatory gap in the USA, when, after a USFDA enforcement action, it relaunched its products in early 2021, claiming that its use of synthetic nicotine exempted it from regulation as a tobacco product *(80)*.

A key question for policy-makers is therefore whether products containing synthetic nicotine (or other, non-tobacco-derived nicotine alternatives) are covered by existing regulatory frameworks for tobacco products. This depends on the definitions of the terms used in relevant laws and whether those definitions are specific to (and limited to) tobacco-derived products. The WHO Framework Convention on Tobacco Control (WHO FCTC) defines "tobacco products" as "products entirely or partly made of the leaf tobacco as raw material which are manufactured to be used for smoking, sucking, chewing or snuffing", which would appear to exclude non-tobacco synthetic nicotine products *(81)*. The WHO FCTC language does not, however, prevent Member States from including synthetic nicotine products in the definition of "tobacco products" in their national laws or from otherwise including products containing synthetic nicotine in national (or subnational) tobacco control laws. Notably, the Conference of the Parties to the WHO FCTC has requested the Secretariat of the WHO FCTC "to advise, as appropriate, on the adequate classification of novel and emerging tobacco

1 Lavery MS. Tobacco synthetic nicotine bursts on to the scene. 2021 (available on request from the authors).

products such as heated tobacco products to support regulatory efforts and the need to define new product categories" *(81)*.

To better understand the global legal landscape for synthetic nicotine products and how Member States might revise existing regulatory definitions, including to comply with international obligations *(81)*, we surveyed the tobacco control laws of 210 countries and the EU to determine whether and how those laws apply to synthetic nicotine products.

3.3.1 Methods

Our review covered tobacco control regulations for market entry requirements (e.g. registration before marketing), sales restrictions (e.g. age restrictions for sales or restrictions on where retailers place tobacco products), packaging and labelling requirements (e.g. requirements for certain warning statements or images), and advertising regulations (e.g. restrictions on television advertisements). We excluded other kinds of tobacco-related laws, such as tax laws, smoke-free laws and regulation of flavours.

Most of the laws were found on the Tobacco Control Laws website *(82)*, which contains the laws of 210 countries and the EU. The amendment to the definition of "tobacco products" by the USA in March 2022 was not yet available on the website and was accessed elsewhere *(83,84,85)*. Accordingly, laws from a total of 211 jurisdictions were included in the analysis.

Of the 211 jurisdictions, 21 either did not have any laws available or did not have a version in English. Three English-speaking individuals with training in US law (two of the authors, MLB and PJZ, and Annamarie Beckmeyer) reviewed the relevant laws for the remaining 190 jurisdictions (189 countries and the EU). The EU directives are not binding law but are instead "legislative act[s] that set ... out a goal that all EU countries must achieve ..., [leaving] individual countries to devise their own laws on how to reach these goals". We coded the EU as a separate jurisdiction because of the importance of its Tobacco Products Directive to tobacco policy-making in Europe *(86)*.

MonQcle, legal research software *(87)*, was used to code laws for their application to any synthetic nicotine products. If the laws did apply to synthetic nicotine products, they were then coded for whether they applied to any synthetic nicotine product in addition to e-cigarettes and whether the covered synthetic nicotine products subject to market entry requirements, sales restrictions, packaging and labelling requirements and advertising restrictions.

3.3.2 Results

The phrasing of laws in some countries is broad enough to cover certain synthetic nicotine products or to cover such products more broadly. Tobacco control laws in many countries, however, do not clearly apply to such products (Table 3).

Table 3. Applicability of tobacco control laws in 211 jurisdictions to control of products containing synthetic nicotine

Coverage of products containing synthetic nicotine	Number of jurisdictions	Characteristics
No coverage	92	"Tobacco products" defined as products containing elements made from tobacco plants
Clear coverage of certain products	52	E-cigarettes (and other specific product types) defined to include nicotine derived from any source, but "tobacco product" otherwise limited to products made from tobacco plants
Broader coverage	29	"Tobacco products" defined to explicitly include synthetic nicotine or nicotine derived from any source
Unclear coverage	17	Product definitions refer to tobacco plant or smoke without expressly requiring that the products be made or derived from tobacco
Not available	21	

Laws that do not cover synthetic nicotine products

Of the 190 laws coded, 92 did not apply to any type of synthetic nicotine product. Many of the jurisdictions in this category had laws that define the products covered according to their tobacco content. For example, before March 2022, US law defined "tobacco products" (for the purposes of federal regulation) as products "made or derived from tobacco" *(84,85,88)*.

Some laws in this category did not expressly include the relevant terms, but the terms themselves suggested that synthetic nicotine products are probably not covered. For example, in some laws, the term "tobacco product" was used without a definition. Countries in the WHO African Region were the most likely to have laws that did not apply to any type of synthetic nicotine product.

Laws that clearly cover only certain synthetic nicotine products

In 52 jurisdictions, the laws include definitions broad enough to cover certain synthetic nicotine products –usually e-cigarettes – but not other currently marketed synthetic nicotine products such as pouches, toothpicks and chewing-gums. Many of these laws define "tobacco products" according to the tobacco content (as in the category above) but then separately define "electronic cigarettes" or similar terms without specifying the source or content of nicotine.

Other laws in this category do not define the relevant terms but include terms that can encompass e-cigarettes that contain synthetic nicotine. For example, in some laws, terms such as "electronic cigarettes" or "electronic nicotine delivery systems" are used without defining their limits. These terms are therefore probably broad enough to cover e-cigarettes that contain synthetic, rather than tobacco-derived, nicotine.

The laws of some jurisdictions apply to a limited extent to products other than e-cigarettes. For instance, some laws also cover herbal smoking products that contain no tobacco, which could leave room to include combustible products infused with synthetic nicotine.

Of the laws that cover only certain synthetic nicotine products, a few completely ban the sale and distribution of e-cigarettes. There is no clear geographical pattern of jurisdictions that have implemented laws covering only certain synthetic nicotine products. There does, however, appear to be a temporal trend, as most laws in this category were enacted in 2017 or later.

Laws that cover synthetic nicotine products more broadly

The laws of 29 jurisdictions are drafted broadly enough to cover all or most synthetic nicotine products that are currently marketed and also products that may emerge (Table 4). Only a few of these laws completely ban a broader range of synthetic nicotine products.

Table 4. Examples of national laws that include product definitions that cover all synthetic nicotine products

Country	Date	Comments and definitions
Republic of Moldova	2015	The Republic of Moldova adopted comprehensive revisions to its Tobacco Control Act to comply with its obligations under the WHO FCTC and to align its policies with those of the EU pursuant to the Moldova–EU Association Agreement. In addition to regulating "tobacco products", the revised law regulates "related products", defined as including "products made of plants for smoking *and* products that contain nicotine, including electronic cigarettes" [emphasis added] *(89)*.
Singapore	2010	Singapore's law has included the regulation of "tobacco substitutes" since 2010. Although the definition has been amended over time, it has consistently been used as a catch-all term to regulate products that contain nicotine (regardless of source) but are not included in the other defined categories in the Tobacco Act *(90)*.
USA	2022	In response to the introduction of synthetic nicotine products that claimed to be outside the reach of the US Tobacco Control Act, the USA amended the definition of "tobacco product" in the Act to include "any product...containing nicotine from any source, that is intended for human consumption" *(91)*.

Countries in the WHO European Region are most likely to have laws that cover a broader range of synthetic nicotine products. Most of the laws were passed after 2016.

For example, the law in the Republic of Moldova differentiates "related products" from "tobacco products" to include "products made of plants for smoking and products that contain nicotine, including electronic cigarettes", which provides broad coverage of existing and emerging synthetic nicotine products *(89)*. The law in Singapore includes a definition that provides broad coverage of synthetic nicotine products, "tobacco product" being defined as including "tobacco substitute", which in turn is defined as "any article, object or thing that contains nicotine", with no requirement that nicotine be derived from tobacco. The law expressly excludes from "tobacco substitute" "(a) a cigarette or cigar, or any other form of tobacco; (b) a tobacco derivative; (c) a mixture containing any form of tobacco or a tobacco derivative; (d) a therapeutic product registered under the Health Products Act" *(90)*.

The USA now regulates but does not explicitly ban synthetic nicotine products *(92)*. In March 2022, the Federal Food, Drug, and Cosmetic Act was amended to include synthetic nicotine products within USFDA tobacco product authorities. The definition of "tobacco product" now covers "any product made or derived from tobacco *or containing nicotine from any source* that is intended for human consumption, including any component, part, or accessory of a tobacco product" [emphasis added]. This definition thus covers all or most currently marketed synthetic nicotine products as well as products that may emerge. Under this law, synthetic nicotine products now require premarketing authorization from the USFDA before they can be legally sold. No synthetic nicotine products have yet received such authorization, but the USFDA reported that > 1 million marketing applications from > 200 companies had been received *(92)*. The USFDA refused 925 000 of the applications and accepted 8600 for further review.

Laws with unclear coverage of synthetic nicotine products

The laws of 17 jurisdictions were not clear about whether the definitions cover synthetic nicotine products. For example, some of the laws referred to tobacco when defining nicotine products but did not state whether application of the law was limited to nicotine derived from tobacco. The Tobacco Control Laws website did not have sufficient information in English on the laws in some countries for the authors to be able to determine whether the laws applied to products containing synthetic nicotine.

3.3.3 Discussion

Various legal adjustments could include synthetic nicotine products in the scope of tobacco control regulations. The approaches that some countries have adopted cover only e-cigarette or e-liquid synthetic nicotine products. These approaches do not include potential regulation of other kinds of synthetic nicotine products that are currently marketed or that may emerge, which will undermine comprehensive regulation of novel tobacco and nicotine products. As approaches in countries such as the Republic of Moldova, Singapore and, most recently, the USA show, however, legal adjustments could include the full range of currently marketed synthetic nicotine products and products that may emerge under tobacco control regulations.

Although our analysis was limited to synthetic nicotine, it provides an example of how the tobacco industry may seek to exploit gaps or uncertainty in laws to market new products or to evade tobacco-related regulations. Further work on the legal landscape of nicotine analogues may be useful to help countries in developing appropriate regulatory approaches *(73)*.

The description of the global legal landscape for synthetic nicotine products has several limitations. Only laws available in English and only tobacco control laws that cover market entry, sales restrictions, packaging and labelling and advertising were coded. The laws that were not available in English and other types of tobacco laws, such as tax laws, may include synthetic nicotine products. The coded laws generally did not include subnational jurisdictions, where the laws may define terms differently. The Tobacco Control Laws website may not be complete, as it may not include the most recently adopted laws or court decisions that affect the interpretation or enforceability of laws. Furthermore, the English versions of laws may not faithfully reflect the original versions.

Importantly, the laws we coded did not include laws to regulate products other than tobacco products. Even if synthetic nicotine products are not covered by a country's regulatory scheme for tobacco products, they may be subject to regulation as drugs (or drug–device combination products) or to other laws for consumer protection. Such laws may provide countries with opportunities to regulate synthetic nicotine products without changing their tobacco control laws *(76)*. For example, before the US law was amended in March 2022 to bring synthetic nicotine products under the law, public health groups urged the USFDA to regulate synthetic nicotine products as drugs *(93)*. Synthetic nicotine manufacturers such as Hangsen and NGL boast that their products "provide the same satisfaction smokers are seeking from their nicotine", which implicitly acknowledges that the products they sell are used for their effects as addictive drugs *(48)*. Most product websites include warnings or disclaimers, acknowledging that the nicotine in their products is addictive and may be hazardous. Additionally, the Australian Therapeutic Goods Administration requires a prescription for buying e-cigarettes containing nicotine *(94)*. Although this requirement appears to cover synthetic nicotine products, it is not imposed through Australia's tobacco control laws and is thus outside the scope of this analysis.

Finally, this analysis did not allow assessment of whether the requirements imposed through tobacco control laws or other kinds of law are enforced. Enforcement may vary within and between countries.

3.4 Recommendations for consideration by policy-makers

1. Countries in which there is a regulatory gap for synthetic nicotine products (as compared to products containing nicotine derived from tobacco) should consider amending their tobacco control laws to ensure that they include synthetic nicotine products.
2. Countries that choose to amend their tobacco control laws to cover synthetic nicotine products should consider legal adjustments that extend the coverage of the laws to the full range of synthetic nicotine

products that are currently marketed as well as products that may emerge. These may include products that contain synthetic nicotine analogues, other chemicals with similar properties or chemical systems that generate nicotine or analogues in situ.

3. Countries are advised to enforce standards for the purity of synthetic nicotine in products, preferably those of the European and US pharmacopoeias. Regulators should consider implementation of product standards to ban the mixing of tobacco-derived nicotine with synthetic nicotine in marketed products.
4. Policy-makers are advised to enforce uniform labelling rules for products containing nicotine, either natural or synthetic, and to declare the content of *S*-nicotine and, separately, the content of *R*-nicotine and any other nicotine analogue or any other chemical with similar properties.
5. Countries should consider banning synthetic nicotine products that contain *R*-nicotine, or any nicotine analogue apart from *S*-nicotine, at levels that exceed those in tobacco-based products, until the safety of consumption of these chemicals in such products is established.
6. Regulators should consider restricting marketing practices for promotion of synthetic nicotine as generally "tasteless and odourless", "purer" or "healthier" than purified tobacco-derived nicotine, unless scientific evidence to support such claims is provided.

3.5 Conclusions

Companies are marketing an increasingly wide range of synthetic nicotine products, which, if not regulated, may undermine work to reduce use of tobacco and nicotine addiction and the work of WHO Member States to regulate tobacco and nicotine products comprehensively. Knowledge about the effects on human health of synthetically derived nicotine in different types of consumer products is still incomplete. Although synthetic nicotine products are not clearly regulated under current tobacco control legislation in many countries, the laws in some countries have been updated to cover these products. The information presented above shows that countries could make various legal adjustments to close regulatory gaps for synthetic nicotine products, including adjustments that cover both the broad range of products currently on the market and those that might emerge.

References

1. What you need to know about new synthetic nicotine products. Washington DC: Truth Initiative; 2021 (https://truthinitiative.org/research-resources/harmful-effects-tobacco/what-you-need-know-about-new-synthetic-nicotine-products).
2. Ramamurthi D, Chau C, Lu Z, Rughoobur I, Sanaie K, Krishna P et al. Marketing of "tobacco-free" and "synthetic nicotine" products (white paper). Palo Alto (CA): Stanford University; 2022 (https://tobacco-img.stanford.edu/wp-content/uploads/2022/03/13161808/Synthetic-Nicotine-White-Paper-3-8-2022F.pdf).
3. Schmid T. A real up and comer: synthetic nicotine. Tobacco Asia, 14 February 2021 (https://www.tobaccoasia.com/features/a-real-up-and-comer-synthetic-nicotine/).
4. Ray K, Schuman V. Next Generation Labs CEO Vincent Schuman announces ZIA™ gum with TFN® synthetic nicotine. Chicago (IL): Cision® PR Web; 2017 (https://www.prweb.com/releases/2017/11/prweb14861999.htm).
5. Lucy TM Gum. Midletown (PA): Lucy Goods; 2023 (https://lucy.co/products/chewpark).
6. NicotinePicks. Kirksville (MO); 2023 (https://nicotinepicks.com/).
7. Crave Nicotine ToothPicks. Stockton (CA): Crave Nicotine; 2023 (https://web.archive.org/web/20210609132039/https://podalifestyle.com/).
8. Poda and our flagship Beyond Burn™ Poda Pods are set to revolutionize the heat-not-burn industry. Vancouver (BC): Poda Lifestyle and Wellness Ltd; 2021 (https://web.archive.org/web/20210609132039/https://podalifestyle.com/).
9. Outlaw Dip. Deer Park (NY): Outlaw Dip Co.; 2023 (https://outlawdip.com/faq/).
10. Ronin Smokes. Cambridge (Ont): Ronin Smokes; 2023 (https://www.roninsmokes.com/).
11. The Puff Bar. New & improved. Gendale (CA): Puff Bar; 2021 (https://web.archive.org/web/20210726184612/https://puffbar.com/collections/puff-bar).
12. Juice Head pouches. Huntington Beach (CA): Juice Head; 2023 (https://juicehead.co/collections/juice-head-pouches).
13. Pachamama SYN Vape Disposable – 1500 puffs. Louiville (KY): VaporFi; 2023 (https://www.vaporfi.com/pachamama-synthetic-disposable-vape-pen/).
14. BidiTM Vapor launches BidiTM Pouch, a nicotine delivery product in a tin pack. Jacksonville Beach (FL): QRX Digital; 2020 (https://www.newswire.com/news/biditm-vapor-launches-biditm-pouch-a-nicotine-delivery-product-in-a-21245979).
15. Can't light up? ZIA™. Boca Raton (FL): Next Generation Labs; 2018 (https://www.nextgenerationlabs.com/wp-content/uploads/2017/07/ZIA-gum-presentation-final.pptx).
16. VAPORX. Seoul: VAPORX; 2022 (http://vaporx.co.kr/22).
17. Zhang H, Pang Y, Luo Y, Li X, Chen H, Han S et al. Enantiomeric composition of nicotine in tobacco leaf, cigarette, smokeless tobacco, and e-liquid by normal phase high-performance liquid chromatography. Chirality. 2018;30(7):923–31. doi: 10.1002/chir.22866.
18. Pictet A, Rotschy A. Synthese des Nicotins [Synthesis of nicotines]. Ber Dtsch Chem Gesells. 1904;37(2):1225–35. doi:10.1002/CBER.19040370206.
19. Wagner FF, Comins DL. Recent advances in the synthesis of nicotine and its derivatives. Tetrahedron. 2007;63(34):8065–82. doi:10.1016/j.tet.2007.04.100.
20. Anderson H. Manufacture of nicotine. In: British American Tobacco Records; 1964. p. 100048807-8 (https://www.industrydocuments.ucsf.edu/docs/lgcy0212).
21. Moates RF. Feasibility of synthetic nicotine production. In: RJ Reynolds Records; Master Settlement Agreement 1967. p. 500613486-9 (https://www.industrydocuments.ucsf.edu/docs/srhn0096).
22. Southwick E. Synthesis of nicotine. In: Myers L, editor. Liggett & Myers Records 1978. p. lg0292754-lg5 (https://www.industrydocuments.ucsf.edu/docs/gpyw0014).

23. Ye X, Zhang Y, Song X, Liu Q. Research progress in the pharmacological effects and synthesis of nicotine. Chem Select. 2022;7(12):e202104425. doi:10.1002/slct.202104425.
24. Arnold M, inventor; Next Generation Labs, LLC, assignee. Process for the preparation of (R,S)-nicotine. US patent 9,556,142. 2017 (https://patents.google.com/patent/US20160115150A1/en?oq=20160115150).
25. Arnold M, inventor; Next Generation Labs LLC, Kaival Labs LLC, assignee. Nicotine replacement therapy products comprising synthetic nicotine. US patent 10,610,526. 2020.
26. Hellinghausen G, Lee JT, Weatherly CA, Lopez DA, Armstrong DW. Evaluation of nicotine in tobacco-free-nicotine commercial products. Drug Test Anal. 2017;9(6):944–8. doi:10.1002/dta.2145.
27. Puff Bar. Tobacco free. Los Angeles (CA): Cool Clouds Distribution Inc.; 2021 (https://puffbar.com/pages/about-puff-bar).
28. Duell AK, Kerber PJ, Luo W, Peyton DH. Determination of (R)-(+)- and (S)-(–)-nicotine chirality in Puff Bar e-liquids by (1)H NMR spectroscopy, polarimetry, and gas chromatography–mass spectrometry. Chem Res Toxicol. 2021;34(7):1718–20. doi:10.1021/acs.chemrestox.1c00192.
29. Weber B, Pan B, inventors; Siegfried AG, Contraf-Nicotex-Tobacco GmbH, assignee. Enantiomeric separation of racemic nicotine by addition of an o,o′-disubstituted tartaric acid enantiomer patent US20200331883A1. 2019 06/27/2019 (https://patents.google.com/patent/WO2019121649A1/en).
30. Weber BT, Lothschütz C, Pan B, inventors; Siegfried AG. Contraf-Nicotex-Tobacco GmbH assignee. Preparation of racemic nicotine by reaction of ethyl nicotinate with n-vinylpyrrolidone in the presence of an alcoholate base and subsequent process steps USA 2020 (https://patents.google.com/patent/US20200331884A1).
31. S-isomer tobacco free nicotine. Irvine (CA): Five Pawns; 2020 (https://web.archive.org/web/20200811035816/https://fivepawns.com/blogs/five-pawns-news-events/s-isomer-tobacco-free-nicotine).
32. Synthetic nicotine. Plainville (CT): Tea Time Eliquid Co.; 2021 (https://web.archive.org/web/20211028035558/https://teatimeliquid.com/pages/synthetic-nicotine).
33. Towards a clean nicotine future. London: Zanoprima; 2023 (https://www.zanoprima.com/).
34. McCague R, Narasimhan AS, inventors; Zanoprima Lifesciences Limited (London, GB), assignee. Process of making (S)-nicotine. USA patent 10,913,962. 2021 02/09/2021 (https://patents.google.com/patent/US10913962B2).
35. What is MOTiVO™? Seoul: Hangsen International Group Ltd; 2023 (https://perma.cc/BE6K-P3US).
36. Method for preparing nicotine of high optical purity. (https://patents.google.com/patent/WO2022105482A1/en).
37. Altria announces definitive agreement to acquire NJOY Holdings, Inc. Richmond (VA): Altria Group; 2023 (https://investor.altria.com/press-releases/news-details/2023/Altria-Announces-Definitive-Agreement-to-Acquire-NJOY-Holdings-Inc/default.aspx).
38. Willis B, Ahmed MM, Freund W, Sawyer D, inventors; NJOY, LLC, assignee. Synthesis and resolution of nicotine. USA patent US10759776B2. 2020 09/01/2020 (https://image-ppubs.uspto.gov/dirsearch-public/print/downloadPdf/10759776).
39. SyNic Pure Nicotine 1000mg/mL. Newbury Park (CA): Nicotine River; 2023 (https://nicotineriver.com/collections/synic%E2%84%A2-nicotine/products/synic-pure-nicotine-1000mg-ml).
40. Rossel S. Synthetic nicotine is gaining acceptance. Tobacco Reporter, 1 December 2019 (https://tobaccoreporter.com/2019/12/01/mirror-image/).

41. tfn® Nicotine. Not derived from tobacco leaf, stem, or waste dust. Phoenix (AZ): Liquid Nicotine Wholesalers; 2023 (https://liquidnicotinewholesalers.com/tfn-pure-liquid-nicotine.html).
42. CNT Nicotine. Phoenix (AZ): Liquid Nicotine Wholesalers; 2023 (https://liquidnicotinewholesalers.com/cnt-nicotine.html).
43. Cultra™ Pure Nicotine. Phoenix (AZ): Liquid Nicotine Wholesalers; 2023 (https://liquidnicotinewholesalers.com/cultra-pure-liquid-nicotine.html).
44. PurNic Pure Nicotine 1000mg/mL. Thousand Oaks (CA): Nicotine River; 2023 (https://nicotineriver.com/collections/purnic%E2%84%A2-nicotine/products/purnic-pure-nicotine-1000mg-ml).
45. Next Generation Labs LLC has been granted a Notice of Allowance from China for its Process for The Preparation of (R-S) Synthetic Nicotine – Patent #201580069647.2. Chicago (IL): Cision PRWeb; 2021 https://www.prweb.com/releases/next_generation_labs_llc_has_been_granted_a_notice_of_allowance_from_china_for_its_process_for_the_preparation_of_r_s_synthetic_nicotine_patent_201580069647_2/prweb17809747.htm).
46. Zanoprima Lifesciences Ltd v. Hangsen International Group Ltd (6:22-cv-00268) District Court, W.D. Texas; 2022 (https://perma.cc/SV6B-VYDA).
47. Hangsen releases synthetic nicotine. Tobacco Reporter, 11 Setember 2020 (https://tobaccoreporter.com/2020/09/11/hangsen-and-geek-vape-release-synthetic-nicotine-product/).
48. What is TFN®. Boca Raton (FL): Next Generation Labs; 2021 (http://www.nextgenerationlabs.com/).
49. Specific nicotine isomers ratios could potentially offer nicotine use at satisfying but non-addictive levels as revealed by Next Generation Labs CEO Vincent Schuman. Chicago (IL): Cision PRWeb; 2017 (https://www.prweb.com/releases/2017/11/prweb14911138.htm).
50. Shimada A, Iizuka H, Kawaguchi T, Yanagita T. [Pharmacodynamic effects of d-nicotine – Comparison with l-nicotine]. Nihon Yakurigaku Zasshi. 1984;84(1):1–10. PMID:6489864.
51. Ikushima S, Muramatsu I, Sakakibara Y, Yokotani K, Fujiwara M. The effects of d-nicotine and l-isomer on nicotinic receptors. J Pharmacol Exp Ther. 1982;222(2):463–70. PMID:7097565.
52. Martin BR, Aceto MD. Nicotine binding sites and their localization in the central nervous system. Neurosci Biobehav Rev. 1981;5(4):473–8. doi:10.1016/0149-7634(81)90017-8.
53. Zhang X, Gong ZH, Nordberg A. Effects of chronic treatment with (+)- and (–)-nicotine on nicotinic acetylcholine receptors and N-methyl-D-aspartate receptors in rat brain. Brain Res. 1994;644(1):32–9. doi:10.1016/0006-8993(94)90343-3.
54. Meltzer LT, Rosecrans JA, Aceto MD, Harris LS. Discriminative stimulus properties of the optical isomers of nicotine. Psychopharmacology (Berl). 1980;68(3):283–6. doi:10.1007/BF00428116.
55. Clarke PB, Kumar R. Characterization of the locomotor stimulant action of nicotine in tolerant rats. Br J Pharmacol. 1983;80(3):587–94. doi:10.1111/j.1476-5381.1983.tb10733.x.
56. Kumar R, Pratt JA, Stolerman IP. Characteristics of conditioned taste aversion produced by nicotine in rats. Br J Pharmacol. 1983;79(1):245–53. doi:10.1111/j.1476-5381.1983.tb10518.x.
57. Romano C, Goldstein A, Jewell NP. Characterization of the receptor mediating the nicotine discriminative stimulus. Psychopharmacology (Berl). 1981;74(4):310–5. doi:10.1007/BF00432737.
58. Rosecrans JA, Meltzer LT. Central sites and mechanisms of action of nicotine. Neurosci Biobehav Rev. 1981;5(4):497–501. doi:10.1016/0149-7634(81)90020-8.

59. Goldberg SR, Risner ME, Stolerman IP, Reavill C, Garcha HS. Nicotine and some related compounds: effects on schedule-controlled behaviour and discriminative properties in rats. Psychopharmacology (Berl). 1989;97(3):295–302. doi:10.1007/BF00439441.
60. Jacob P 3rd, Benowitz NL, Copeland JR, Risner ME, Cone EJ. Disposition kinetics of nicotine and cotinine enantiomers in rabbits and beagle dogs. J Pharm Sci. 1988;77(5):396–400. doi:10.1002/jps.2600770508.
61. Nwosu CG, Godin CS, Houdi AA, Damani LA, Crooks PA. Enantioselective metabolism during continuous administration of S-(–)- and R-(+)-nicotine isomers to guinea-pigs. J Pharm Pharmacol. 1988;40(12):862–9. doi:10.1111/j.2042-7158.1988.tb06289.x.
62. Nwosu CG, Crooks PA. Species variation and stereoselectivity in the metabolism of nicotine enantiomers. Xenobiotica. 1988;18(12):1361–72. doi:10.3109/00498258809042260.
63. Crooks PA, Godin CS. N-Methylation of nicotine enantiomers by human liver cytosol. J Pharm Pharmacol. 1988;40(2):153–4. doi:10.1111/j.2042-7158.1988.tb05207.x.
64. Yang J, Chen YK, Liu ZH, Yang L, Tang JG, Miao MM et al. Differences between the binding modes of enantiomers S/R-nicotine to acetylcholinesterase. RSC Adv. 2019;9(3):1428–40. doi:10.1039/c8ra09963d.
65. Saareks V, Mucha I, Sievi E, Vapaatalo H, Riutta A. Nicotine stereoisomers and cotinine stimulate prostaglandin E2 but inhibit thromboxane B2 and leukotriene E4 synthesis in whole blood. Eur J Pharmacol. 1998;353(1):87–92. doi:10.1016/s0014-2999(98)00384-7.
66. Thuerauf N, Kaegler M, Dietz R, Barocka A, Kobal G. Dose-dependent stereoselective activation of the trigeminal sensory system by nicotine in man. Psychopharmacology (Berl). 1999;142(3):236–43. doi:10.1007/s002130050885.
67. Mao H, Wang H, Hu X, Zhang P, Xiao Z, Liu J. One-pot efficient catalytic oxidation for bio-vanillin preparation and carbon isotope analysis. ACS Omega. 2020;5(15):8794–803. doi:10.1021/acsomega.0c00370.
68. Zhang L, Kujawinski DM, Federherr E, Schmidt TC, Jochmann MA. Caffeine in your drink: natural or synthetic? Anal Chem. 2012;84(6):2805–10. doi:10.1021/ac203197d.
69. Cheetham AG, Plunkett S, Campbell P, Hilldrup J, Coffa BG, Gilliland S 3rd et al. Analysis and differentiation of tobacco-derived and synthetic nicotine products: addressing an urgent regulatory issue. PLoS One. 2022;17(4):e0267049. doi:10.1371/journal.pone.0267049
70. Nicotine (USP 29-NF 24). North Bethesda (MD): United States Pharmacopeial Convention; 2020 (https://online.uspnf.com/uspnf/document/1_GUID-3D851985-2C16-408D-99E1-F241A9767168_4_en-US).
71. Liu B, Chen Y, Ma X, Hu K. Site-specific peak intensity ratio (SPIR) from 1D (2)H/(1)H NMR spectra for rapid distinction between natural and synthetic nicotine and detection of possible adulteration. Anal Bioanal Chem. 2019;411(24):6427–34. doi:10.1007/s00216-019-02023-6.
72. Han S, Cui L, Chen H, Fu Y, Hou H, Hu Q et al. Stable isotope characterization of tobacco products: a determination of synthetic or natural nicotine authenticity. Rapid Commun Mass Spectrom. 2023;37(3):e9441. doi:10.1002/rcm.9441.
73. Vagg R, Chapman S. Nicotine analogues: a review of tobacco industry research interests. Addiction. 2005;100(5):701–12. doi:10.1111/j.1360-0443.2005.01014.x.
74. Clemens KJ, Caillé S, Stinus L, Cador M. The addition of five minor tobacco alkaloids increases nicotine-induced hyperactivity, sensitization and intravenous self-administration in rats. Int J Neuropsychopharmacol. 2009;12(10):1355–66. doi:10.1017/S1461145709000273.
75. Hall BJ, Wells C, Allenby C, Lin MY, Hao I, Marshall L et al. Differential effects of non-nicotine tobacco constituent compounds on nicotine self-administration in rats. Pharmacol Biochem Behav. 2014;120:103–8. doi:10.1016/j.pbb.2014.02.011.

76. Harris AC, Tally L, Muelken P, Banal A, Schmidt CE, Cao Q et al. Effects of nicotine and minor tobacco alkaloids on intracranial-self-stimulation in rats. Drug Alcohol Depend. 2015;153:330–4. doi:10.1016/j.drugalcdep.2015.06.00.
77. Zettler PJ, Hemmerich N, Berman ML. Closing the regulatory gap for synthetic nicotine products. Boston Coll Law Rev. 2018;59(6):1933–82. PMID:30636822.
78. Jordt SE. Synthetic nicotine has arrived. Tob Control. 2023;32(e1):e113–7. doi:10.1136/tobaccocontrol-2021-056626.
79. Merkley JA, Kaine T, Warren E, Brown S, Markey EJ, Baldwin T et al. [Letter 16 November 2021]. Washington DC: United States Senate; 2021 (https://www.merkley.senate.gov/imo/media/doc/21.11.16%20Signed%20Synthetic%20Nicotine%20Letter%20to%20FDA.pdf).
80. Maloney J. Puff Bar defies FDA crackdown on fruity e-cigarettes by ditching the tobacco. The Wall Street Journal, 2 March 2021 (https://www.wsj.com/articles/puff-bar-defies-fda-crackdown-on-fruity-e-cigarettes-by-ditching-the-tobacco-11614681003).
81. FCTC/COP8(22) Novel and emerging tobacco products. Geneva: World Health Organization; 2018 (https://fctc.who.int/who-fctc/governance/conference-of-the-parties/eight-session-of-the-conference-of-the-parties/decisions/fctc-cop8(22)-novel-and-emerging-tobacco-products).
82. Tobacco control laws. Washington DC: Campaign for Tobacco-Free Kids; 2023 (https://www.tobaccocontrollaws.org/).
83. H.R.2471 – Consolidated Appropriations Act, 2022. Washington DC: Congress.gov; 2022 (https://www.congress.gov/bill/117th-congress/house-bill/2471/text).
84. 21 USC 321: Definitions; generally. In: United States Code. Washington DC: Office of the Law Revision Counsel;2022 (https://uscode.house.gov/view.xhtml?req=granuleid:USC-prelim-title21-section321&num=0&edition=prelim).
85. Section 101 of the Tobacco Control Act – Amendment of Federal Food, Drug, and Cosmetic Act (FDCA). Silver Spring (MD): Food and Drug Administration; 2023 (https://www.fda.gov/tobacco-products/rules-regulations-and-guidance/section-101-tobacco-control-act-amendment-federal-food-drug-and-cosmetic-act-fdca).
86. Types of legislation. Brussels: European Union; 2023 (https://european-union.europa.eu/institutions-law-budget/law/types-legislation_en).
87. About LawAtlas.org. Philadelphia (PA): Temple University, Beasley School of Law, Center for Public Health Law Research; 2023 (https://lawatlas.org/page/lawatlas-about).
88. Family Smoking Prevention and Tobacco Control Act and Federal Retirement Reform. Public law 111-31 – June 22, 2009. WashingtonDC: Government Printing Office Act 2009 (https://www.govinfo.gov/content/pkg/PLAW-111publ31/pdf/PLAW-111publ31.pdf).
89. Parlamentul Lege Nr. 278 din 14–12-2007 privind controlul tutunului [Parliamentary law no. 278 of 14 12 2007 on tobacco]. Chişinău; Republic of Moldova; 2023 (https://www.legis.md/cautare/getResults?doc_id=128322&lang=ro).
90. Tobacco (Control of Advertisements and Sale) Act 1993. 2020 revised edition. Singapore: Legislation Division of the Attorney-General's Chambers of Singapore; 2023 (https://sso.agc.gov.sg/Act/TCASA1993).
91. Lipstein A, Zeller M. E-cigarette companies found a loophole in synthetic nicotine – it won't stop the FDA. The Hill, 7 April 2022 (https://thehill.com/opinion/healthcare/3260879-e-cigarette-companies-found-a-loophole-in-synthetic-nicotine-it-wont-stop-the-fda/).
92. Regulation and enforcement of non-tobacco nicotine (NTN) products. Silver Spring (MD): Food and Drug Administration; 2023 (https://www.fda.gov/tobacco-products/products-ingredients-components/regulation-and-enforcement-non-tobacco-nicotine-ntn-products).

93. American Academy of Pediatrics, American Cancer Society Cancer Action Network, American Heart Association, American Lung Association, Campaign for Tobacco Free Kids, Parents Against Vaping E-cigarettes et al. Letter to Dr Janet Woodcock, Acting Commissioner, US Food and Drug Administration. Re: Synthetic nicotine and Puff Bar. Washington DC: Tobacco Free Kids; 2021 (https://www.tobaccofreekids.org/assets/content/what_we_do/federal_issues/fda/regulatory/2021_03_18_puff-bar-synthetic-nicotine.pdf).
94. TGA confirms nicotine e-cigarette access is by prescription only. Canberra (ACT): Department of Health and Aged Care, Therapeutic Goods Administration; 2020 (https://www.tga.gov.au/news/media-releases/tga-confirms-nicotine-e-cigarette-access-prescription-only).

4. Nicotine pouches: characteristics, use, harmfulness and regulation

Charlotte GGM Pauwels, National Institute for Public Health and the Environment, Centre for Health Protection, Bilthoven, Netherlands (Kingdom of the)

Reinskje Talhout, National Institute for Public Health and the Environment, Centre for Health Protection, Bilthoven, Netherlands (Kingdom of the)

Jennifer Brown, Johns Hopkins Bloomberg School of Public Health, Baltimore (MD), USA

Rula Cavaco Dias, Healthier Populations Division, Health Promotion Department, No Tobacco Unit, World Health Organization, Geneva, Switzerland

Ranti Fayokun, Healthier Populations Division, Health Promotion Department, No Tobacco Unit, World Health Organization, Geneva, Switzerland

Contents

Key findings, challenges and regulatory implications

- Nicotine pouches have recently become available in many markets worldwide, and their sales are growing rapidly.
- Nicotine pouches deliver sufficient nicotine to induce and sustain nicotine addiction.
- Nicotine pouches have attractive properties, such as appealing flavours, and can be used discretely without the stigma of smoking.
- Nicotine is harmful to health, including to the nervous and cardiac systems.

- There are few data on nicotine pouches because they have been on the market for only a short time. A cautionary approach is warranted, given their similarities to conventional oral tobacco products, in particular snus.
- Nicotine pouches are not regulated or not specifically regulated in several jurisdictions. Some countries had already made their regulations and laws "future-proof" and resilient, so that nicotine pouches are regulated under existing laws. Others have recently updated their laws, while some retain definitions that refer only to conventional tobacco products.

Keywords: nicotine products, nicotine pouches, characteristics, harmfulness, regulation, regulatory mechanisms

4.1 Introduction

In the past decade, novel and emerging nicotine and tobacco products, such as electronic nicotine delivery systems (ENDS), non-nicotine delivery systems (ENNDS) and heated tobacco products (HTPs) have proliferated on markets globally. Some of these products, such as ENDS, have also been marketed and promoted to children and adolescents by tobacco and related industries *(1,2)*. Since about 2018, another category of products, commonly known as nicotine pouches, has been introduced onto several markets, as the tobacco industry continues to expand its portfolio of novel and emerging nicotine and tobacco products *(3)*. Other names used to describe these products include "tobacco-free nicotine pouches", "tobacco leaf-free pouches" and "tobacco-derived nicotine pouches"; in this paper, they are referred to as "nicotine pouches". In some jurisdictions, such as the United States of America (USA), they are referred to as "white pouches".

Synthetic nicotine is becoming increasingly popular, although most nicotine-containing products on the market in the USA reportedly contain tobacco-derived nicotine *(4)*. Nicotine pouches are pre-portioned pouches that contain nicotine. They are similar to conventional smokeless tobacco products such as snus in some respects, including appearance, inclusion of nicotine and manner of use (placing them between the gum and lip); however, unlike snus, which contains tobacco, nicotine pouches reportedly do not contain tobacco but rather cellulose powder and some other ingredients. The nicotine may, however, have been extracted from tobacco and may therefore contain substances originally present in the tobacco, as in ENDS. They have been promoted as "tobacco-free", which could be misleading if the nicotine was extracted from tobacco.

The pouches are available in flavours similar to those in, for example, ENDS, ENNDS and conventional smokeless tobacco products. These flavours can enhance the effects of nicotine by sustaining use and can improve palatability and increase their appeal to adults and especially to young people, including nicotine naive adolescents. Some ingredients in nicotine pouches, such as alkaline agents which increase pH, may increase the delivery of nicotine *(5)*. Some nicotine pouches are marketed as containing synthetic nicotine, usually a racemic mixture of *S*- and *R*- nicotine isomers; a few are stereoselective, containing more of the more potent *S*-isomer, which predominates in the tobacco plant. (See also Paper 2.) Little is known about the pharmacological and metabolic effects of *R*-nicotine in humans *(6)*. Until 2022, the definition of tobacco products of the US Food and Drug Administration (USFDA) included tobacco-derived nicotine, and products containing synthetic nicotine were not legally considered tobacco products. The definition was changed in 2022, since when the USFDA has regulatory authority over tobacco products containing nicotine from any source that are not used for therapeutic purposes *(7)*. In some other jurisdictions, nicotine products are not considered tobacco products unless they are explicitly included in the tobacco law. (See e.g. section 6.2.)

Nicotine pouches were first introduced in Europe but are now available in other countries, such as Indonesia, Kenya, Pakistan and the USA and some countries in the WHO Western Pacific Region. Some of these countries have sought technical assistance from WHO to address these products. Sales of nicotine pouches are increasing rapidly in many parts of the world *(5,8)*, including Denmark, Norway, Sweden and the USA. For example, the sales of nicotine pouches increased from US$ 642 000 in 2016 to US$ 52 million in 2018 in the USA *(5)*, and sales are expected to increase in European countries such as Austria, Croatia, Germany and the United Kingdom *(8)*. Euromonitor International reported an estimate that, globally, 6.8 billion units of nicotine pouches had been sold in 2021, representing more than a 2000% increase over its estimated retail volume in 2018. It was further estimated that, by the end of 2023, projected global retail volume sales will amount to more than 11 billion units *(9)*.

Introduction of new products that closely resemble traditional tobacco and nicotine products poses serious regulatory challenges in all WHO regions. Many manufacturers and retailers promote them as "healthy alternatives", and these products are often advertised with themes that appeal to young people *(10,11)*. A large collection of nicotine pouch advertisements is available online *(11)*.

Manufacturers have attempted to persuade regulators to classify nicotine pouches as non-tobacco products, as it is sometimes unclear whether these products are included in tobacco regulations or whether they occupy a regulatory "grey area" *(6)*. For example, they are often promoted as "non-tobacco" products, "white pouches" and "tobacco-free products". In some countries, especially low- and middle-income countries, manufacturers of these products claim that the nicotine contained in them is not derived from tobacco and therefore claim that the products fall outside the scope of tobacco control law *(6)*. Tobacco manufacturers also seek regulatory exemption for newer nicotine and tobacco products, including nicotine pouches, as in the case of Lyft in Kenya. The Lyft product was approved for sale in Kenya by the country's drug regulatory authority and has been available since July 2019 *(12)*. Several health advocacy groups in Kenya submitted a petition to the national Cabinet Secretary for Health, urging him to ban the Lyft nicotine pouches, arguing that they had been allowed onto the Kenyan market illegally. Sales were suspended once the rationale for approval of the pouches as a drug was questioned by the Cabinet Secretary, who subsequently informed British American Tobacco (BAT) that Lyft had to adhere to Kenya's requirements for tobacco products. Health advocacy groups now insist these products should not be available at all *(12)*.

Member States have sought technical assistance from WHO on defining nicotine pouch products and the knowledge and evidence available on these products, which includes the potential and actual risks associated with the products, their characteristics and how they are regulated in countries. This paper summarizes the known characteristics of nicotine pouches, the users of the product, the potential risks of their use and mechanisms for regulating nicotine pouches. This information, from the scientific literature, internet searches, the web pages of manufacturers and market data on nicotine products, will improve regulators' understanding of these products, country experience and regulatory challenges. The paper also provides guidance on regulatory options for nicotine pouches and some recommendations for consideration by countries.

4.2 Methods section

A search was conducted in the bibliographic database PubMed and other sources (e.g. general web search, specialized search on Euromonitor International, ECigIntelligence and Tobacco Intelligence, web pages of manufacturers and market data on nicotine products). Peer reviewed publications up to March 2023 were included. These were initially screened on title and abstract, and then further considered for full review. Keywords that were searched included "nicotine products", "nicotine pouches", "tobacco leaf-free pouches", "tobacco-derived nicotine pouches", "non-tobacco products", "white pouches", "tobacco-free products", "characteristics", "harmfulness", "regulation" and "regulator mechanisms".

Further, in 2020, WHO distributed a questionnaire to WHO regional advisors in all six regions to elicit country experiences with nicotine pouches, the regulatory mechanisms in place and difficulties found in regulation. A further questionnaire was disseminated to the WHO Global Tobacco Regulators Forum in 2021, who were given three weeks to complete the questionnaire and return to WHO. A subsequent questionnaire was formulated, and data collection done in 2022. Information was informally sought from regulators in European Union Member States via email and a review was conducted on tobacco control laws in the Tobacco Control Laws Databases of WHO and the Campaign for Tobacco Free Kids. 124 national laws were reviewed in total.

4.3 Characteristics of the products

Different brands of nicotine pouches and different products within a brand differ in weight, nicotine concentration and bioavailability, flavour and pouch size. Like traditional tobacco-containing snus, 20–25 nicotine pouches are typically packed in a pocket-sized tin (Fig. 1). In some countries, the tins may have a compartment for discarding used pouches (e.g. Zyn, Ace) *(13)*. The brand names, flavours and pouch sizes (e.g. "slim") are often listed on the lid. In some countries, the nicotine content is described as "strength" on a dot or a numerical scale (e.g. 4 out of 5; Fig. 1, right). The nicotine content varies from brand to brand and may be expressed in mg or mg/pouch. The lack of a requirement for standardized labelling and the resulting variety of expression of nicotine concentration probably confuses consumers. A warning is often placed on the lid or the bottom of the tin. Although such warnings are not required if the product is not regulated as a tobacco product, it may give the false impression that the company is abiding by the provisions under tobacco product regulations.

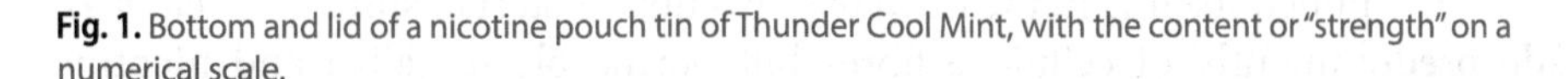

Fig. 1. Bottom and lid of a nicotine pouch tin of Thunder Cool Mint, with the content or "strength" on a numerical scale.

Photo credit: WHO

A single nicotine pouch weighs 149–800 mg *(5,14)* and generally contains nicotine at 3–50 mg/g, equal to a nicotine dose of 2–32.5 mg per pouch *(3,15,16)*, whereas a traditional portion of snus weighs 0.3–1.13 g, with a nicotine dose of 6.81–20.6 mg/g wet tobacco *(17)*. Some nicotine pouches with exceptionally high nicotine levels have been reported, e.g. pouches with up to 120 mg/g nicotine have been reported on the Estonian market, the strongest products coming from the Russian Federation *(8)*. A nicotine content of 1.29–6.11 mg per pouch was measured in 37 brands (2–6 mg/pouch) from six manufacturers, with 1.12–47.2% moisture content, pH 6.9–10.1 and 7.7–99.2% free-base nicotine, which is more bio-available than protonated nicotine (see section 5.2) *(5)*. The nicotine in these products is either derived from tobacco or synthesized. The brands offer a variety of flavours, such as fruity and sweet, but also coffee and menthol. In some cases, the flavours are combined and described for example as "a balanced combination of sweet and tart pineapple with creamy coconut and a nutty undertone" (Lyft) and given "concept" names, such as "tropic breeze" (Velo). Mint flavours appear to be used most widely in the USA, representing 54.6% of the total US nicotine pouch market in 2019 *(18)*. An increase in sales of fruit-flavoured nicotine pouches was observed between January 2019 and June 2020 *(19)*. Cooling and fruit categories dominate the market, representing almost 70% of the flavours in these products *(20)*. Nicotine pouches are available in several sizes, the smaller ones promoted as "slim" or "mini" on the package.

The pouch itself is made of water-insoluble material, similar to tea bags, made predominantly of cellulose fibres but permeable to saliva and nicotine *(21)*. The pouch contains an off-white or white powder containing either salts consisting of nicotine and an acid or free-base nicotine (Fig. 2). Other ingredients in addition to nicotine include cellulose, water, salt and other additives, such as pH-adjusting agents, filler, noncaloric sweeteners, a stabilizer (hydroxypropyl cellulose) and flavourings *(3,22)*.

The user places one nicotine pouch under the upper lip, where nicotine and flavours are released. Brand websites and web shops advise retaining the pouch for a minimum of 5 min to up to 1 h *(23)*. Shortly after the pouch is placed in the mouth, a tingling sensation is felt (due to the nicotine) that can last for up to 15 min *(22)*. Data from surveys and websites indicate that Zyn users consume 10–12 portions *(22)* daily *(24)*.

Fig. 2. Nicotine pouch package labelled "slim", a nicotine pouch and its content

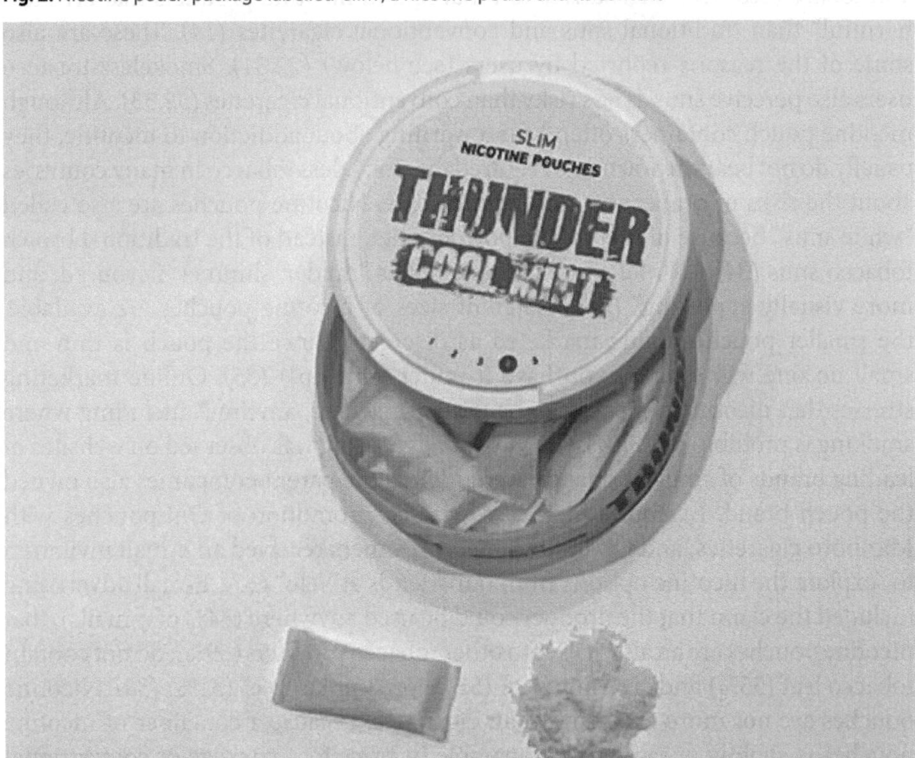

Photo credit: WHO

4.4 Marketing

The nicotine pouches currently on the market are produced mainly by large tobacco manufacturers such as BAT (Lyft, Velo, Zonnic) *(25)*, Altria Group Inc. (On!), Swedish Match (Zyn, G.4) *(21)*, Imperial Brands (Skruf, Knox, ZoneX) *(26,27)*, Philip Morris International (Shiro, Sirius), Swisher (Rogue), and Japan Tobacco International (Nordic Spirit) *(3,28)*. In the USA, the Federal Trade Commission reported on nicotine pouches in 2021 for the first time *(29)*, when the companies sold 140.7 million units of such products in the USA, for US$ 420.5 million. Nicotine pouch sales increased from 163 178 packages of 15–20 pouches (US$ 709 635) in July 2016 to 45 965 455 units (US$ 216 886 819) in June 2020 *(19)*. The highest US market share in 2020 was that of Swedish Match (78.7%), followed by Altria (10%) and BAT (7.6%). Small companies also manufacture nicotine pouches, such as Ace Superwhite by the Ministry of Snus (Denmark) or N!Xs by Microzero AB (Sweden and other European countries) *(30)*.

Nicotine pouches are marketed online and by tobacconists as smoke- and tobacco leaf-free alternatives to tobacco and nicotine products that are "less harmful" than traditional snus and conventional cigarettes *(14)*. These are also some of the reasons reported by users (see below) *(22,31)*. Smokeless-tobacco users also perceive snus as less risky than conventional cigarettes *(32,33)*. Although nicotine pouch containers often bear a warning about addiction to nicotine, they usually do not bear the warnings required for smokeless tobacco in many countries about the risks of oral cancer and gum disease. Nicotine pouches are also called "white snus", because of their white powder filler, instead of the traditional brown tobacco snus *(34)*. "White snus" is marketed as "milder, slimmer, flavoured, and more visually appealing" *(26)*. Different sizes of nicotine pouches are available, the smaller pouches being marketed as discreet ("Since the pouch is thin and small no one will see that you have it under your lip") *(35)*. Online marketing stresses that nicotine pouches can be used "anywhere, anytime", including where smoking is prohibited *(23,34)*. Cross-over advertising was observed on websites of leading brands of conventional cigarettes when the parent companies also owned the pouch brand. Examples include Altria's co-promotion of On! pouches with Marlboro cigarettes, and On! and Camel consumers received an e-mail invitation to "explore the nicotine options from our friends at Velo" *(37)*. E-mail advertising included the claim that the product could be used anywhere (84% of e-mails), that nicotine pouches are an alternative to other tobacco products (69%), do not contain tobacco leaf (55%) and are "spit-free" (52%) or "smoke-free" (31%) *(38)*. Nicotine pouches are not more expensive than cigarettes, because a container of nicotine pouches is slightly cheaper or comparable in price to a package of conventional cigarettes *(3)*. It has also been claimed by companies that, in contrast to e-cigarettes, batteries and charging devices are not necessary *(35,39)* and that, in contrast to traditional snus, the white pouches look cleaner and do not stain the teeth *(34,36)*.

According to a Euromonitor report *(40)*, "use of influencer marketing and social media platforms, such as Instagram, has been embraced by modern oral manufacturers". An article by the Bureau of Investigative Journalism summarizes marketing tactics for new BAT products, including heated tobacco and oral nicotine, presenting nicotine products as "cool" and "aspirational" in a glossy youth-focused advertising campaign; paying social media influencers to promote nicotine pouches; sponsoring music and sporting events; and an international offer of free samples of nicotine pouches, which appears to have attracted underage people and non-smokers *(10)*. Nicotine pouches are promoted on the social media accounts of musicians, football players and influencers in many countries *(10)*.

4.5 User profile

A few studies have reported the prevalence of nicotine pouch use. The overall prevalence among 10 296 adult current cigarette smokers or recent ex-smokers in Australia (0.1%), Canada (0.9%), England (1.1%) and the USA (0.7%) was 0.8%.

Among the few current and ex-smokers, more males (1.1%) than females (0.5%) used nicotine pouches. In all the countries studied, the prevalence was highest among those aged 18–24 years (2.3%) (25–39 years, 1.4%; 40–54 years, 0.4%; ≥ 55 years, 0.1%) *(41)*.

In a survey of 3883 smokers, vapers, dual users and recent ex-users in the United Kingdom in 2019, 15.9% had heard of nicotine pouches, and 3.1% had seen them for sale; 4.4% had ever used nicotine pouches, and 2.7% were current users *(42)*. In a survey of a sample of 5805 people representative of the Dutch population, only 6.9% were aware of nicotine pouches, mainly because they knew someone who used them (33%) *(31)*. Of the respondents, 0.6% had ever used a nicotine pouch, and 0.06% were current users. Current smokers had higher than average ever use (1.91%), especially those who preferred menthol cigarettes (6.26%). Awareness among adolescents (13–17 years) was relatively high (9.1%), but only 0.3% had ever used a nicotine pouch, and none reported current use.

In an online repeat cross-sectional survey in 2019 of 11 714 young people aged 16–19 years in Canada, 11 170 in England and 11 838 in the USA, 1% in Canada, 1.3% in England and 1.5% in the USA had used nicotine pouches in the previous 30 days *(43)*. Data from the 2021 National Youth Tobacco Survey in the USA indicated that 1.9% of middle- and high-school students (age 11–18 years) had ever used nicotine pouches *(44)*. Further, 0.8% of the students reported current use (past 30 days) of nicotine pouches. Of the students who reported current use of nicotine pouches, most (63.5%) reported having used them on 1–5 days in the past 30 days, and 17.2% reported use on 20–30 days in the past 30 days. Additionally, 61.6% of current users reported having used flavoured nicotine pouches in the past 30 days, mint and menthol being the most commonly reported flavours.

A study conducted in the USA in early 2021 from a web-based survey of US adults who were current, established smokers (had smoked at least 100 cigarettes in their lifetime and now smoked every day or on some days) found that 29.2% had ever seen or heard of nicotine pouches, 5.6% had ever used them, and 16.8% expressed interest in using them in the next 6 months *(45)*. Younger adult smokers were more likely to have ever seen or heard of nicotine pouches than older adult smokers. Among adults who smoked, those with more education had lower odds of ever using nicotine pouches, while those who had attempted to quit before using traditional methods or had ever used smokeless tobacco had higher odds of ever use.

The demographics of Zyn users and the patterns and reasons for use were investigated in a study based on data from Swedish Match North America *(22)*. The average Zyn user was about 33 years old, male, white, had finished high school and earned more than US$ 50 000 per year (i.e. middle income). The majority were current smokeless tobacco users and former tobacco users (mostly former dual cigarette–smokeless tobacco users). Two other studies reported similar profiles of nicotine pouch users: male, 25–34 or 44 years and had formerly smoked and/or vaped *(31,42)*. Zyn users found nicotine pouches moderately to extremely appealing. The reasons for use were "less harmful to my health than other tobacco products" (62%), "ease of use" (53%), "no one can tell when using it" (50%), "less harmful to my health than cigarettes" (49%), and "no smell like smoke/tobacco and to avoid spitting" (48%). Interestingly, 40% of the never users were "curious to see what it was like" *(22)*. Among the Dutch respondents, nicotine pouches were used mainly "at a party" (38%), "with friends" (38%) or "at home" (26%) *(31)*. The main reasons for using nicotine pouches were "out of curiosity" (72%) and "it is tasty and/or pleasant" (23%), but also because they considered "it is less unhealthy than cigarettes" (23%). Only 8% indicated the "availability of different flavours" an important reason for use.

4.6 Evaluation of potential harmfulness of the products

4.6.1 Attractiveness

The Partial Guidelines on Articles 9 and 10 of the WHO Framework Convention on Tobacco Control (WHO FCTC) *(46)* recommend regulation of attractive product characteristics, in particular to decrease uptake by young never users. Nicotine pouches have many attractive features, as mentioned above. For example, they are available in a variety of fruit, mint and other flavours (e.g. cinnamon, liquorice and coffee) *(3,5,22)* and contain sweeteners *(22)*. The cost of the product in the USA is slightly lower than or comparable to that of a pack of conventional cigarettes *(3,47)*, which might be a barrier to use for some but not all potential users. Further attractive features are, for example, the perception that the product is effective for quitting smoking, less harmful than other tobacco products and easy and discreet to use, in places where smoking is banned *(22)*.

4.6.2 Addictiveness

Nicotine pouches contain sufficient nicotine to sustain addiction *(3,47)*: Zonnic 4-mg delivers 2 mg of nicotine *(47)*, similar to the levels delivered by cigarettes. Release of nicotine from On! pouches into artificial saliva released equivalent levels of nicotine with all flavours *(48)*.

Two studies have addressed the pharmacokinetics of nicotine pouches *(16,49)*. In a study by Lunell et al. *(16)*, which was funded and designed by Swedish Match, pouches with a concentration of 3 mg, 6 mg or 8 mg nicotine were tested. After 1 h of use, 1.6 mg (56%), 3.5 mg (60%) and 3.8 mg (50%) nicotine respectively were released from the bags, respectively, and the amount of nicotine in the users' blood gradually increased during use, with peak concentrations of 7.7 ng/mL, 14.7 ng/mL and 18.5 ng/mL. The authors reported that Zyn (6 and 8 mg) delivered nicotine as quickly and to a similar extent as smokeless tobacco products. In a study by Rensch et al. *(49)*, funded by Altria Client Services LLC, nicotine pouches with various flavours and 4 mg nicotine were tested. The amount of nicotine in venous blood from participants increased gradually during the 30 min of use and for 10 min afterwards, to a peak concentration of 9.6–12.1 ng/mL. Use of the pouches reduced the urge to smoke or craving for a cigarette. All the nicotine pouches were considered pleasant but not as much as one's own brand of cigarette. Flavour did not appear to influence the pharmacokinetics of nicotine or the subjective responses.

In comparison, 1–2 mg of nicotine are inhaled from one tobacco cigarette over about 5 min *(50)*. The peak plasma nicotine concentration in venous blood after smoking one cigarette is 10–30 ng/mL and is reached within 5–8 min of the first puff of the cigarette *(50,51)*. Thus, the amount of nicotine to which users of nicotine pouches are exposed, especially from pouches with ≥ 4 mg per pouch), is in the same range as that to which smokers are exposed. One difference is that peak concentrations are reached in a very short time during smoking, while there is a slower, more gradual increase with use of nicotine pouches. A similar observation was made for snus, nicotine plasma levels rising less rapidly than during smoking a cigarette *(51)*.

The slower release of nicotine is an important difference, because it is precisely the fast peak that makes smoking so addictive. The faster a drug is absorbed and reaches the brain, the greater the "rush" it causes and the stronger the rewarding effect. In addition, a short interval between the act and the "reward" provides strong conditioning of behaviour *(52)*. Nicotine replacement therapy products, such as nicotine chewing-gum, which is absorbed in the stomach and intestines, and patches, which are absorbed through the skin, result in very slow absorption of nicotine and are therefore much less addictive *(50)*. The rate at which nicotine is absorbed from nicotine pouches appears to be closer to that of nicotine chewing-gum than that of inhalable nicotine-containing products *(16,50)*.

Nicotine pouches such as Zyn contain pH adjusters *(22)*, which probably increase the addictive potential of the product, since a higher pH results in more so-called "free" nicotine *(5)*, which makes the products harsher but is more easily absorbed in the mouth than other forms of nicotine. Tobacco snus products with a higher pH deliver more nicotine to the user *(53)*. Nicotine pouch products vary in pouch content mass, moisture content (1.12–47.2%), alkalinity (pH 6.86–10.1), and percentage of free nicotine (7.7–99.2%). The total nicotine content ranges from 1.29 to 6.11 mg/pouch and that of free nicotine from 0.166 to 6.07 mg/pouch *(5)*.

4.6.3 **Toxicity**

Nicotine pouches do not have the chemical by-products of burning or smouldering tobacco leaf, and they are not inhaled. The ground tobacco leaves of conventional smokeless brands emanate toxic chemicals that are not present in nicotine pouches. Indeed, tobacco-free nicotine pouches may have the fewest harmful constituents of all tobacco and nicotine products *(3,54)*. The term "tobacco-free" may, however, be misleading, as many pouches contain nicotine extracted from tobacco and may thus be not entirely free from tobacco residues. Product constituents, exposure and biomarkers of harm have not been investigated independently *(3,55)*. The main risk factor is nicotine, a known toxicant *(56)* registered under the European Union REACH regulations *(57)*. It is classified as acutely toxic (category 2) after oral, dermal or inhalation exposure and is subject to hazard statements H300: fatal if swallowed, H310: fatal in contact with skin, and H330: fatal if inhaled *(55)*. The higher the nicotine dose of tobacco-containing snus, the larger the increase in heart rate and systolic blood pressure when used by never-tobacco users *(58)*. A concern associated with synthetic nicotine is that the pharmacological and metabolic effects of *R*-nicotine are largely unknown *(6)*. Nicotine extracted from tobacco may be contaminated with tobacco-specific nitrosamines, which are carcinogenic. Most of the other ingredients are also used in food and can therefore be assumed to be relatively safe by the oral route *(212)*, although this has not been addressed, and these ingredients should be studied in the context of nicotine pouches, particularly for local effects. A study by BAT on toxicants such as metals, aldehydes and tobacco-specific nitrosamines in nicotine pouches showed low levels of chromium and formaldehyde in some but not all samples *(59)*.

Dentists have warned of the harmful effects of nicotine pouches *(60,61)*, while a tobacco industry publication shows minimal enamel staining *(62)*. Chaffee et al. *(63)*, reviewed the literature on oral and periodontal effects and concluded that evidence was lacking.

A study of screening assays in vitro by BAT showed toxicological responses to reference cigarette extract in most, while a snus extract had minimal-to-moderate effects and a nicotine pouch extract gave little or no response in all the assays *(64)*.

4.7 Population effects and related factors

For tobacco product users, nicotine pouches might be perceived as a less harmful alternative to conventional cigarettes, heated or smokeless tobacco products, and it would be best to refrain from use of tobacco and nicotine completely. Uptake by never-tobacco or -nicotine users, however, results in exposure to nicotine, which may cause addiction and may even be a gateway to use of other nicotine and tobacco products. Unfortunately, limited data are available on these effects. Data from Swedish Match showed that most Zyn users were former tobacco users (43%), and only a few were never users (4%); most used Zyn every day *(22)*. Zyn appealed most to dual cigarette-smokeless tobacco users (76%), smokeless tobacco users (52%) and smokers (36%), while never and former tobacco users showed much less but still some interest (11–12%) *(22)*. Novel nicotine products can, however, be taken up rapidly by adolescents and young adults, as seen, for example, with Juul e-cigarettes in the USA *(3)*. Uptake by never smokers, in particular young people, can be stimulated by several factors, such as marketing and design, the variety of flavours and discretion *(3)*. The likelihood of progression to the use of tobacco products, as has been reported for e-cigarettes *(65,66)*, is not known, but the nicotine levels are sufficiently high to sustain addiction, which is generally ≤ 50 mg/g nicotine per pouch, although nicotine pouches with up to 120 mg/g of nicotine have been reported on certain markets.

Nicotine pouches may undermine tobacco control policies such as bans on flavours in tobacco products, including conventional cigarettes. Nicotine pouches can be used discreetly in places where smoking is not allowed and may lead to dual use with conventional cigarettes, which would undermine the beneficial effects of tobacco-free policies. Discontinuation of nicotine exposure imposed by non-smoking rules (e.g. in workplaces, on transport systems, in restaurants and bars) helps tobacco users to quit. Thus, sustaining nicotine dosing in places where smoking is not allowed exacerbates addiction and makes quitting less likely.

Another concern is that nicotine pouches blur the distinction between nicotine replacement therapy and smokeless tobacco products *(47)*, as some manufacturers promote these products as nicotine replacement therapy or tools for stopping use of smokeless and smoked tobacco products. Pouch advertisements make both explicit and implicit promises of their usefulness in tobacco cessation: "Designed with smokers in mind" (On!), "I can breathe again" (Zyn) and "never going back" (Zyn).

A study funded by both the manufacturer of Zonnic pouches and the New Zealand National Heart Foundation showed that, for smokers, a Zonnic pouch is as effective as nicotine chewing-gum in relieving craving but subjectively more attractive *(67)*. Overall, nicotine pouches are not proven tools for cessation, and it is unknown how their availability will affect overall smoking cessation rates; there may be competition with proven cessation tools *(47)*.

4.8 Regulation and regulatory mechanisms

According to ECigIntelligence *(68)* and WHO Member States, nicotine pouches are currently available in more than 30 countries, and the market is set to expand in coming years. Between 2020 and 2022, WHO disseminated questionnaires and elicited information from various tracks to capture country experiences with nicotine pouches, the regulatory mechanisms in place and challenges found in regulation. In total, 71 countries provided information on nicotine pouches, and 124 national laws were reviewed. The majority of Member States that provided information reported that nicotine pouches had entered their market between 2018 and 2020, and many reported that the sale of these products was becoming an issue in their countries.

Various regulatory approaches have been adopted, including regulation of nicotine pouches as consumer products, poisons, medical or pharmaceutical products, nicotine pouches (in their own class), nicotine-containing products and tobacco products. These classifications have resulted in bans on nicotine pouches in 12 countries, including Australia and the Russian Federation, regulation in some other countries, and application of existing tobacco control regulations in others.

While the WHO review identified 22 countries in which nicotine pouches are regulated, these products appear not to be regulated in 161 countries, albeit with general consumer laws applying. Very few tobacco control laws cover nicotine pouches, and very few countries regulate or ban nicotine pouches specifically. The majority of countries that regulate nicotine pouches do so through non-tobacco control laws, such as laws on pharmaceutical products, poisons, food and general consumer protection. Table 1 presents examples of regulation of nicotine pouches through various approaches.

Table 1. Examples of approaches taken by countries to regulate nicotine pouches

Regulatory approach	Countries	Law or regulation	Description
Consumer product	Austria, Bulgaria, Croatia, Cyprus, Dominican Republic, Greece, Iceland, Luxembourg, Malta, Poland, Portugal	Consumer laws apply	In Austria, nicotine pouches are classified as both "consumer products" and "medicines". As long as no claims are made about smoking cessation aids, they are classified as "consumer products". In other situations, nicotine pouches are classified as "medicines".
Food	Germany, Netherlands (Kingdom of the)	Commodity law and Article 14 Regulation (EC) No. 178/2002 of the European Parliament and of the Council of 28 January 2002 (Netherlands, Kingdom of the)	In Germany, nicotine pouches are banned because they contain nicotine, an unauthorized novel food ingredient. In Netherlands (Kingdom of the), nicotine pouches containing ≥ 0.035 mg of nicotine per pouch may no longer be sold or traded, as they are classified as harmful foods.
Poison	Brunei Darussalam, Ireland	Poisons Act	In Brunei Darussalam, nicotine pouches are classified as both a "poison" and an "imitation tobacco product". They are listed as a "poison" under the Poisons Act; importation and sale of poisons require a license. (See note on "imitation tobacco product" below).
Medicine or pharmaceutical product	Austria, Canada, Chile, Finland, Hungary, Japan, Malaysia, South Africa	Canadian Food and Drugs Act Finnish Medicinal Products Act (section 3)	In Austria, nicotine pouches are classified as "medicines" if smoking cessation claims are made. Otherwise, they are classified as a "consumer product". Pouches that deliver < 4 mg of nicotine per dose are exempt from prescription, are regulated as "natural health products" and are subject to the Natural Health Products Regulations in Canada and as a licensed self-medication product in Finland. Pouches that deliver > 4 mg per dose are considered a prescription drug and subject to the requirements of the Food and Drug Regulations (Canada) and the Medicinal Products Act (Finland). The pouch that delivers the drug is considered a Class I medical device. No nicotine pouch has yet been granted authorization for sale as a drug in Canada.

Nicotine pouch Nicotine-containing product Tobacco-free products Tobacco alternatives Imitation tobacco	Belgium, Brunei Darussalam, Estonia, New Zealand, Republic of Moldova	Tobacco Order 2005 (Brunei Darussalam) Smoke free Environments and Regulated Products Amendment Act 2020 (New Zealand) Law No. 278-XVI on Tobacco and Tobacco Products, as amended in 2015 (Republic of Moldova)	In Belgium, nicotine pouches are classified as "similar to tobacco products". In Brunei Darussalam, nicotine pouches are classified as both a "poison" and as an "imitation tobacco product". Nicotine pouches may be considered an "imitation tobacco product" and are therefore prohibited. Nicotine pouches are not currently sold in the country. See note on "poisons". In Estonia, nicotine pouches are considered "snus imitation products" and taxed as "alternative tobacco products". See Table 2. In New Zealand, the Government prohibits the import for sale, packaging and distribution of oral nicotine products (unless approved as medicines). A significant change to the legislation by amending the definition of "tobacco product" was avoided; instead, oral nicotine pouches are directly prohibited, consistent with regulation of snus and chewing tobacco under New Zealand law. See Table 2 for more details on regulation in the Republic of Moldova.
Tobacco product	USA	Code of Federal Regulations – Title 21, Volume 8	See Table 2.

As these products are relatively new on many markets, they are unregulated in several countries. In some, current tobacco control or other laws include no measures that could be applied to this product. In other countries, the products are not specifically regulated, although general consumer protections laws apply. Some countries are exploring ways in which to address nicotine pouches, such as considering them as nicotine chewing gum and imposing excise duties. The definitions applied to nicotine pouches under existing tobacco control laws and their legal interpretation are listed in Table 2.

Table 2. Examples of legal definitions in tobacco control laws applied to nicotine pouches, and legal interpretations

Country	Relevant regulations or law	Relevant definition(s)	Interpretation
Estonia	Tobacco Act (RT I 2005, 29, 210), as amended in 2018	"Products related to tobacco products" are defined as "products used similarly to tobacco products which imitate consumption of tobacco products and products used to replace tobacco products, including electronic cigarette, herbal products for smoking, different materials to replace waterpipe tobacco and tobacco-free snus, regardless of the nicotine yield of such products".	The provisions of the Act apply to tobacco products and products related to tobacco products. As such, nicotine pouches are considered snus imitation products and therefore regulated under the Tobacco Act; bans on advertising, sales to minors and point-of-sale display apply, and pouches are taxed as alternative tobacco products
Republic of Moldova	Law No. 278-XVI on Tobacco and Tobacco Products, as amended in 2015	"Nicotine-containing products" are defined as "any product consumed by inhalation, ingestion or otherwise, to which nicotine is added during the production process or is added by the consumer himself before or during consumption"; "tobacco-related products" are defined as "products made of plants for smoking and products which contain nicotine, including electronic cigarettes".	Under the Law, nicotine-containing products are regulated and Articles 23a and 23e specifically apply to nicotine pouches, stipulating that "a) the nicotine content does not exceed 2 mg per unit or product" and "e) the product does not contain additives specified in paragraph (3) of Article 11". This includes nicotine that is added by the consumer before or during consumption and tobacco-related products that meet the specified definition.

Russian Federation	Federal Law No. 303-FZ of July 31, 2020 "On Amendments to Certain Legislative Acts of the Russian Federation on the Protection of Citizens' Health from the Consequences of Consuming Nicotine-containing Products	Nicotine-containing products are defined as "products that contain nicotine (including those obtained by synthesis) or its derivatives, including nicotine salts, intended for the consumption of nicotine and its delivery by sucking, chewing, sniffing or inhaling, including products with heated tobacco, solutions, liquids or gels containing liquid nicotine in a volume of at least 0.1 mg/mL, nicotine-containing liquid, powders, mixtures for sucking, chewing, sniffing, and are not intended for consumption (except for medical products registered in accordance with the legislation of the Russian Federation, food products containing nicotine in natural form, and tobacco products)".	The law prohibits the wholesale and retail trade of nicotine-containing products intended for chewing and sucking, effectively banning nicotine pouches. This applies to any form of nicotine, including synthetic nicotine.
USA	Code of Federal Regulations – Title 21, Volume 8	Tobacco products are defined as "any product made or derived from tobacco that is intended for human consumption, including any component, part, or accessory of a tobacco product (except for raw materials other than tobacco used in manufacturing a component, part, or accessory of a tobacco product".	All tobacco products, including nicotine pouches that meet the definition of a "tobacco product", are subject to the USFDA's regulatory authority. Tobacco products are covered under the Federal Food, Drug and Cosmetic Act and its implementing regulations. Under the Act, the USFDA's regulatory authority covers the manufacture, sales, distribution, labelling, advertising, promotion and marketing of cigarettes, cigarette tobacco, roll-your-own tobacco, smokeless tobacco and other tobacco products that the Agency, through regulation, deems subject to the law.

A complementary but independent study was conducted by the Johns Hopkins Institute for Tobacco Control as part of its biannual survey to collect information from countries on tobacco and nicotine product regulation in 2021 *(69)*. A total of 67 countries in all six WHO regions in various income categories provided information. The policy scan identified 34 countries that regulate nicotine pouches, of which 23 regulate both tobacco-derived and synthetic nicotine, while the other 11 regulate only tobacco-derived nicotine pouches. Representatives of 38 countries that do not regulate synthetic nicotine pouches cited the wording of their legislation as the main barrier to regulating synthetic nicotine. Of 33 countries that reported that nicotine pouches were sold on their markets, 20 had regulations, while 14 countries that reported that nicotine pouches were not sold on their markets nevertheless had regulatory policies in place.

4.8.1 Regulatory considerations

Countries that are interested in taking regulatory action with respect to nicotine pouches have two possible regulatory pathways: ban or regulate. In banning or regulating these products, countries that are Parties to the WHO FCTC should take into account their obligations under the Convention in formulating or adopting policies with regard to these products. Countries should also consider existing relevant national laws (on food, consumers, drugs and tobacco), trade classification and classification according to product constituents or characteristics. The Harmonized Commodity Description and Coding System, administered by the World Customs Organization *(70)*, is a standardized system adopted by many countries for classifying traded products. The classification may affect application of domestic laws, and countries may consider whether the Harmonized System codes apply to nicotine pouches.

In regulating nicotine pouches, Parties may introduce measures to prevent nicotine addiction, in line with Article 5 (2b) of the WHO FCTC. This provides that Parties shall, in accordance with their capabilities,

> adopt and implement effective legislative, executive, administrative and/or other measures and cooperate, as appropriate, with other Parties in developing appropriate policies for preventing and reducing tobacco consumption, nicotine addiction and exposure to tobacco smoke.

Countries may also consider banning these products, in line with WHO FCTC Article 2.1, which encourages Parties to implement measures beyond those required by the WHO FCTC. If a country opts to regulate rather than imposing a ban, it may also consider prohibiting or restricting ingredients that may be used to increase the palatability of these products (such as flavours), as recommended in paragraph 3.1.2.2 of the Partial Guidelines on Articles 9 and 10 of the WHO FCTC *(46)*, in order to reduce uptake by young people or never users.

In regulating nicotine pouches, their classification is an important consideration, as it determines to a large extent how a product is regulated. In some jurisdictions, such as the European Union, nicotine is classified among chemicals; however, nicotine pouches are not regulated in a harmonized manner, as they are currently not covered by the Tobacco Products Directive. Countries may also consider whether domestic laws can be applied to these products, including consumer, food and tobacco control laws.

4.8.2 Country case study: Netherlands (Kingdom of the)

A report from the National Institute for Public Health and the Environment (RIVM) *(14)* described nicotine products available on the Dutch market and reported that nicotine pouches were becoming increasingly popular. The report included information that nicotine is addictive and harmful to health, such as to the nervous system and can cause cardiac arrhythmia, particularly at high dosages. The RIVM therefore advised the Dutch Ministry of Health, Welfare and Sport to discourage use of nicotine pouches by imposing stricter regulations and organizing public information campaigns. Nicotine pouches currently fall under the Commodities Act. The RIVM considered the question of which existing legislation could apply to nicotine products without tobacco. According to the Ministry of Health, these products are not presently within the scope of the Tobacco Act, as they do not contain any tobacco. Policy-makers could consider adding nicotine products without tobacco to the list of products covered by the legislation, for example, by broadening the definition of tobacco and related products. In view of the harmful and addictive effects of nicotine pouches, people, particularly young people, should be prevented from starting their use. On 9 November 2021, the Dutch State Secretary of Health declared that he intended to include nicotine products without tobacco in the law on tobacco and smoking products and to prohibit nicotine pouches in particular *(71)*. Until that is done, these products remain under the Commodities Act. RIVM also proposed that nicotine pouches containing ≥ 0.035 mg of nicotine per bag be considered harmful foods *(72)*, and these products were prohibited under the Commodities Act in November 2021.

4.9 Discussion

Although nicotine pouches are relatively new on many markets, tobacco manufacturers appear to be expanding their markets and are lobbying governments to classify and license nicotine pouches as non-tobacco products. Manufacturers are also seeking to ensure that more lenient regulations are applied to these products than to conventional tobacco products. A key strategy is conflation of product categories (i.e. blurring the line between different product categories) to create confusion in order to penetrate global markets, maximize profits and "get a seat at the table" with regulators. Regulators sometimes lack information on the harm caused by these products and on the regulatory options for addressing the challenges they pose. One of the challenges faced by some regulators is the claim by some manufacturers that, as the products do not contain tobacco and/or that the nicotine contained in the products is not derived from tobacco, they should not be regulated under tobacco laws. Some manufacturers also attempt to bypass ministries of health and have the products registered by other ministries in order to evade strict regulations. Further, these products are pitched as "less harmful" or "smoke free" alternatives to conventional products and sold with a variety of flavours, which could undermine tobacco control policies, such as bans on flavours and smoke-free laws.

Nicotine pouches contain significantly fewer ingredients and toxicants than conventional cigarettes and are being marketed as and perceived by consumers as "less harmful". While the pouches may present fewer risks than conventional tobacco products, manufacturers should not make such claims, unless they are proven and authorized by regulators. Governments can use their policies and regulatory frameworks to decide to educate their populations according to the available evidence. Regulations that distinguish between nicotine and tobacco products and between nicotine derived from tobacco and synthetic nicotine open the possibility for discussions on whether a product is "tobacco-free" or not and should be regulated more leniently. From the perspective of public health, such a distinction is not fruitful, and regulators should therefore consider widening their regulatory frameworks to include non-therapeutic nicotine products in general, irrespective of whether they contain tobacco or whether the nicotine is derived from tobacco.

While there are limited data on the prevalence of use, the available evidence suggests that strategies similar to those used to market conventional tobacco products are used to market nicotine pouches. These products are similar in appearance to conventional smokeless tobacco products, such as snus, contain nicotine and are used similarly. The attractiveness of these products, including the flavours, suggests that they could sustain use through improved palatability. This is a public health concern, especially in relation to young people and non-users of tobacco. The nicotine content of some of these products, which may be as high as or higher than that of conventional tobacco products, suggests reasonable concern about nicotine addiction. Although limited data are available on these products because of their recent introduction, a cautionary approach is warranted in view of their similarities to conventional products. These products are sold online and by tobacconists, and their sale is largely uncontrolled or unrestricted in many countries, especially on the Internet, including sales to the USA of very high-strength nicotine pouches from Europe. Some of these products are difficult to distinguish from conventional smokeless tobacco products *(8)*. In view of their intense marketing and use of flavours that are attractive to young people, countries are encouraged to protect their existing policies or formulate new policies, as appropriate. In addition, they should broaden their regulatory requirements to cover the wide range of nicotine and tobacco products that are appearing on several markets around the world.

Our preliminary analysis of national laws and the results of the survey indicate that nicotine pouches are unregulated or not specifically regulated in several jurisdictions, and manufacturers have exploited the regulatory vacuum. Other countries have, however, previously made their regulations and laws "future-proof" and resilient to ensure that nicotine pouches are regulated under existing laws. Some have recently updated their laws, whereas some still use the definitions that cover conventional products. The industry might use the latter case to its advantage, using strategies to "get a seat at the table" and present themselves as part of the solution to reducing tobacco use, despite fuelling widespread use of nicotine. A few countries have nevertheless designated nicotine pouches as tobacco products, and other countries may consider acting similarly. Parties to the WHO FCTC interested in banning or regulating nicotine pouches can use certain provisions of the Convention to protect their populations. Some countries in the WHO European Region that have made legislative amendments to include these products have met opposition from tobacco manufacturers. It is urgent to harmonize regulation of new tobacco and nicotine products to ensure strong protection of health, as required in the Treaty on the Functioning of the European Union *(8)*. It is particularly important to regulate access and promotion to young people.

4.10 Research gaps, priorities and questions

Currently, limited information is available on nicotine pouches, including on their abuse potential, harm, user profiles and population effects. Furthermore, there is no information on long-term dependence of these products, given the short time they have been on the market. More data, preferably studies independent of the tobacco industry, are required on:

- prevalence of use and user profiles, including tobacco use status;
- whether nicotine pouches can help tobacco users to quit tobacco use;
- whether these products are used in addition to cigarettes or other nicotine and tobacco products (dual use);
- monitoring of product use to ensure that nicotine pouches do not promote nicotine addiction among non-smokers, especially young people;
- the possibility that these products are a gateway to use of conventional tobacco products and addiction, especially for young people;
- the potential for increasing attractive features, such as flavour profiles, and the effect of factors such as marketing on perception and use;
- the precise content of nicotine, flavourings, other additives and contaminants;
- short and long-term health effects of nicotine and other substances in nicotine pouches, including synthetic nicotine;
- the effects of switching completely from use of tobacco products to nicotine pouches on exposure and health; and
- the actual outcomes of people who smoke, smokeless tobacco users, never and ex-users who initiate use of nicotine pouches.

4.11 Policy recommendations for product regulation and information dissemination

Policy-makers should adopt common regulatory principles that have been applied successfully to tobacco and related products in many jurisdictions to:

- minimize product appeal and uptake by young people,
- increase product safety and
- minimize false health beliefs.

According to the legal definition of tobacco products or definitions in other relevant laws, countries could explore use of existing tobacco control or other relevant laws to regulate nicotine pouches. Any decision should be in accordance with the country's domestic regulatory context and should ensure maximum protection of the health of its citizens, especially children and young people.

Recommendations for policy-makers, particularly to protect young people and non-users, are as follow.

- Establish or extend surveillance of the product and of users, including their demographics; use of other tobacco and related products; the brands and types; and the flavours used in nicotine pouches in order to assess prevalence and user profiles.
- Regulate all forms of marketing of nicotine pouches and take all other action necessary to minimize access, appeal and initiation by young people.
- Inform the general public about the risks for toxicity and addiction associated with the nicotine in these pouches.
- Require health warning on packages of nicotine pouches, for example on the effects of nicotine, which could include effects on users, the detrimental effects on fetal development in pregnant women, and the damaging effects on brain development in young people, including on learning.
- Prohibit health-related claims by manufacturers, including their potential effectiveness as cessation products, unless the products are licensed and approved as such by regulators.
- Set an upper limit on nicotine to reduce the addictiveness of the products and harm from inadvertent ingestion.
- Protect existing and formulate new policies, as appropriate, to broaden the regulatory requirements to cover the wide range of nicotine and tobacco products appearing on several markets around the world.
- Regulate nicotine pouches in the same manner as products of similar appearance, content and use.
- Ensure that nicotine pouches are not classified as pharmaceutical products unless they are proven to act as nicotine replacement therapy and undergo stringent pharmaceutical registration for licensing as such by the appropriate national regulatory authority.

- Regulate nicotine pouches to prevent all forms of marketing, and take all other action necessary to minimize access and appeal to and initiation by young people.
- Protect tobacco control activities from all commercial and other vested interests related to nicotine pouches, including the direct and indirect interests of the tobacco industry, and ban all forms of marketing and promotional activities.
- Fully implement Article 5.3 of the WHO FCTC to protect policies against undue influence by the tobacco and related industries.

4.12 Conclusions

Nicotine pouches have recently become available on many markets worldwide. They contain sufficient nicotine to induce and sustain nicotine addiction and have many attractive properties, such as appealing flavours and packaging and discreet use. They contain fewer toxicants and therefore expose users to fewer harmful and potentially harmful constituents than conventional tobacco products; however, no use of non-therapeutic nicotine and of tobacco products is recommended for maximum protection of health, as the benefits of quitting tobacco use are apparent almost immediately. Uptake of nicotine pouches results in exposure to toxic nicotine, which may cause nicotine addiction and subsequently lead to use of other nicotine and tobacco products. Nicotine pouches are not regulated or not specifically regulated in several jurisdictions, whereas other countries have made their regulations and laws "future-proof" and resilient such that nicotine pouches are regulated under existing laws. Other countries maintain definitions that refer only to traditional products.

References

1. St Claire S, Fayokun R, Commar A, Schotte K, Prasad VM. The World Health Organization's World No Tobacco Day 2020 campaign exposes tobacco and related industry tactics to manipulate children and young people and hook a new generation of users. J Adolesc Health. 2020;67:334–7. doi:10.1016/j.jadohealth.2020.06.026.
2. WHO study group on tobacco product regulation. Report on the scientific basis of tobacco product regulation: eighth report of a WHO study group (WHO Technical Report Series, No. 1029). Geneva: World Health Organization; 2021. (https://www.who.int/publications/i/item/9789240022720).
3. Robichaud MO, Seidenberg AB, Byron MJ. Tobacco companies introduce "tobacco-free" nicotine pouches. Tob Control. 2019;29(E1):e145–6. doi:10.1136/tobaccocontrol-2019-055321.
4. The rise of products using synthetic nicotine. Washington DC: Campaign for Tobacco-Free Kids; 2022 (https://www.tobaccofreekids.org/assets/factsheets/0420.pdf).
5. Stanfill S, Tran H, Tyx R, Fernandez C, Zhu W, Marynak K et al. Characterization of total and unprotonated (free) nicotine content of nicotine pouch products. Nicotine Tob Res. 2021;23:1590–6. doi:10.1093/ntr/ntab030.
6. Jordt SE. Synthetic nicotine has arrived. Tob Control. 2023;32(e1):e113–7. doi:10.1136/tobaccocontrol-2021-056626.

7. Requirements for products made with non-tobacco nicotine take effect April 14. Silver Spring (MD): Food and Drig Administration. 2022 (https://www.fda.gov/tobacco-products/ctp-newsroom/requirements-products-made-non-tobacco-nicotine-take-effect-april-14).
8. Salokannel M, Ollila E. Snus and snus-like nicotine products moving across Nordic borders: Can laws protect young people? Nordic Stud Alcohol Drugs. 2021;38(6):540–4. doi:10.1177/145507252199570.
9. Market sizes. London: Euromonitor International; 2022.
10. Chapman M, Okoth E, Törnkvist A, Margottini L, Irfan A, Cheema U. New products, old tricks? Concerns big tobacco is targeting youngsters. London: The Bureau of Investigative Journalism, 21 February 2021 (https://www.thebureauinvestigates.com/stories/2021-02-21/new-products-old-tricks-concerns-big-tobacco-is-targeting-youngsters).
11. Pouches & gums. SRITA collection of pouch advertisements. Palo Alto (CA): Stanford Research into the Impact of Tobacco Advertising; 2023. (https://tobacco.stanford.edu/pouches_gums/).
12. Burki TK. Petition to ban nicotine pouches in Kenya. Lancet Oncol. 2021;22:756. doi:10.1016/S1470-2045(21)00267-9.
13. Put the used pouch in the garbage. Stockholm: Swedish Match, 21 November 2016 (https://web.archive.org/web/20210617145948/https://www.swedishmatch.com/Media/Pressreleases-and-news/News/put-the-used-pouch-in-the-garbage/).
14. Pauwels CGGM, Bakker-'t Hart IME, Hegger I, Bil W, Bos PMJ, Talhout R. Nicotineproducten zonder tabak voor recreatief gebruik [Nicotine products without tobacco for recreational use]. In. Bilthoven: National Institute for Public Health and the Environment; 2021 (https://open.overheid.nl/documenten/ronl-fa15b269-76a3-466d-a513-fe457e3cf6f4/pdf).
15. GrantSnus. Berlin: Kordula UAB. (https://grantsnus.com/).
16. Lunell E, Fagerström K, Hughes J, Pendrill R. Pharmacokinetic comparison of a novel non-tobacco-based nicotine pouch (ZYN) with conventional, tobacco-based Swedish snus and American moist snuff. Nicotine Tob Res. 2020;22:1757–63. doi:10.1093/ntr/ntaa068.
17. Lawler TS, Stanfill SB, Tran HT, Lee GE, Chen PX, Kimbrell JB et al. Chemical analysis of snus products from the United States and northern Europe. PLoS One. 2020;15:e0227837. doi:10.1371/journal.pone.0227837.
18. Delnevo CD, Hrywna M, Miller Lo EJ, Wackowski OA. Examining market trends in smokeless tobacco sales in the United States: 2011–2019. Nicotine Tob Res. 2021;23:1420–4. doi:10.1093/ntr/ntaa239.
19. Marynak KL, Wang X, Borowiecki M, Kim Y, Tynan MA, Emery S. Nicotine pouch unit sales in the US, 2016–2020. JAMA. 2021;326:566–8. doi:10.1001/jama.2021.10366.
20. Market report. Barcelona: Tobacco Intelligence; 2022.
21. Market development. Smokefree products in the US. Stockholm: Swedish Match; 2021 (https://web.archive.org/web/20210617151128/https://www.swedishmatch.com/Our-business/smokefree/Market-development/smokefree-products-in-the-us/).
22. Plurphanswat N, Hughes JR, Fagerstrom K, Rodu B. Initial information on a novel nicotine product. Am J Addict. 2020;29:279–86. doi:10.1111/ajad.13020.
23. Q&A. Amsterdam: Snussie.com; 2019 (https://web.archive.org/web/20210118194301/https://www.snussie.com/en/service/qanda/).
24. How many nicotine pouches per day? London: Haypp.com; 2020 (https://www.haypp.com/uk/nicopedia/how-many-nicotine-pouches-per-day/).
25. Modern and traditional oral products. London: British American Tobacco; 2022 (https://web.archive.org/web/20210614154017/https://www.bat.com/snus).
26. Brand portfolio. Bristol: Imperial Brands; undated (https://web.archive.org/web/20210617202521/https://www.imperialbrandsplc.com/about-us/brand-portfolio.html).

27. Tobacco-free innovations ushers in new era SKRUF. Bristol: Imperial Brands; 2018 (https://web.archive.org/web/20191118113541/https:/www.imperialbrandsplc.com/sustainability/case-studies/reduced-harm-ngps/tobacco-free-innovation-ushers-in-new-era-skruf.html).
28. JTI Sweden and Nordic snus. Stockholm: Japan Tobacco International; undated (https://web.archive.org/web/20210617153749/https://www.jti.com/europe/sweden).
29. Kôp all white och nikotinfri portion [Buy all white and nicotine-free portion]. Göteborg: Microzero AB; undated (https://web.archive.org/web/20210303173503/https://nixs.se/).
30. FTC report finds annual cigarette salesincreased for the first time in 20 years. Washington DC: Federal Trade Commission; 2021 (https://www.ftc.gov/news-events/press-releases/2021/10/ftc-report-finds-annual-cigarette-sales-increased-first-time-20).
31. Havermans A, Pennings JLA, Hegger I, Elling JM, de Vries H, Pauwels CGGM et al. Awareness, use and perceptions of cigarillos, heated tobacco products and nicotine pouches: A survey among Dutch adolescents and adults. Drug Alcohol Depend. 2021;229(Pt_B):109136. doi:10.1016/j.drugalcdep.2021.109136.
32. Pillitteri JL, Shiffman S, Sembower MA, Polster MR, Curtin GM. Assessing comprehension and perceptions of modified-risk information for snus among adult current cigarette smokers, former tobacco users, and never tobacco users. Addict Behav Rep. 2020;11:100254. doi:10.1016/j.abrep.2020.100254.
33. Wackowski OA, Rashid M, Greene KL, Lewis MJ, O'Connor RJ. Smokers' and young adult non-smokers' perceptions and perceived impact of snus and e-cigarette modified risk messages. Int J Environ Res Public Health. 2020;17(18):6807. doi:10.3390/ijerph17186807.
34. Everything about nicotine pouches and the difference between the tobacco pouch snus. London: Haypp.com; undated (https://web.archive.org/web/20210617203157/https://www.haypp.com/eu/whats-a-nicotine-pouch).
35. Nicotine pouches. Stockholm: Zyn.com (https://web.archive.org/web/20210616130922/https://www.zyn.com/international/en/).
36. The history of the tobacco pouch and nicotine pouches. London: Haypp; 2021 (https://web.archive.org/web/20210617203006/https://www.haypp.com/eu/the-history-of-nicotine-pouches).
37. Talbot EM, Giovenco DP, Grana R, Hrywna M, Ganz O. Cross-promotion of nicotine pouches by leading cigarette brands. Tob Control. 2021. doi:10.1136/tobaccocontrol-2021-056899.
38. Czaplicki L, Patel M, Rahman B, Yoon S, Schillo B, Rose SW. Oral nicotine marketing claims in direct-mail advertising. Tob Control. 2021;31(5):663–66. doi:10.1136/tobaccocontrol-2020-056446.
39. Dryft. Moorpark (CA): Kretek International Inc.; 2023 (https://www.kretek.com/company/brands/dryft.)
40. Nicotinepouches,aviablealternativetosmoking?London:EuromonitorInternational;2020(https://www.euromonitor.com/nicotine-pouches-a-viable-alternative-to-smoking-/report).
41. Li L, Borland R, Cummings KM, Gravely S, Quah ACK, Fong GT et al. Patterns of non-cigarette tobacco and nicotine use among current cigarette smokers and recent quitters: findings from the 2020 ITC Four Country Smoking and Vaping Survey. Nicotine Tob Res. 2021;23:1611–6. doi:10.1093/ntr/ntab040.
42. Brose LS, McDermott MS, McNeill A. Heated tobacco products and nicotine pouches:a survey of people with experience of smoking and/or vaping in the UK. Int J Environ Res Public Health. 2021;18(16):8852. doi:10.3390/ijerph18168852.
43. East KA, Reid JL, Rynard VL, Hammond D. Trends and patterns of tobacco and nicotine product use among youth in Canada, England, and the United States from 2017 to 2019. J Adolesc Health. 2021;69:447–56. doi:10.1016/j.jadohealth.2021.02.011.
44. Gentzke AS, Wang TW, Cornelius M, Park-Lee E, Ren C, Sawdey MD et al. Tobacco product use

and associated factors among middle and high school students – National Youth Tobacco Survey, United States, 2021. MMWR Surveill Summ. 2022;71(5):1–29. doi:10.15585/mmwr.ss7105a1.

45. Hrywna M, Gonsalves NJ, Delnevo CD, Wackowski OA. Nicotine pouch product awareness, interest and ever use among US adults who smoke, 2021. Tob Control. 2022. doi:10.1136/tobaccocontrol-2021-057156.
46. Partial guidelines for implementation of Articles 9 and 10 of the WHO Framework Convention on Tobacco Control. Geneva: World Health Organization; 2010 (https://www.who.int/fctc/guidelines/Guideliness_Articles_9_10_rev_240613.pdf?ua=1).
47. Kostygina G, England L, Ling P. New product marketing blurs the line between nicotine replacement therapy and smokeless tobacco products. Am J Public Health. 2016;106:1219–22. doi:10.2105/AJPH.2016.303057.
48. Aldeek F, McCutcheon N, Smith C, Miller JH, Danielson TL. Dissolution testing of nicotine release from OTDN pouches: product characterization and product-to-product comparison. Separations. 2021;8(1):7. doi:10.3390/separations8010007.
49. Rensch J, Liu J, Wang J, Vansickel A, Edmiston J, Sarkar M. Nicotine pharmacokinetics and subjective response among adult smokers using different flavors of on!® nicotine pouches compared to combustible cigarettes. Psychopharmacology (Berl). 2021;238(11):3325–34. doi:10.1007/s00213-021-05948-y.
50. Hukkanen J, Jacob P 3rd, Benowitz NL. Metabolism and disposition kinetics of nicotine. Pharmacol Rev. 2005;57:79–115. doi:10.1124/pr.57.1.3.
51. Digard H, Proctor C, Kulasekaran A, Malmqvist U, Richter A. Determination of nicotine absorption from multiple tobacco products and nicotine gum. Nicotine Tob Res. 2013;15:255–61. doi:10.1093/ntr/nts123.
52. Benowitz NL. Clinical pharmacology of nicotine: implications for understanding, preventing, and treating tobacco addiction. Clin Pharmacol Ther. 2008;83:531–41. doi:10.1038/clpt.2008.3.
53. Pickworth WB, Rosenberry ZR, Gold W, Koszowski B. Nicotine absorption from smokeless tobacco modified to adjust pH. J Addict Res Ther. 2014;5:1000184. doi:10.4172/2155-6105.1000184.
54. Palmer AM, Toll BA, Carpenter MJ, Donny EC, Hatsukami DK, Rojewski AM et al. Reappraising choice in addiction: novel conceptualizations and treatments for tobacco use disorder. Nicotine Tob Res. 2022;24(1):3–9. doi:10.1093/ntr/ntab148.
55. A summary of data on the bioavailability of nicotine and other ingredients from the use of oral nicotine pouches and assessment of risk to users. London: HM Government, Committee on Toxicity of Chemicals in Food, Consumer Products and the Environment. 2021 (https://cot.food.gov.uk/sites/default/files/2021-05/TOX-2021-22%20Nicotine%20pouches.pdf).
56. Price LR, Martinez J. Cardiovascular, carcinogenic and reproductive effects of nicotine exposure: A narrative review of the scientific literature. F1000Res. 2019;8:1586. doi:10.12688/f1000research.20062.2
57. REACH (Registration, Evaluation, Authorisation and Restriction of Chemicals) (EC 1907/2006). Brussels: European Commission, Environment; 2006 (https://ec.europa.eu/environment/chemicals/reach/reach_en.htm).
58. Ozga JE, Felicione NJ, Elswick D, Blank MD. Acute effects of snus in never-tobacco users: a pilot study. Am J Drug Alcohol Abuse. 2018;44:113–9. doi:10.1080/00952990.2016.1260581.
59. Azzopardi D, Liu C, Murphy J. Chemical characterization of tobacco-free "modern" oral nicotine pouches and their position on the toxicant and risk continuums. Drug Chem Toxicol. 2022;45(5):2246–54. doi:10.1080/01480545.2021.1925691.
60. Keogh A. Nicotine pouches. Br Dent J. 2021;230:61–2. doi:10.1038/s41415-021-2622-y.

61. Sahni V. Blurred line on nicotine. Br Dent J. 2021;230:325. doi:10.1038/s41415-021-2855-9.
62. Dalrymple A, Bean EJ, Badrock TC, Weidman RA, Thissen J, Coburn S et al. Enamel staining with e-cigarettes, tobacco heating products and modern oral nicotine products compared with cigarettes and snus: an in vitro study. Am J Dent. 2021;34:3–9. PMID:33544982.
63. Chaffee BW, Couch ET, Vora MV, Holliday RS. Oral and periodontal implications of tobacco and nicotine products. Periodontology. 2000;87(1):241–53. doi:10.1111/prd.12395.
64. Bishop E, East N, Bozhilova S, Santopietro S, Smart D, Taylor M et al. An approach for the extract generation and toxicological assessment of tobacco-free "modern" oral nicotine pouches. Food Chem Toxicol. 2020;145:111713. doi:10.1016/j.fct.2020.111713.
65. Khouja JN, Suddell SF, Peters SE, Taylor AE, Munafò MR. Is e-cigarette use in non-smoking young adults associated with later smoking? A systematic review and meta-analysis. Tob Control. 2020;30(1):8–15. doi:10.1136/tobaccocontrol-2019-055433.
66. Yoong SL, Hall A, Turon H, Stockings E, Leonard A, Grady A et al. Association between electronic nicotine delivery systems and electronic non-nicotine delivery systems with initiation of tobacco use in individuals aged < 20 years. A systematic review and meta-analysis. PLoS One. 2021;16(9):e0256044. doi:10.1371/journal.pone.0256044.
67. Thornley S, McRobbie H, Lin RB, Bullen C, Hajek P, Laugesen M et al. A single-blind, randomized, crossover trial of the effects of a nicotine pouch on the relief of tobacco withdrawal symptoms and user satisfaction. Nicotine Tob Res. 2009;11:715–21. doi:10.1093/ntr/ntp054.
68. ECigIntelligence.com.
69. Duren M, Atella L, Welding K, Kennedy RD. Nicotine pouches: a summary of regulatory approaches across 67 countries. Tob Control. 2023. doi:10.1136/tc-2022-057734.
70. Harmonized Commodity Description and Coding System. Brussels: World Customs Organization; 2023 (https://www.wcoomd.org/en/topics/nomenclature/overview.aspx#:~:text=What%20is%20the%20Harmonized%20System,World%20Customs%20Organization%20(WCO)).
71. Kamerbrief tabak en alcohol [Parliamentary brief tobacco and alcohol]. The Hague: State Secretary of Health, Ministry of Health, Well-being and Sport; 2021 (https://www.rijksoverheid.nl/documenten/kamerstukken/2021/11/09/verzamelbrief-tabak-en-alcohol).
72. Beoordeling van het nicotinegehalte in nicotinezakjes waarbij de Acute reference Dose niet overschreden wordt [Assessment of the nicotine in nicotine pouches not exceeding the acute reference dose]. Bilthoven: National Institute for Public Health and the Environment; 2021 (https://www.rivm.nl/sites/default/files/2021-11/FO_nicotinezakjes%20tox_20211101_def_anon.pdf).

5. Biomarkers of exposure, effect and susceptibility for assessing electronic nicotine delivery devices and heated tobacco products, and their possible prioritization

Irina Stepanov, McKnight Distinguished University Professor and Director, Institute for Global Cancer Prevention Research, University of Minnesota, Minneapolis (MN), USA

Stephen S. Hecht, PhD, Wallin Professor of Cancer Prevention, American Cancer Society Professor, Masonic Cancer Center, University of Minnesota, Minneapolis (MN), USA

Contents

Abstract

Biomarkers have been used extensively in studies of cigarettes and other conventional tobacco products, providing valuable data on harmful exposures, biological effects, and the disease susceptibility of users and non-users exposed to second-hand smoke. This report provides an evaluation of the published literature on use of such biomarkers in studies of electronic nicotine delivery systems (ENDS) and heated tobacco products (HTPs) and an assessment of the potential utility and limitations of biomarkers in tobacco control. The reviewed evidence indicates that switching from smoking conventional cigarettes to exclusive ENDS use is associated with reductions in biomarkers of exposure to several toxicants and carcinogens that play key roles in smoking-induced diseases. The levels of many such biomarkers are, however, higher in dual users (people who continue to use cigarettes and ENDS at the same time), which is much more common than switching completely. In addition, the health effects of the changes in exposure are not yet well understood, and biomarkers of biological effects suggest that ENDS pose certain risks to users – particularly dual users and when compared with non-use of any tobacco or nicotine product. The review of the published literature underscores the lack of independent, non-industry research on exposure and effects resulting from HTP use. The report proposes a panel of priority biomarkers for tobacco control, identifies relevant research gaps, notes the need for industry-independent research, and recommends regulatory priorities.

Keywords: biomarker, exposure, biological effect, toxicity, electronic cigarette, electronic nicotine delivery system, heated tobacco product, health effect

5.1 Background

This section was commissioned to provide evidence-based recommendations on the use of biomarkers for assessing the nicotine and tobacco products that have emerged in the 21st century, in particular electronic nicotine delivery devices (ENDS) and heated tobacco products (HTPs), and to propose policy options to achieve the objectives and measures outlined in the relevant decision (FCTC/COP8(22)). This document serves as a background paper for the ninth technical report of the WHO Study Group on Tobacco Product Regulation (TobReg).

Biomarkers are powerful tools for objective assessment of human exposure to chemical toxicants and carcinogens in tobacco products, ENDS and HTPs and the resulting disease-related biological effects. Such objective assessment is crucial, because product analysis alone is insufficiently informative for predicting constituent uptake, which is significantly affected by users′ behaviour *(1,2)*. Furthermore, chronic diseases associated with use of nicotine and tobacco products, such as cancer, chronic obstructive pulmonary disease (COPD) and cardiovascular disease, take a long time to develop, and their monitoring – while essential for long-term tobacco control policies – is not suitable for regulatory decisions when new products are introduced onto the market. Therefore, biomarkers can serve as surrogate indicators for assessing such health risks. Nevertheless, use of biomarkers for regulatory purposes has been limited. A contributing factor is that countries with limited resources for tobacco control must prioritize their allocation effectively to reduce the public health harm that results from tobacco use, and, in many cases, other activities are prioritized before measuring biomarkers. Furthermore, the challenge of distinguishing differences in the levels of biomarkers due to variations among products from the differences due to user behaviour was the basis for TobReg′s conclusion in 2008 that measurement of biomarkers is not a suitable regulatory strategy for monitoring differences among cigarette products *(3)*. Substantial new research has, however, been conducted during the past 15 years, with new biomarkers, new technologies and new evidence. Further, manufacturers continue to introduce a constant stream of new nicotine and tobacco product types onto markets worldwide, some of which differ significantly in their chemical composition not only from traditional products such as cigarettes but also from other relatively recent products. For example, HTPs and, more recently, tobacco-free nicotine pouches are significantly different from e-cigarettes in their chemical composition and mode of use; therefore, knowledge on the health effects associated with e-cigarettes cannot be used directly to guide regulatory decisions on such emerging products. Therefore, advances in biomarker research must be summarized in the context of the current product landscape, in order to reassess their potential use as proxies in tobacco control.

In this paper, we review the current literature on biomarkers of exposure, biomarkers of biological effects, including those associated with specific diseases, and biomarkers of susceptibility that have been used in studies of electronic cigarettes and other ENDS and HTPs. In particular, it covers:

- biomarkers of exposure used for ENDS and HTPs;
- biomarkers of biological effects that are part of the pathophysiology of various relevant diseases;
- biomarkers of susceptibility;
- a discussion of the state of biomarker research and implications for tobacco control, including research gaps and limitations of biomarkers;
- recommendations on possible prioritization of biomarkers for tobacco control;
- recommendations on addressing research gaps and priorities; and
- relevant policy recommendations.

The literature search was conducted primarily in the PubMed database and the SciFinder search tool, which retrieves data from the Medline and CAplus databases. Important relevant articles cited in publications obtained in the database research were also included. In addition, the websites of the US Centers for Disease Control and Prevention, the US Food and Drug Administration and other relevant websites that contain information on exposures and effects associated with ENDS and HTPs were used.

5.2 Biomarkers of exposure

5.2.1 Definition and overview of biomarkers of exposure commonly used in studies of tobacco and nicotine products

A biomarker is defined by Oxford Languages as a measurable substance in an organism the presence of which is indicative of some phenomenon such as environmental exposure. Within this general definition, a biomarker of exposure is an entity that can be reliably quantified and is related to a specific exposure. In the context of this report, a biomarker of exposure can confirm use of, or exposure to, specific nicotine or tobacco products, or indicate changes in exposure to specific chemical compounds when individuals switch between products. The structures of the biomarkers discussed in this report and their sources are presented in Fig. 1.

Fig. 1. Structures of some constituents and biomarkers discussed in this report

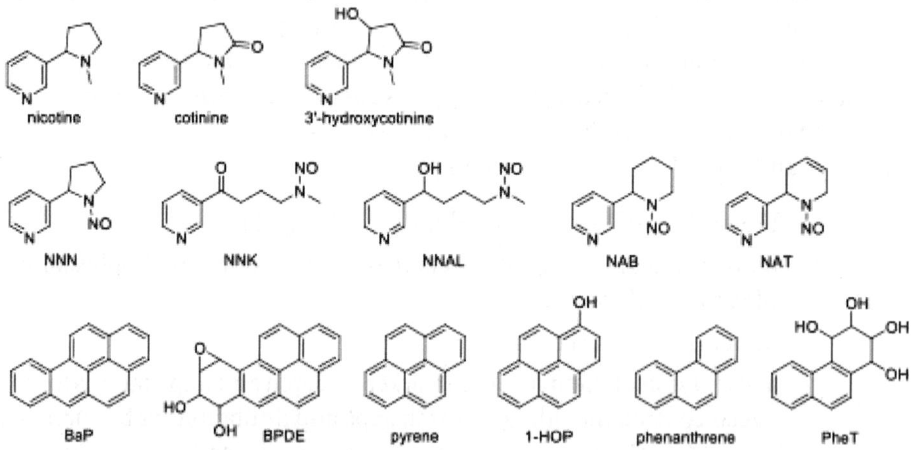

Carbon monoxide (CO). CO is a product of incomplete combustion of organic matter. Exhaled CO is a useful, widely applied biomarker of exposure to all tobacco products the use of which involves combustion; these include cigarettes, cigars, pipes and hookah, but can also include HTPs because of the evidence of some level of combustion when such products are used. Marijuana smoking can also increase exhaled CO. CO is not produced in significant amounts during use of ENDS, HTPs or smokeless tobacco if no combustion is involved. Exposure to CO is associated with blood carboxyhaemoglobin (COHb) but is more commonly measured as CO in exhaled breath, as this test can be performed easily with commercially available devices. Various cut-off points of exhaled CO have been proposed to distinguish smokers from non-smokers, as other factors, such as high levels of environmental pollution, can affect measurements. A cut-off point of 5–6 parts per million (ppm) CO in exhaled breath has been suggested to distinguish users of smoked tobacco products from users of other tobacco products or non-users of any tobacco product *(4)*. CO binds rapidly to haemoglobin in the blood, which can lead to various health effects by diminishing its oxygen-carrying ability. Such effects can be a particular problem for people with underlying cardiovascular or pulmonary disease. Cigar smoking is an especially rich source of CO exposure *(5)*.

Nicotine and its metabolites. Addiction to nicotine is the single most important reason why people continue to use products that efficiently deliver this substance, despite the known adverse health effects of tobacco product use. All tobacco and nicotine products deliver nicotine, with varying pharmacokinetics, resulting in binding to nicotinic cholinergic receptors in the brain and the release of dopamine, which mediates the pleasurable sensations associated with use of these products *(4,6)*. The time between inhaling tobacco smoke and the release of dopamine is only a few seconds, which helps to explain smokers′ addiction *(4,6)*. As the half-life of nicotine in the body is only about 2 h, it is not a very useful quantitative biomarker of nicotine exposure. The major metabolite of nicotine – cotinine – has been widely used as a biomarker of nicotine uptake due to its longer half-life of approximately 16 h (range, 8–30 h, depending on individual characteristics). Thus, cotinine has been quantified in serum, plasma, whole blood, saliva and urine as a biomarker of nicotine uptake. While cotinine is a good general biomarker of nicotine exposure, individual differences in enzymes involved in its formation and further metabolism, including CYP2A6 and UGT2B10, can affect cotinine measurements. Thus, the gold standard biomarker of nicotine exposure is "total nicotine equivalents", which comprise urinary nicotine, cotinine and 3′-hydroxycotinine and their glucuronides. This biomarker is strongly correlated with urinary metabolite measurements that include these compounds and also with several minor nicotine metabolites such as nicotine *N*-oxide *(4,6)*.

Tobacco-specific nitrosamines and metabolites. Tobacco-specific nitrosamines are a group of carcinogens formed during tobacco curing and processing by reactions of tobacco alkaloids such as nicotine, nornicotine, anabasine and anatabine with nitrite in tobacco *(7–10)*. All tobacco-containing products contain tobacco-specific nitrosamines, including *N*′-nitrosonornicotine (NNN), *N*′-nitrosoanabasine, *N*′-nitrosoanatabine and 4-(methylnitrosamino)-1-(3-pyridyl)-1-butanone (NNK), as well as some minor products *(11)*. Tobacco-specific nitrosamines are present at lower levels in HTP emissions than in the smoke of conventional cigarettes but are generally not found, or are present in very low quantities, in the emissions from ENDS, as discussed below. As the name implies, the occurrence of these carcinogens is specific to tobacco products, including smoked and smokeless tobacco products *(12)*. NNN and NNK are powerful carcinogens, inducing tumours at relevant sites in laboratory animals, such as the oral mucosa, oesophagus, and lung *(9)*. Tumours are observed in animals treated chronically with low doses of these compounds *(13,14)*. It has been clearly demonstrated that users of tobacco products take up NNN and NNK *(11)*. Thus, NNN and NNK are widely regarded as important causes of cancer in people who use smokeless tobacco or smoked products; they were classified as "carcinogenic to humans" by the International Agency for Research on Cancer (IARC) *(11)*. NNK is metabolized in laboratory animals and humans to 4-(methylnitrosamino)-1-(3-pyridyl)-1-butanol (NNAL), which has carcinogenic activity similar to that of NNK *(9)*. Urinary NNAL has been widely used as a biomarker of exposure to NNK *(10)*. Its tobacco specificity and carcinogenic activity combined make it an important biomarker of exposure to tobacco carcinogens and of cancer risk. Prospective epidemiological studies of cigarette smokers have demonstrated a significant association between relatively high levels of urinary NNAL and lung cancer risk *(15)*. Urinary NNN has similarly been used as a biomarker of NNN exposure and carcinogenicity and was significantly related to the incidence of oesophageal cancer in a prospective study of cigarette smokers *(16)*. Thus, NNAL and NNN are considered potentially useful biomarkers of relevant cancer risks in people who use tobacco-containing products and may be useful in predicting cancer risk; however, further studies are required. NNN has also been identified in the saliva of e-cigarette users as a result of endogenous formation in the oral cavity from nicotine and/or its metabolite nornicotine *(17)*.

Polycyclic aromatic hydrocarbon (PAH) metabolites. PAH, like CO, are products of incomplete combustion of organic matter. Thus, mixtures of PAH are found in the smoke of cigarettes, cigars, pipes and hookah but in far lower quantities in smokeless tobacco that does not contain fire-cured tobacco, where their presence is due in part to environmental pollution *(18–20)*. Similarly, the levels of PAH are consistently lower than in conventional smoked products or not detected at all in ENDS or HTPs *(21)*. PAH have been known since the 1970s to contribute significantly to tobacco-smoke carcinogenesis from studies in many animal models of the carcinogenic activity of selected subfractions and individual compounds in tobacco smoke condensate, including benzo[*a*]pyrene (BaP), chrysene, methylchrysenes, benzofluoranthenes, benz[*a*]anthracenes and others *(22)*. More than 500 PAH have been at least partially identified in tobacco smoke, and BaP as a representative PAH is classified as "carcinogenic to humans" by IARC *(18,23)*. 1-Hydroxypyrene (1-HOP) and hydroxyphenanthrenes, urinary metabolites of the non-carcinogenic PAH pyrene and phenanthrene, which are components of all PAH mixtures, have been widely used as biomarkers of exposure to PAH *(14)*. The population-based National Health and Nutrition Examination Survey (NHANES) showed significantly higher levels of these metabolites in cigarette smokers than non-smokers in the USA, the exposure of the latter group resulting from inhaling polluted air or consuming charbroiled food *(24)*. Cigarette smoking has consistently been shown to be a major source of exposure to PAH.

A pathway of metabolism of PAH that leads to carcinogenesis is formation of diol epoxides *(25)*. This important metabolic pathway can be quantified by analysis of a urinary BaP-tetraol, an end-product of metabolic BaP-diol epoxide hydrolysis; however, a more practical approach is use of phenanthrene tetraol (PheT), as its concentration in urine is more than 1000 times higher than that of BaP tetraol *(10,26)*. The levels of PheT in cigarette smokers were significantly associated with lung cancer in the Shanghai Cohort study *(15)*. There is no doubt that PAH contribute to cancer risk in cigarette smokers, although the relative extent of their contribution to the etiology of specific cancers versus those of other toxicants and carcinogens discussed here is presently unknown.

Volatile toxicants and carcinogens and their metabolites. Numerous volatile toxicants and carcinogens are produced during the combustion of tobacco. These include the IARC Group 1 (carcinogenic to humans) compounds formaldehyde, ethylene oxide, benzene and 1,3-butadiene; Group 2A (probably carcinogenic to humans) compounds acrolein, acrylamide, dimethylformamide and styrene; and Group 2B (possibly carcinogenic to humans) compounds propylene oxide, acrylonitrile, crotonaldehyde, ethyl benzene and propylene oxide, among others *(27)*. Other volatile compounds, such as methacrolein and methyl vinyl ketone, have well-established toxic effects similar to those of acrolein *(28–30)*. Acrolein is considered one of the most toxic compounds in tobacco smoke, and its non-cancer hazard index is the highest among common smoke constituents *(31)*. Acrolein and related compounds are implicated as causes of COPD *(32)*. Most of these compounds or their metabolites, mainly mercapturic acids, are detectable in blood or urine of all humans due to endogenous processes, inflammation and environmental or dietary exposure, but cigarette, cigar, pipe, hookah and marijuana smoking usually result in significantly higher levels than in non-smokers *(33–46)*. Cyanoethyl mercapturic acid (CEMA), a metabolite of acrylonitrile, which is not an endogenous compound and is seldom encountered in significant quantities in the general environment except in tobacco smoke, is a particularly useful biomarker for distinguishing users of smoked tobacco products from non-smokers. Thus, a cut-off point of 27 pmol/mL urine of CEMA differentiated cigarette smokers from nonsmokers with a sensitivity and specificity greater than 99% *(47)*. ENDS and HTPs also generate volatile toxicants and carcinogens but generally at much lower levels than conventional cigarettes *(48,49)*.

Metals. The occurrence of various metals, including arsenic, beryllium, cadmium, hexavalent chromium, cobalt, lead, nickel and radioactive polonium, has been reported in tobacco *(20)*. The highest mean concentrations in total particulate matter of cigarette smoke were those of cadmium and lead, at 40.2 ± 5.4 and 11.0 ± 1.1 ng/cigarette (ISO conditions), respectively *(50)*. These results were consistent with those obtained in other studies *(20)*. In agreement with these data, the NHANES study demonstrated that blood and urinary cadmium levels and blood lead levels were higher in smokers than in non-smokers *(51)*. Similar results were found for cadmium in blood and urine of a German population *(52)*. Cadmium and its compounds are carcinogenic to humans, causing lung cancer and possibly kidney and prostate cancers *(53)*. Lead is toxic to the neurological, renal, cardiovascular, haematological, immunological, reproductive and developmental systems *(54)*.

Some studies have reported the presence of metals in ENDS aerosols. Chromium and lead were reliably measured in e-cigarette aerosol *(55,56)*. Other studies have reported the presence of cadmium, copper, nickel, manganese, aluminium and tin and shown that product design is an important factor in the levels of metals *(57,58)*. Because they remain for a long time in exposed people, cadmium and lead may serve as long-term markers of cumulative exposure *(59)*.

5.2.2 Application of biomarkers of exposure in studies of ENDS and HTPs

Carbon monoxide (CO)

Several investigations, including randomized clinical trials and cross-sectional studies, have included quantification of exhaled CO or blood COHb and demonstrated significantly lower levels in ENDS users than in cigarette smokers, and most studies did not find elevated CO or COHb in exclusive ENDS users (reviewed in *21* and *60*). Some examples are cited here. Oliveri et al. *(61)* reported a 47% lower concentration of COHb in adult exclusive ENDS users than in cigarette smokers. Hatsukami et al. *(62)* observed a significant 60% reduction in expired CO when cigarette smokers switched to ENDS for 8 weeks, although not all study participants switched to exclusive ENDS use. McRobbie et al. *(63)* reported a significant 80% decrease in expired CO when cigarette smokers switched to ENDS for 4 weeks. Czoli et al. *(64)* found a significant 41% reduction in expired CO when subjects switched from 7 days of dual use of cigarettes and ENDS to 7 days of exclusive ENDS use. O´Connell et al. *(65)* found a significant 88–89% reduction in expired CO and an 84–86% reduction in COHb when cigarette smokers switched to ENDS for 5 days. Cravo et al. *(66)* reported rapid decreases in expired CO and blood COHb in subjects who switched from conventional cigarettes to ENDS. Expired CO decreased from 20.3 ppm to 7.4 ppm after 1 week of ENDS use and was 7.6–9.0 ppm from week 2 until the end of the study (12 weeks), and COHb decreased from 6.79% to 4.06–4.37% after 1 week of ENDS use until the end of the study. Morris et al. *(67)* reported a 79% reduction in COHb when subjects switched from conventional cigarettes to ENDS use for 9–14 days.

A review by Akiyama and Sherwood (tobacco industry researchers) *(60)* provides comparative biomarker results after cigarette smoking and in 30 clinical trials of HTPs, with a median intervention period of 8 days. Reductions of 80–90% in expired CO and of 50–90% in COHb were observed within 1 week in most studies. These results are consistent with significantly less combustion in HTPs than during cigarette smoking. For example, in one study in which subjects switched from conventional cigarettes to an HTP, expired CO decreased by approximately 80% within 6–7 days, reaching levels similar to those achieved after smoking cessation *(68)*. In a comparison of a menthol HTP with menthol cigarette smoking, average COHb was reduced by 62% within 5 days of switching from smoking to the HTP, similar to that achieved in 5 days of abstinence *(69)*.

Nicotine and its metabolites

Randomized clinical trials and cross-sectional studies of the levels of biomarkers of nicotine and its metabolites in cigarette smokers and ENDS users were reviewed by Akiyama and Sherwood *(60)* and by Scherer et al. *(21)*. Some studies indicated lower levels of urinary total nicotine equivalents in ENDS users than in cigarette smokers, while others reported no difference. For example, Round et al. *(70)* conducted a randomized, parallel-group clinical study of smokers who switched to an ENDS product for 5 days. Total nicotine equivalents measured in 24-h urine samples decreased by 38.3% ($P < 0.05$), and plasma cotinine and nicotine were similarly statistically significantly decreased. Shahab et al. *(71)* conducted a cross-sectional study and found that urinary total nicotine equivalents were not significantly different in cigarette smokers and ENDS users. In theory, the levels of total nicotine equivalents should be similar in smokers and ENDS users, as both products are designed to deliver nicotine efficiently, and there is likely to be some self-titration, although differences in ENDS product characteristics and use patterns may lead to the different results. It is important to note that use of ENDS with non-salt liquids, in which most of the nicotine is present in unprotonated form, leads to predominantly oral absorption of nicotine. This results in slower nicotine pharmacokinetics and may therefore have lower abuse liability than conventional cigarettes. Many currently marketed ENDS contain nicotine in the form of salts, however, which makes ENDS aerosols easy to inhale and results in faster nicotine absorption.

In the review by Akiyama and Sherwood *(60)*, the levels of total nicotine equivalents in HTP users were similar to those in cigarette smokers in most studies, not differing by more than 20%. For example, in a three-arm, parallel-group study, 160 Japanese adult smokers were randomized to a menthol HTP ($n = 78$) or a menthol cigarette ($n = 42$) for 5 days in a confined setting and 85 days in ambulatory settings. No significant differences in the levels of total nicotine equivalents were found between the HTP and conventional cigarette users in either setting *(72)*, although substantial differences in the characteristics of different HTPs may be related to differences in nicotine delivery.

Tobacco-specific nitrosamines and metabolites

Significant reductions in urinary NNAL were reported in all the randomized clinical trials in which cigarette smokers switched to ENDS or HTPs, and its levels were also significantly lower in ENDS users than in cigarette smokers in cross-sectional studies, including the Population Assessment of Tobacco and Health (PATH) study *(60,73,74)*. NNAL is barely detected in the urine of ENDS users because it is a metabolite of NNK, which occurs only in tobacco-containing products. The low levels that are occasionally detected in ENDS users may be due in part to carryover from use of tobacco products (due to the long half-life of NNAL) or exposure to second-hand tobacco smoke *(75–77)*. NNN levels were either extremely low or not detected in the urine of ENDS users *(60,78)*. Bustamante et al. *(17)* presented evidence for the presence of NNN in the saliva of ENDS users (14.6 ± 23.1 pg/mL) and concluded that it was formed endogenously, as it was not detected above trace amounts in ENDS liquids. Scherer et al. *(79)* did not find statistically significantly higher levels of tobacco-specific nitrosamines or their metabolites in the urine or saliva of ENDS or HTP users than in non-users of tobacco products.

PAH metabolites

A randomized clinical trial of cigarette smokers who switched to ENDS for 5 days showed significant 63.5% and 63.8% reductions in urinary 1-HOP and 3-hydroxyBaP, respectively, as well as significant reductions in fluorene and naphthalene metabolites *(70)*. The results were similar when mentholated products were used. A similar 5-day switching trial showed a 70.5% reduction in urinary 1-HOP *(65)*. In another trial in which cigarette smokers switched to ENDS for 8 weeks, significant 20% reductions were found in PheT *(62)*. A comparison of urinary 1-HOP levels in ENDS users with those reported in three studies of cigarette smokers showed significant 57–61% reductions in ENDS users *(80)*. As reviewed by Akiyama and Sherwood *(60)*, many randomized clinical trials have shown a reduction in 1-HOP after switching from conventional cigarettes to HTPs. The reductions were frequently greater than 60%, although some trials reported 15–30% reductions. Similar results were observed in the PATH study *(74)*. The results summarized here are consistent with substantial decreases in exposure to combustion products in users of both ENDS and HTPs.

Volatile toxicants and carcinogens and their metabolites

Consistently, randomized clinical trials of cigarette smokers who switched to ENDS found significant decreases in biomarkers of exposure to volatile toxicants and carcinogens, including acrolein, acrylamide, acrylonitrile, benzene, 1,3-butadiene, crotonaldehyde, and ethylene oxide *(60)*. In a study in which participants were randomized to 8 weeks of instructions for complete substitution of cigarettes with e-cigarettes, significant decreases in urinary biomarkers of acrylamide (32%), acrolein (47%), acrylonitrile (66%) and crotonaldehyde (47%) were observed *(62)*. In a study in which smokers were randomized to 5 days of ENDS use, significant decreases were found in the levels of mercapturic acids of acrolein (70.5%), acrylonitrile (85.9%), benzene (89.7%), 1,3-butadiene (55.5%), crotonaldehyde (77.5%) and ethylene oxide (62.3%) *(71)*. Cross-sectional studies, including the PATH study *(60,74)* and a recent study in which cigarette smokers or ENDS users were confined for 3 days *(81)*, gave similar results.

A number of studies, most of which were based on one or two times, have shown higher levels of urinary biomarkers of exposure to volatile agents such as acrylonitrile, acrolein, crotonaldehyde and propylene oxide in ENDS users than non-users of any tobacco or nicotine product (reviewed in *82*). One study in which urine samples were obtained monthly for 4–6 months found significantly higher levels of 3-hydroxypropyl mercapturic acid, a major metabolite of acrolein, in the urine of ENDS users than in non-users of any tobacco or nicotine product *(82)*.

Several randomized clinical trials of cigarette smokers who switched to HTPs, conducted by industry researchers, also showed large decreases in the mercapturic acids of volatiles. The results were consistent across all the published industry trials *(60)*.

A clinical study in which 10 subjects per group were confined for 3 days and used only their specified product (cigarettes, ENDS, HTPs, oral tobacco, nicotine replacement therapy or non-users of any tobacco or nicotine product) showed slight increases in mercapturic acids related to acrolein, acrylamide, and crotonaldehyde in HTP users than in users of other non-cigarette products *(81)*.

Metals

Cadmium and lead are the toxic metals to which cigarette smokers are exposed at the highest levels, as noted above. Data from Wave 1 of the PATH study (2013–2014) also indicated significantly higher levels of urinary cadmium and lead in ENDS users than in never users of any tobacco product or ENDS, by 23% and 19%, respectively *(73)*. The authors noted that the long half-lives of biomarkers of metal exposure were possible confounding factors, as some ENDS users may have been former smokers or were exposed in other ways. Prokopowicz et al. *(83)* reported that the blood levels of cadmium decreased significantly in cigarette smokers who switched to ENDS, while there was no significant difference in blood lead levels. Smokers had significantly higher levels of both biomarkers than non-smokers. A cross-sectional study of urinary elements including chromium, nickel, cobalt, silver, indium, manganese, barium, strontium, vanadium and antimony, in addition to cadmium and lead, showed no differences in the levels of these elements in ENDS users and non-smokers *(84)*. A review found inconsistent results with respect to biomarkers of lead, chromium, nickel, selenium and strontium in ENDS users as compared with non-users *(85)*.

No biomarkers of exposure to metals have been reported in HTP users. A search in PubMed for "metal exposures heated tobacco products" and a similar Google search produced only unrelated articles or monographs.

Salivary propylene glycol as a novel biomarker for ENDS

Propylene glycol is a major constituent of e-cigarette aerosol. An assay was developed recently for measuring propylene glycol in saliva,[2] showing that the average concentrations of propylene glycol in the saliva of ENDS users were approximately 100 times higher than those in non-smokers and 30 times higher than those in smokers. Therefore, salivary propylene glycol could be used as a novel biomarker to validate ENDS use.

Biomarkers of exposure and dual use

A significant number of smokers who adopt ENDS continue to smoke conventional cigarettes (referred to as "dual users"), with varying degrees of substitution *(86)*. Studies have shown that dual users generally have similar or higher levels of many biomarkers of exposure compared to those of exclusive smokers and that complete switching to ENDS is necessary to achieve meaningful reductions in exposure *(87–90)*. In a recent study, Anic et al. *(90)* used biomarker data from 2475 adults in the PATH Study who were smokers in Wave 1 (2013–2014) and who transitioned to exclusive or dual ENDS use or quit tobacco products in Wave 2 (2015). Cigarette smokers who became dual users of cigarettes and ENDS did not have significant reductions in most of the assessed biomarkers. Table 1 gives examples of data from that study, with the levels of some of the biomarkers discussed above for smokers who continued exclusive smoking, became dual users, or quit any tobacco or nicotine use.

Table 1. Biomarker levels in PATH study participants in Wave 2, by product use status

	Product use status			
Biomarker (source)	Exclusive smoking	Dual smoking and ENDS use	Exclusive ENDS use	No use of tobacco or nicotine
TNE, µmol/g creatinine (nicotine)	31.2 [28.0 ; 34.8]	38.5 [30.3 ; 48.9]	9.1 [3.6 ; 22.9]	0.1 [0.0 ; 0.1]
NNAL, ng/g creatinine (NNK)	218.1 [199.2 ; 238.8]	231.9 [187.0 ; 287.5]	12.5 [5.7 ; 27.3]	5.0 [3.6 ; 7.0]
1-HOP, ng/g creatinine (pyrene)	316.8 [298.3 ; 336.4]	308.1 [277.4 ; 342.2]	113.4 [93.0 ; 138.4]	167.9 [148.9 ; 189.4]
CEMA, µg/g creatinine (acrylonitrile)	131.7 [121.1 ; 143.2]	128.1 [104.7 ; 156.6]	8.6 [4.9 ; 14.9]	3.9 [2.9 ; 5.3]
3HPMA, µg/g creatinine (acrolein)	1342.2 [1247.8 ; 1443.7]	1531.6 [1321.6 ; 1774.9]	303.8 [228.7 ; 403.7]	299.9 [255.8 ; 351.5]
4HBMA, µg/g creatinine (1,3-butadiene)	31.8 [29.6 ; 34.2]	33.9 [28.9 ; 39.8]	5.1 [3.9 ; 6.8]	5.4 [4.6 ; 6.4]

Source: Anic et al. *(90)*
Each cell shows the geometric mean and [95% CI]. All participants included in these analyses were exclusive smokers at wave 1.

2 Tang MK, Carmella SG, unpublished data; 2022.

5.3 Biomarkers of biological effects (harm or disease)

5.3.1 Definitions and overview of biomarkers of biological effects commonly used in studies of tobacco and nicotine products

Various definitions of biomarkers of biological effect have been used in the literature *(91)*. These biomarkers are commonly referred to as "biomarkers of potential harm", which have been defined as "the measurement of an effect due to exposure; these include early biological effects, alterations in morphology, structure, or function, and clinical symptoms consistent with harm, including preclinical changes." *(92–94)*. It should be noted that this definition encompasses (i) a continuum of biological effects and (ii) a spectrum of relevant diseases. In the context of tobacco and nicotine product use, the predominant health outcomes of relevance include cancer and cardiovascular and respiratory diseases. Time is also an important variable. Interpretation of the implications of an acute change as opposed to a long-term change may depend on the biomarker. A definition that is tailored to 'combustible' tobacco product use was proposed by tobacco industry researchers, which specifies that a biomarker of biological effect is "a significant, objective, measurable alteration in a biological sample, after smoking a tobacco product, ... which is altered in a proportion of smokers and is reversible on cessation of smoking" *(95)*. This definition includes the notion of reversibility, which is relevant for studies of potential changes in biological effects when users of traditional tobacco products (e.g. conventional cigarettes) switch to tobacco or nicotine products with different harmful constituent yields (e.g. ENDS). The consequences of biomarker reversibility should be further investigated in studies of changes in health effects.

DNA adducts. DNA **add**ition prod**ucts**, commonly called adducts, are produced by reactions with DNA of certain organic or inorganic intermediates formed during cellular metabolism of inhaled toxicants or carcinogens as well as by reactions of intermediates formed from some endogenous compounds. Adducts to DNA bases or phosphates are central to the carcinogenic process because they can cause miscoding in DNA and the consequent mutations observed in many critical growth control genes involved in cancer. Cells have DNA repair systems to mend the damage, but, when the repair systems are inefficient or error prone, mutations can occur in DNA when adducted bases are misread and the wrong base is inserted by DNA polymerases. The result is a permanent mutation, which may occur in critical genes involved in growth control, leading to cancer initiation. Many mutations are produced in the DNA of various tissues during the metabolism of tobacco carcinogens *(96–101)*. Numerous studies with a variety of methods, including ^{32}P-postlabelling, immunoassays and mass spectrometry, have examined specific types of DNA adducts in various tissues of cigarette smokers and non-smokers *(102–108)*. Many of the studies show higher levels of certain DNA adducts in tissues of smokers than in those of non-smokers, but interpretation was complicated as, in some cases, the numbers of subjects were small or the methods lacked appropriate validation.

Cytokines, chemokines and reactive proteins. Inflammation and oxidative damage are significant factors in various diseases caused by cigarette smoking, including cancer, cardiovascular disease and COPD. The processes involve infiltration of lymphocytes, macrophages and neutrophils into tissues under stress and secretion of pro- and anti-inflammatory cytokines and other factors. Such biomarkers are commonly measured in plasma or serum, and some – such as interleukin (IL)-6 and C-reactive protein (CRP) – are significantly higher in smokers than in nonsmokers *(107)*. The levels of these biomarkers have been directly linked to various relevant diseases. For example, CRP, BCA-1/CXCL13, MDC/CCL22 and IL-1RA have been prospectively associated with the risk of lung cancer *(108)*. In relation to cardiovascular disease (CVD), oxidized LDL is a representative indicator of lipid profiles *(109)*, and CRP, IL-6, fibrinogen and soluble ICAM-1 are indicators of thrombosis and endothelial dysfunction, which play a basic role in the initiation and progression of atherosclerosis, vasoconstriction and coronary heart disease *(110–114)*. COPD is associated with elevated IL-8, TNF-α, IL-6 and RANTES ("regulated on activation, normal T cell expressed and secreted"), reflecting a predominance of macrophages and T cells, which are correlated with the degree of airflow obstruction and emphysema and appear to play a predominant role in apoptosis, leading to lung destruction *(115–124)*. COPD is also characterized by increases in fibrinogen, which is associated with reduced lung function *(118–121,125–131)*.

Urinary prostaglandin metabolites. Two urinary biomarkers, prostaglandin E_2 metabolite (PGEM) and (*Z*)-7-[1*R*,2*R*,3*R*,5*S*)-3,5-dihydroxy-2-[(*E*,3*S*)-3-hydroxyoct-1-enyl]cyclopentyl]hept-5-enoic acid (8-*iso*-$PGF_{2α}$) are considered to be biomarkers of inflammation and oxidative damage, respectively. PGEM is a metabolite of prostaglandin E_2, while 8-*iso*-$PGF_{2α}$ is a product of lipid peroxidation *(132)*. 8-*iso*-$PGF_{2α}$ has also been quantified in blood. Inflammation and oxidative damage are clearly associated with cigarette smoking, and they play a significant role in cancer induction by enhancing the activity of cigarette smoke carcinogens through mechanisms involving co-carcinogenesis or tumour promotion *(102)*. They also have established roles in cardiovascular disease and COPD *(133)*. The Shanghai Cohort Study found an independent association between urinary levels of 8-*iso*-$PGF_{2α}$ in cigarette smokers and the risk of lung cancer, after adjustment for smoking intensity and duration and other possible confounding factors *(134)*. A significant association was also observed in former smokers but not in never smokers, indicating a probable interaction between tobacco smoke carcinogens and oxidative damage *(134)*. The levels of urinary 8-*iso*-$PGF_{2α}$ decrease more slowly than biomarkers of exposure after cessation of cigarette smoking. The PATH study found that more than 6 months were required for the geometric mean levels of 8-*iso*-$PGF_{2\alpha}$ to return to non-smoker levels, while another study reported a 27% decrease after 12 weeks of cessation *(135,136)*.

5.3.2 Application of biomarkers of biological effect in studies of ENDS and HTPs

DNA adducts

Few studies have been published on DNA damage by ENDS in oral cells (reviewed in *137)*. Mixed results were obtained in a variety of in-vitro studies in which cultured oral cells were exposed to ENDS aerosol or liquid, some studies indicating possible DNA damage while others did not.

A clinical study was conducted of the acrolein–DNA adduct (8*R*/*S*)-3-(2′-deoxyribos-1′-yl)-5,6,7,8-tetrahydro-8-hydroxypyrimido[1,2-*a*]purine-10(3*H*)-one (γ-OH-Acr-dGuo) in ENDS users and non-users of any tobacco product in oral cells of 20 people per group who visited the clinic once a month for 3 months. The levels of γ-OH-Acr-dGuo were significantly nine times higher in ENDS users than in non-users and lower than in cigarette smokers *(106,138)*. These results demonstrate specific DNA adduct formation in the oral mucosa of ENDS users rather than non-users, signalling a possible carcinogenic effect. In a study of apurinic/apyrimidinic sites in DNA, a type of endogenous DNA damage common in all human tissues, the levels in ENDS users were significantly 45% and 42% lower than in non-smokers and smokers, respectively, based on data from a single clinic visit (30–35 subjects per group). The direct relation between apurinic/apyrimidinic sites and ENDS use or cigarette smoking is unclear *(139)*.

No studies were found on the effects of HTP use on DNA damage.

Cytokines, chemokines and reactive proteins

Table 2 summarizes data on some circulating and urinary biomarkers of biological effect commonly measured in studies of tobacco and nicotine product use, with geometric mean ratios for users of various product types *(135,140)*. In reviewing data on such biomarkers, it is important to note that they are not specific to exposure to a particular tobacco or nicotine product and are likely to be influenced by pre-existing sub-clinical disease from previous smoking.

Table 2. Geometric mean ratio (GMR) and range, by product use status, for commonly measured biomarkers of biological effects in studies of tobacco and nicotine product use

Biomarker	Matrix	Indicative of	GMR in ENDS vs smokers	GMR in ENDS vs nonsmokers	Population[a]
IL-6	Serum	Inflammation	0.84 (0.71–0.98)	0.98 (0.82–1.18)	PATH
hs-CRP	Serum or plasma	Inflammation, cardiovascular risk	0.73 (0.57–0.93)	0.86 (0.66–1.11)	PATH
Fibrinogen	Plasma	Inflammation, coagulation, cardiovascular risk	0.96 (0.92–1.01)	0.99 (0.94–1.04)	PATH
sICAM	Serum	Inflammation, cardiovascular risk	0.82 (0.75–0.89)	1.02 (0.95–1.1)	PATH
LDL	Plasma	Cardiovascular risk	0.52 (0.24, 1.14)	0.60 (0.31, 1.16)	NHANES[b]
HDL-C	Plasma	Cardiovascular risk	1.00 (0.50, 2.00)	1.82 (0.95, 3.49)[c]	NHANES[b]
TGL	Plasma	Cardiovascular risk	0.26 (0.06, 1.02)	0.42 (0.12, 1.51)	NHANES[b]
8-iso-PGF2a	Urine	Oxidative stress	0.75 (0.68–0.83)	1.10 (0.98–1.22)	PATH

[a] US cohorts. NHANES: National Health and Nutrition Examination Survey; PATH: Population Assessment of Tobacco and Health Study.
[b] Data for exclusive ENDS users with no prior history of smoking
[c] Lower levels of high-density lipoprotein cholesterol (HDL-C) are associated with higher cardiovascular risk.

ENDS users. Analysis of data from Wave 1 of the PATH study shows that the levels of IL-6, hs-CRP, and sICAM-1 in former smokers who switched to exclusive ENDS use are significantly lower than those in current exclusive cigarette users and comparable to those in former smokers who did not use any tobacco or nicotine product *(135)*. Fibrinogen levels were, however, similar in ENDS users and smokers (GMR, 0.96; 95% CI, 0.92 ; 1.01). The analysis also showed that the levels of these biomarkers did not differ by frequency of ENDS use by current exclusive users, and there was no association with the time since smoking cessation. A recent study of data on HDL-C, low-density lipoprotein cholesterol, triglycerides and fasting blood glucose in 8688 adults in two National Health and Nutrition Examination Survey cycles (2015–2016 and 2017–2018) found no statistically significant effect of exclusive ENDS use on these measures *(140)*. Despite common reports of lower levels of cytokines and other circulating inflammatory biomarkers in exclusive ENDS users than in smokers, the results are not consistent across studies, biomarkers or device types *(135,141–144)*. The availability of definitive data on these biomarkers in ENDS users is important, because many studies in vitro, in vivo and in humans indicate that ENDS aerosols may induce inflammation and cause respiratory and cardiovascular effects *(145–148)*. For example, in a study by Mohammadi et al. *(149)*, endothelial function was measured in chronic ENDS users, chronic cigarette smokers and nonusers by assessing the effects of participants' sera on release of nitric oxide (NO) and hydrogen peroxide and cell permeability in cultured endothelial cells. Sera from ENDS users had effects similar to those in smokers in reducing vascular endothelial growth factor-induced NO secretion by endothelial cells, release of hydrogen peroxide, greater permeability and changes in circulating biomarkers of inflammation, thrombosis and cell adhesion. These results suggest that ENDS use may induce changes in endothelial function. A study of salivary inflammatory biomarkers conducted in India showed that the levels of salivary CRP, TNF-α and IL-1b were significantly higher in ENDS users than in non-users and similar to those in smokers *(150)*.

HTP users. Data on circulating inflammatory biomarkers in HTP users are primarily from tobacco industry-conducted studies. For example, Philip Morris International (PMI) published several reports on such biomarkers in study participants recruited in various countries who switched from smoking to a prototype HTP *(151)*. A study in Japanese smokers who switched to an HTP for 6 days included measurement of serum club cell 16-kDa protein (CC16), which is an indicator of lung epithelial injury; no change was observed *(152)*. Another PMI study, in 316 Polish smokers randomized to an HTP or continued smoking condition for 1 month, included assessment of a broad panel of biomarkers associated with cardiovascular risk *(153)*. An increase (i.e. an improvement) in HDL-C was reported in the HTP group; however, reductions in red blood cell count, haemoglobin and haematocrit were observed. Most other biomarkers did not change after a switch from smoking to HTP use for 1 month. A longer switching study was conducted by PMI in the USA, in which 984 adult smokers were randomized to an HTP device or continued smoking for 6 months *(154)*. Reductions were reported in four biomarkers of effect (HDL-C, white blood cell count, forced expiratory volume in 1 min ($FEV1\%_{pred}$) and COHb) in smokers who switched to an HTP as compared with those randomized to continued smoking. In all these studies, statistically significant decreases in exposure to smoke constituents and in urinary mutagenicity were reported in participants randomized to HTP use.

British American Tobacco researchers reported on a longer (12 months) ambulatory clinical study in which smokers were randomized to an HTP, continued smoking or abstinence *(155)*. Statistically significant positive changes were observed in white blood cell count (reduction) and FeNO (increase) after 6 months of HTP use as compared with continuous smoking. The levels of 11-dTX B2 were also reduced after 6 months of HTP use, but the difference between HTP use and continuous smoking did not reach statistical significance. Further, no substantial effect of switching on sICAM-1 or HDL was observed (only descriptive statistics were provided). In an updated report from this study, the levels of most biomarkers at 12 months were similar to those at 6 months *(156)*. While some biomarkers of biological effect changed in a positive direction after switching to HTP (suggesting less harm than smoking), the outcomes were worse than those of participants who quit.

A real-world, post-marketing study of HTP conducted by researchers from Japan Tobacco *(157)* involved measurement of a panel of inflammatory markers, including HDL-C, triglycerides, sICAM-1, white blood cell count, 11-DHTXB2 and 2,3-d-TXB2 (biomarkers of platelet activation). Reductions were reported in biomarkers of effect in users of HTP (average, 1.2 years of use), although urinary 2,3-d-TXB2 was worse than in non-smokers.

Prostaglandin metabolites and other related urinary biomarkers

Prostaglandin metabolites. Analysis of data from the PATH study indicated that former smokers who currently exclusively used ENDS products had levels of urinary 8-*iso*-$PGF_{2\alpha}$ similar to those of former smokers who did not use ENDS products and to those of participants who had never used tobacco *(135)*. It was not clear, however, whether current ENDS users who were not former smokers also had elevated levels of 8-*iso*-$PGF_{2\alpha}$, and the relatively slow decrease in urinary 8-*iso*-$PGF_{2\alpha}$ upon smoking cessation probably compounds the lack of clarity *(135,136)*. It is important to note, however, that 8-*iso*-$PGF_{2\alpha}$ was significantly higher among dual users of smoking and e-cigarettes than in exclusive smokers in that study (GMR, 1.09; 95% CI 1.03 ; 1.15). Minimal, non-significant changes in urinary isoprostanes, including 8-*iso*-$PGF_{2\alpha}$ were observed in a study in which cigarette smokers switched to an HTP *(158)*. When 20 smokers switched to ENDS or HTPs after 1 week of not-using any tobacco product, significant increases in blood 8-*iso*-$PGF_{2\alpha}$ were reported *(159)*. Several industry-sponsored clinical trials addressed the effects of HTP use on urinary 8-*iso*-$PGF_{2\alpha}$. For example, when healthy adult smokers were randomized to a menthol HTP or a smoking abstinence group for 91 days, the levels of urinary 8-*iso*-$PGF_{2\alpha}$ decreased by 13% ($P < 0.05$) and were similar to those in the smoking abstinence group *(160)*. In the PMI study of 984 US cigarette smokers, a 6.8% reduction in urinary 8-*iso*-$PGF_{2\alpha}$ was observed in those who switched to HTP use for 6 months as compared with those who continued to smoke cigarettes *(154)*. In the post-marketing study of real-world HTP use in Japan, the level of this biomarker was somewhat higher in HTP users than in non-users, albeit at borderline significance (P=0.0646) *(157)*.

Indicators of preclinical changes and symptoms

Other urinary biomarkers of oxidative stress. In DNA, guanine is the major target for direct oxidation by inflammation-induced radicals. The most abundant product of such oxidation is 8-oxo-7,8-dihydro-2′-deoxyguanosine, which can cause chromosomal aberrations and induce mutations and is widely used as a biomarker of oxidative stress *(115)*. Some studies have found significantly higher levels of 8-oxo-7,8-dihydro-2′-deoxyguanosine in ENDS users than in nonsmokers *(161,162)*. Sakamaki-Ching et al. *(162)* found no difference in the levels of this biomarker between ENDS users and smokers.

Direct assessment of chronic disease outcomes due to the use of ENDS and HTPs requires prospective cohort studies, which last many years and involve large numbers of participants. As this is not feasible in most studies and not suitable for time-sensitive regulatory decisions, indicators of preclinical changes and symptoms have commonly been used as surrogate measures of respiratory and cardiovascular risk in users of tobacco and nicotine products. Preclinical indicators and symptoms of cardiovascular disease may include such measures as blood pressure, heart rate, arterial stiffness, platelet reactivity and other cardiovascular outcomes. For respiratory diseases such as COPD, commonly measured preclinical indicators and symptoms include respiratory function (FEV1), forced vital capacity (FVC), FEV1/FVC and diffusing capacity of the lung for CO (DLCO), as well as coughing, wheezing, shortness of breath and other symptoms *(163–166)*. Interpretation of cardiopulmonary preclinical indicators is often based on lipid profile, fibrinogen, D-dimer and hs-CRP. There are no robust preclinical indicators of cancer. Reports of large population-based epidemiological studies of the association between ENDS use and disease outcomes are becoming available, most of which have been published since 2020. Such emerging data will play a key role in future assessments of the predictive value of biomarkers of exposure and biological effects in studies of novel and emerging tobacco products.

ENDS users. Several studies have shown that ENDS use increases blood pressure, heart rate, arterial stiffness, platelet reactivity and other cardiovascular outcomes as compared with no tobacco or nicotine product use *(141,147,148,167,168)*. Respiratory effects and symptoms (e.g. resistance to air flow, accumulation of lipid-laden macrophages in lungs) have been also reported in ENDS users, and a longitudinal study of Waves 1 and 2 of the PATH study showed that people who were exclusive ENDS users at baseline had a higher prevalence of subsequent respiratory symptoms than nonusers (33.6% vs 21.7%, respectively) *(168–170)*. Another report based on Waves 1–4 of the PATH study showed that ENDS use was associated with higher risks for respiratory disease (COPD, emphysema, chronic bronchitis and asthma) than in non-users *(171)*. Switching from smoking to ENDS may, however, result in improvements in some of these indicators and outcomes. A systematic review of six population-based studies with samples ranging in size from 19 475 to 161 529 found a lower odds ratio for respiratory outcomes (COPD, chronic bronchitis, emphysema, asthma and wheezing) but no change in cardiovascular outcomes (stroke, myocardial infarction and coronary heart disease) in former smokers who used ENDS as compared with current smokers *(172)*. A randomized crossover study of hookah users showed increased arterial stiffness and higher levels of inflammatory markers with use of e-hookah than with use of conventional tobacco hookah *(173)*. Differences between preclinical indicators and symptoms (i.e. between

COPD and CVD) are not surprising; they could be due to a complex interaction between exposure and specific biological effects elicited by the exposure. For example, the reduction in risk for lung cancer occurs over 20–25 years after smoking cessation and never reaches that of a never smoker *(174)*. In contrast, the CVD risk falls to that of a never smoker in only 1–3 years *(174)*; however, even low levels of exposure, such as fewer than three cigarettes per day or even second-hand smoke in nonsmokers, increase the risk for CVD *(175–177)*.

HTP users. There are few independent reports on indicators of health effects in HTP users. A cross-sectional study of 58 336 students aged 12–18 years in the 2018 Korea Youth Risk Behavior Survey study found an association between HTP use and asthma, allergic rhinitis and atopic dermatitis *(178)*. Two cases of acute eosinophilic supplementation associated with HTP use were reported in Japan. One case involved a 20-year-old man who had used 20 HTP sticks per day for 6 months and had doubled his consumption 2 weeks before hospitalization *(179)*. The second case was in a 16-year-old boy with bronchial asthma who developed cough, shortness of breath and fatigue immediately after smoking an HTP, the symptoms worsening over the course of 2 weeks of HTP use *(180)*. Most tobacco industry reports on preclinical indicators in HTP users are limited to respiratory measures, namely FEV_1. These studies show either no change or a modest improvement in this measure after switching from smoking to HTP use *(155,157)*.

Notable emerging biomarkers of biological effects for studies of ENDS and HTPs

DNA methylation profile. Extensive literature supports use of epigenetic modifications as a measure of the effect of smoking *(181,182)*. Studies of DNA methylation in saliva and bronchoalveolar lavage fluid found that the epigenetic profiles for ENDS use were similar to those of non-users *(183,184)* and that ENDS use did not affect cg05575921, a highly hypomethylated aryl hydrocarbon receptor repressor (AHRR) site in smokers and a sensitive, specific marker of smoking status *(185,186)*; however, hypomethylation of LINE-1 repeat elements and global loss of DNA hydroxymethylation were reported in leukocytes of ENDS users, suggesting systemic effects *(187)*.

The effect of HTP use on DNA methylation was assessed in the Tsuruoka Metabolome Cohort Study in Japan *(188)*, which found that 10 of 17 smoking-associated genes were significantly hypomethylated, and GPR15 expression was markedly upregulated in HTP users as compared with non-smokers, although AHRR expression was significantly lower than in cigarette smokers. These results suggest that HTP use may result in distinct DNA methylation and transcriptome profiles. The implications of such effects should be investigated.

Gene expression. Changes in gene expression induced by smoking and other harmful exposures can indicate disturbances in cellular metabolic pathways, and such changes could serve as biomarkers of biological effects linked to specific health outcomes, including lung cancer *(189–191)*. Cross-sectional studies of ENDS users, smokers and nonsmokers found differential gene expression among the groups. For example, Martin et al. *(192)* found that the nasal epithelium of smokers showed differential downregulation of 53 genes, while that of ENDS users showed differential downregulation of 358 genes as compared with nonsmokers. Upregulation of only one gene – growth response 1 ("*EGR1*") – was the same in smokers and ENDS users, while the remaining overexpressed genes were specific to the two products. In the second cross-sectional study, oral cells were used to assess gene expression in the same two groups, with different results: smokers had more differentially expressed genes than ENDS users *(193)*. The most deregulated pathways in smokers and ENDS users were associated with carcinogenic pathways. Studies of gene expression after an acute exposure to ENDS showed significant changes in oral, blood and respiratory cells in response to the exposure *(194,195)*. More research is necessary to understand the pathophysiological consequences of such findings.

The oral microbiome. The oral microbiome is a complex receptor medium for chemical exposures in the oral cavity. Changes in the oral chemical environment create conditions that may be either detrimental or beneficial to certain bacterial populations, leading to changes in the composition and function of the oral microbiome. The oral cavity is also the gateway for bacteria that colonize the respiratory tract *(196–198)*, and there is accumulating information on the association of the oral microbiome with a variety of chronic diseases, including cancer, CVD and COPD *(196,199–207)*. Cigarette smoking affects the oral microbiome *(208)* through immunosuppressive effects *(209)*, favouring biofilm formation *(210)*, altering oral O_2 tension and pH *(211)* and modifying the chemical environment of the oral cavity *(212)*. The type of tobacco used, frequency of use, and smoking history have been reported to influence the degree and the nature of such changes *(213,214)*.

Recent studies suggest that the oral microbiome signatures in ENDS users are distinct from those in cigarette smokers and former or never smokers, including altered taxonomic composition, increased microbial diversity, a significant increase in the abundance of microbial pathways involved in carbohydrate and amino acid metabolism, and diverse virulence factors *(215–217)*. Some of these traits are favourable (e.g. greater diversity than in smokers), while others suggest inflammatory processes. As in smokers, the relative abundance of *Veillonella* in buccal cells and saliva of ENDS users is significantly higher than in non-users *(215,216)*. *Veillonella* are associated with various infections, including in the mouth, lungs and heart *(218–220)*, and some *Veillonella* species may play a role in endogenous nitrosation through their capacity to reduce nitrate to nitrite *(216,221)*, which might be the main reason for the comparable levels of NNN found in the oral cavities of ENDS users and of smokers found by Bustamante et al. *(17)*.

Only one study was found of the oral microbiome in HTP users *(222)*. The study was conducted in 65 adolescents, aged 14–18, in Ukraine who were HTP or ENDS users or non-users of any tobacco product (control group). The composition of the oral microbiomes of participants who used HTPs was different from that of ENDS users. The findings suggest that both products reduce the number of resident plaque microflora, which leads to the emergence of opportunistic transient streptococci such as *Streptococcus pneumoniae* and *S. pyogenes*.

5.4 Biomarkers of susceptibility

5.4.1 Definition and overview of biomarkers of susceptibility used in studies of tobacco and nicotine products

Individual and population differences in the uptake and/or metabolism of toxicants and carcinogens present in tobacco and nicotine products can contribute to differences in susceptibility to adverse biological effects and the subsequent health outcomes. Biomarkers of susceptibility are predictive indicators of individual characteristics (e.g. gene polymorphisms) that drive such differences. In the context of tobacco control, these biomarkers are useful for interpreting and predicting potential population differences in the levels of biomarkers of exposure or of biological effect among users of the same product type. Furthermore, such biomarkers can potentially be used to identify susceptible populations for targeted cessation interventions.

Nicotine metabolite ratio (NMR). The most commonly used biomarker of susceptibility in studies of tobacco use and disease risk is the ratio of two nicotine metabolites, *trans*-3′-hydroxycotinine and cotinine, referred to as the NMR. This biomarker reflects the activity of CYP2A6, the enzyme primarily responsible for nicotine metabolism, which is mainly defined by the presence or absence of functional polymorphisms in the *CYP2A6* gene *(223)*. Inter-individual differences in the NMR have been associated with smoking behaviour and dose *(224)*, smoking abstinence *(225)*, and the risk of lung cancer *(6)*. Representative values for the NMR in daily tobacco users in the USA, overall and by age, sex and race or ethnicity are available from a recent analysis of Wave 1 of the PATH Study *(226)*.

Urinary metabolites of carcinogens and toxicants. As discussed above, urinary metabolites of tobacco smoke constituents such as nicotine, tobacco-specific nitrosamines, PAH and volatile organic compounds are well established biomarkers of exposure. In addition, analysis of data from the Shanghai Cohort Study and the Singapore Chinese Health Study, two large prospective epidemiological studies, showed that total nicotine equivalents, total NNAL, total NNN and PheT were independently associated with significantly higher cancer risks among smokers (Table 3) *(10,15,227)*. These biomarkers can therefore also be considered biomarkers of susceptibility to disease, namely cancer. In these studies, total nicotine equivalents served as dose monitors for all other constituents of tobacco smoke. Although the effects of total NNAL, total NNN and PheT were still apparent after correction for total nicotine equivalents, this was not the case for mercapturic acid biomarkers of 1,3-butadiene, ethylene oxide, benzene, acrolein and crotonaldehyde *(229)*. The relevance of use of urinary carcinogen and toxicant biomarkers to assess risk in users of ENDS and HTPs requires further research, as users of these products appear to have limited exposure to the relevant parent compounds (NNK, NNN and PAH). As noted above, Bustamante et al. *(17)* presented evidence of the presence of increased levels of NNN in the saliva of ENDS users.

Table 3. Urinary carcinogen and toxicant metabolites that have been prospectively associated with lung cancer risk in smokers

Constituent	Biomarker	Odds ratio	Study population	Reference nos
Nicotine	cotinine	0.85–3.52	Shanghai, Singapore, USA	*229–231*
NNK	Total NNAL	1.57–2.64	Shanghai, Singapore, USA	*229–231*
PAH	PheT	1.23–2.34	Shanghai, USA	*229,231*
Volatile organic compunds	Mercapturic acids	0.97–1.20	Shanghai	*228*

Examples of other potential biomarkers of susceptibility. Certain biomarkers described above, such as DNA methylation, gene expression and the microbiome, are also important individual characteristics that can affect the metabolism of tobacco constituents and/or the protective mechanisms (e.g. immune responses, DNA repair) against their harmful effects. Therefore, DNA methylation, gene expression and the microbiome could serve as biomarkers of susceptibility in studies of tobacco or nicotine products.

5.4.2 Application of biomarkers of susceptibility in studies of ENDS and HTPs

Use of biomarkers of susceptibility in studies of ENDS and HTPs has been limited.

NMR

It is not known whether NMR is predictive of ENDS or HTP use behaviour or of any health outcome resulting from use of these products. The relatively short time since these products have been on the market, their diversity and continuous evolution, and the variation in nicotine delivery are probably the main reasons for lack of data. A study of PATH data (Waves 1 and 2) of the association between NMR and transitions in cigarette smoking and ENDS use *(232)* found a significant two-way interaction, women with higher NMR (i.e. faster nicotine metabolism) being 10 times less likely to quit ENDS use than women with lower NMR. These results indicate that NMR could potentially be used as a biomarker of quitting ENDS use in women.

Other potential biomarkers of susceptibility

Urinary metabolites of carcinogens and toxicants. There is no consistent evidence for substantial increases in metabolites of NNK or PAH in the urine of ENDS or HTP users as compared with non-users of any tobacco or nicotine product *(60,233)*.

DNA methylation, gene expression and the microbiome. Use of these biomarkers in studies of tobacco and nicotine products is relatively recent and limited. The potential associations of these biomarkers with the metabolism of ENDS or HTP constituents or with biological effects in ENDS or HTP users have not been studied.

5.5 Established and validated methods for measuring biomarkers

Well-characterized methods are available for most commonly used biomarkers of exposure and biological effects, particularly those used for large cohorts. Liquid chromatography (LC) or gas chromatography (GC) coupled with mass-spectrometry (MS) are highly sensitive and selective methods of choice for these measurements. Examples of methods, along with their analytical parameters, are illustrated in Table 4. For circulating biomarkers of biological effects, such as cytokines and reactive proteins, immunoassay methods with commercially available kits are typically used. The performance of the kits is validated for quality and specificity by their manufacturers.

Table 4. Examples of validated methods for some biomarkers of exposure and biological effects

Biomarker	Method description	Method characteristics	Reference no(s)
Biomarkers of exposure			
Urinary total nicotine equivalents (TNEs)	LC–MS/MS analysis of nicotine, cotinine, 3′-hydroxycotinine and their glucuronides after enzymatic treatment of urine (to release these biomarkers from their glucuronide conjugates) and solid-phase extraction	Accuracy: 93–96% Intra-day CV: 4.2–7.1% Inter-day CV: 0.4–5%	*234*
Urinary total NNAL	LC–MS/MS analysis of NNAL and its *O*- and *N*-glucuronides after enzymatic treatment of urine and two extraction steps	Accuracy: 94% Intra-day CV: 3.0% Inter-day CV: 5.7%	*235*
Urinary CEMA	LC–MS/MS analysis after a purification step with mixed mode anion exchange on a 96-well plate	Accuracy: 98% Intra-day CV: 6.4% Inter-day CV: 6.6%	*236*
Urinary PheT	GC–NICI–MS/MS analysis after treatment with β-glucuronidase and arylsulfatase and purification on styrene-divinylbenzene plates in a 96-well format	Accuracy: 95% Intra-day CV: 2.9% Inter-day CV: 3.7%	*235*

Biomarkers of biological effect			
Urinary 8-iso-PGF_{2a}	LC–MS/MS analysis after a single purification step	Accuracy: 103% Intra-day CV: 4.0% Inter-day CV: 5.5%	*237*
DNA adducts of acrolein in oral cells	DNA is isolated, and, after hydrolysis and solid-phase extraction, the adducts are quantified by LC-MS/MS.	Accuracy: 96% Intra-day CV: 1.6% Inter-day CV: 3.4%	*238*

CV, coefficient of variation; NICI, negative ion chemical ionization

5.6 Summary of evidence on biomarkers for ENDS and HTPs and implications for public health

Research on biomarkers for evaluating tobacco and nicotine products has proliferated in the past 15 years, extending application of known biomarkers, providing new biomarkers and generating new evidence on their levels in users. The national longitudinal PATH study in the USA and other large, longitudinal cohorts were instrumental platforms for using a broad panel of biomarkers to assess exposure and effects in users of various tobacco product types. A substantial body of published literature supports use of biomarkers of exposure for evaluating ENDS, and new support has become available for use of DNA adducts in oral cells and certain cardiopulmonary biomarkers of biological effects for this purpose. The same biomarkers of exposure and effect are likely to be useful for assessing HTPs; however, most of the research on biomarkers for HTPs to date has been conducted by the tobacco industry. Little information is available on the potential role of NMR, a biomarker of susceptibility, in measuring exposure and effects in ENDS or HTP users.

5.6.1 Summary of available data and implications for public health

The main conclusions of this review of data on biomarkers for ENDS and HTPs and implications for health are outlined below.

Switching from conventional cigarette smoking to exclusive ENDS use is associated with reduced exposure to several toxicants and carcinogens that play key roles in smoking-induced diseases. This conclusion is supported by extensive literature, including the most recent analyses based on Waves 1 and 2 of the PATH study *(74,90)* and a secondary analysis of a Cochrane systematic review of trials of use of ENDS for smoking cessation *(239)*.

Public health implications:

- The extent and nature of the effects on health of these reductions are not yet well understood. Biomarkers of exposure do not account for the potential combined effects of numerous individual constituents and, therefore, have limited capacity to predict changes in disease risk. For example, the levels of many biomarkers of exposure, including those of volatile toxicants such as acrylonitrile and acrolein, are higher in exclusive ENDS users than in non-users of nicotine or tobacco products *(88,142,170,240–242)*, and the effects of these low-level exposures are not well understood. In addition, ENDS users may be exposed to certain organophosphate flame retardants (probable contaminants in ENDS devices) *(243)*, and a study of untargeted chemical profiling showed that ENDS aerosols may contain more than 2000 chemical constituents, many of which are yet to be characterized *(244)*.

- While the long-term effects of ENDS use are poorly understood, smokers who switch completely to ENDS may benefit from reductions in exposure to many tobacco toxicants and carcinogens.

A panel of biomarkers of exposure can be used to determine or validate product use (Table 5). A cut-off point of 5–6 ppm CO will distinguish users of combustible cigarettes from users of ENDS and non-users. ENDS users have urinary TNE (at least 2000 pmol/mg creatinine) *(242)* and lower (< 27 pmol/mL urine) CEMA concentrations than non-users of any tobacco or nicotine product, who will have minimal (essentially zero) TNE *(242)* and CEMA (< 27 pmol/mL urine) *(245,246)*. If the nicotine use status of participants is ambiguous, NNAL (1–2 pmol/mL urine) can be measured as a biomarker of tobacco-specific NNK, which occurs at low levels in ENDS users (0.023 pmol/mL urine) *(242,247)*. Salivary propylene glycol (3.5 μmol/mL in ENDS users and 0.004 μmol/mL in non-users, our unpublished data) can be used to confirm ENDS use. The biomarker half-life should be considered when using these cut-off points, particularly in studies of recent switching from smoking to ENDS use. Anatabine, a minor tobacco alkaloid (which should not be present in exclusive ENDS users), has been used in a few studies.

Table 5. Expected relative values for biomarkers in users of various tobacco and nicotine products

Category of use	CO	Urinary biomarker			Propylene glycol in saliva
		TNE	Total NNAL	CEMA	
Exclusive cigarette smoking	High	High	High	High	Low or ND
Exclusive ENDS use	Low	High	Low or ND	Low or ND	High
Dual use of ENDS and smoking	Variable	High	High	High	Variable
No use of any tobacco or nicotine	Low	Low or ND	Low or ND	Low or ND	Low or ND

ND, not detectable

Public health implications:

- Use of the proposed panel of biomarkers is important for advancing research on exposure and effects in users of ENDS.

Cigarette smokers who become dual users of ENDS and conventional cigarettes do not experience meaningful reductions in most biomarkers of exposure. The amount of smoking appears to be the primary determinant of exposures in dual users.

Public health implications:

- Dual users are not likely to experience improvements in biological effects over those seen with exclusive smoking.
- Dual use exposes users to the same levels of tobacco toxicants and carcinogens as cigarette smoking and also to emissions of ENDS. The health consequences of such mixed exposure are unknown; however, a systematic review suggests that dual use may be associated with the same or a significantly higher risk of self-reported symptoms or disease as exclusive cigarette smoking (248).

DNA adducts in oral cells are useful biomarkers of carcinogen dose and biological effects and can be used to compare the effects of ENDS with those of conventional cigarettes. Such biomarkers should be used more widely, when possible.

Public health implications:

- As stated above, biomarkers of exposure have limited capacity to predict changes in disease risk. As formation of DNA adducts is a key step in chemical carcinogenesis process, these biomarkers might indicate cancer risk.
- In addition, given the reactivity of aldehydes and other inflammatory agents present in ENDS and ENNDS aerosols, urinary biomarkers of exposure may not fully capture important exposures and biological effects at the place of immediate contact of the aerosols with human tissues, such as the oral cavity.

The results of studies of circulating inflammatory cytokines associated with cardiovascular and respiratory effects after ENDS use are inconsistent. One challenge in interpreting the findings is that different types and panels of such biomarkers have been used in different studies.

Public health implications:

- Definitive data on these biomarkers in ENDS users are necessary, as a substantial body of research suggests that ENDS may be a source of inflammatory exposure and thus contribute to respiratory and cardiovascular effects in users.

Studies with indicators of preclinical changes and symptoms suggest that ENDS increase the risks for respiratory and (potentially) cardiovascular effects over that with non-use of any product.

Public health implications:

- While some improvements in respiratory symptoms have been reported in smokers who switch to ENDS, prolonged use of ENDS by former smokers should not be encouraged. Innovative cessation interventions are necessary to help users to quit both smoking and ENDS use.
- ENDS use by never smokers is likely to increase their risks for disease.

Independent studies of biomarkers in HTP users are critically lacking.

Public health implications:

- Although tobacco industry reports indicate significant reductions in biomarkers of exposure and some biomarkers of biological effects in smokers who switch to HTPs, independent academic research is necessary to confirm these findings.
- HTPs are likely to expose users to higher levels of toxicants than ENDS (249); however, no studies with biomarkers are available.

5.6.2 Limitations of biomarkers

The limitations of biomarkers are associated mainly with their specificity, stability and feasibility of measurement.

- Many biomarkers of exposure and biological effect are not specific to a particular tobacco or nicotine product and can be influenced by factors including dietary, environmental and occupational exposures, health status and physical activity. No single biomarker or set of biomarkers can capture all the exposures or effects associated with a tobacco or nicotine product.

- Many biomarkers of biological effect are not specific to one disease. For example, oxidative stress and inflammation are common underlying mechanisms in the pathophysiology of cancer and of cardiovascular and respiratory diseases. Therefore, it is challenging to use such biomarkers to distinguish between the risks for individual diseases.
- Biomarkers have limited, variable half-lives. For example, the half-life of exhaled CO is approximately 8 h or less, the half-life of urinary total NNAL is approximately 3.5 weeks, and the half-life of urinary cadmium can be up to 38 years. Biological stability should be considered when designing and interpreting biomarker studies, particularly in the context of previous smoking and/or the duration of ENDS or HTP product use.
- Measurement of some biomarkers requires highly specialized expertise and instrumentation, which may limit their broad application.

5.6.3 Research gaps

This review of the evidence on use of biomarkers of exposure, effect and susceptibility in studies of ENDS and HTP use indicates the following research gaps and priorities:

- independent (non-tobacco industry) research on ENDS, HTPs and other new and emerging tobacco and nicotine products, such as nicotine pouches;
- research, including untargeted profiling of product emissions and biospecimens, to identify biomarkers specific to ENDS and HTPs, as most biomarkers in current use are based on exposure to cigarette smoke;
- evaluation of biomarkers resulting from use of ENDS and HTPs of different designs and with different ingredients;
- research on biomarkers of biological effects that are specific to exposure to tobacco and nicotine products or to individual health effects (e.g. cardiovascular or respiratory diseases);
- cross-sectional and longitudinal studies to assess and compare ENDS and HTP exposures among populations (including comparisons with non-users of any product) in various countries;
- systematic studies of exposures, biological effects and clinical disease-specific manifestations in ENDS and HTP users, to better characterize the associations with various types of biomarkers and associations of levels of biomarkers with specific disease outcomes; and
- studies to better characterize and communicate the effect of dual and poly-product use, especially with different levels of smoking.

5.7 Recommendations for possible prioritization of biomarkers for tobacco control

Biomarkers are objective measures of harmful exposures and biological effects relevant to disease pathophysiology in users of various tobacco and nicotine products. Therefore, biomarkers of exposure and effects could be useful tools for tobacco control.

Biomarkers of exposure. The advantage of biomarkers of exposure is that they account for the effect of product use patterns on the delivery of harmful constituents to users. Such effects might not be captured by standardized, machine-based product testing in a laboratory. Therefore, biomarkers can be used to more accurately characterize product toxicity and abuse liability. On the basis of the evidence reviewed, the following biomarker panel is recommended for assessing the level of exposure of users of ENDS and HTPs to constituents implicated in smoking-related harm: urinary TNEs (addictiveness), NNAL (exposure to tobacco-derived carcinogens) and CEMA (exposure to combustion). Measurement of exhaled CO and salivary propylene glycol could be added to this panel; however, exhaled CO has a short half-life, and the method for measuring salivary propylene glycol requires validation. Other biomarkers of exposure reviewed in this report can be also used; however, the proposed, limited panel can provide sufficient information on exposure to key classes of harmful constituents and allow classification by product use status.

Biomarkers of biological effects. Biomarkers of biological effects account for interactions among several harmful exposures and potentially for unique exposures that may not be captured by product testing or by urinary biomarkers of exposure. On the basis of the evidence reviewed, the following biological effects are recommended for priority monitoring.

- *DNA adducts in oral cells,* formed by acrolein and potentially by other volatile toxicants, such as formaldehyde. There are clear, striking differences in the levels of these biomarkers between smokers, ENDS users and non-users of tobacco. Given the direct relevance of DNA adduct formation to cancer risk, monitoring of these biomarkers would be valuable for assessing the potential risks of novel products relative not only to conventional cigarettes but also to non-use of any tobacco or nicotine product. DNA adducts in oral cells can also serve as biomarkers of exposure to low but biologically relevant levels of volatile toxicants.
- *Indicators of preclinical changes and symptoms.* Indicators such as blood pressure, heart rate, arterial stiffness, platelet reactivity, respiratory symptoms (coughing, wheezing, shortness of breath) and respiratory function measures (FEV1, FVC, FEV1/FVC and DLCO)

are commonly assessed in general clinical practice and are easy to measure. It is recommended that such measures be included in cross-sectional and longitudinal studies of ENDS and HTP product use, as surrogate measures of respiratory and cardiovascular risk.

Urinary isoprostanes and selected cytokines, for which consistent results have been obtained in comparisons of users of ENDS or HTPs with smokers, could also be used. The absence of differences in the levels of these biomarkers among user groups and the increased levels as compared with non-users of tobacco and nicotine indicate continued systemic oxidative stress and inflammation in ENDS and HTP users. Interpretation of these biomarkers should, however, include recognition that oxidative stress and inflammation are not specific to exposure to these products.

Biomarkers of susceptibility. Anyone exposed to the harmful constituents present in product emissions may be at risk (for addiction or disease). The value of biomarkers of susceptibility is their use for identifying certain population subgroups who are particularly vulnerable to harmful effects. Research on these biomarkers is evolving, and the existing data are insufficient to recommend their direct application in tobacco control. It is recommended, however, that NMR measurements be incorporated into programmes and studies for monitoring TNEs and total NNAL in users of ENDS and HTPs.

Direct epidemiological assessment of disease risk. Emerging epidemiological studies of health outcomes generally show higher risks for ENDS than would be expected from the levels of biomarkers of exposure relevant to cigarette smoke. This could be due in part to the high prevalence of dual use, which may not be captured in such studies, and the yet not well-characterized unique biological effects of ENDS use. Therefore, direct monitoring of health effects resulting from the use of ENDS, HTPs and any new and emerging nicotine or tobacco products should be documented.

5.8 Recommendations for addressing research gaps and priorities

The following strategies are recommended for independent research to address the gaps and priorities that have been identified.

- Conduct research to develop and use biomarkers of exposure, biological effect and susceptibility to better characterize the public health impact of ENDS, HTPs and other new and emerging products.
- Conduct research to compare the relative risks (exposure and biological effects) of use of HTPs and ENDS and comparisons with no tobacco or nicotine product.
- Establish cross-sectional and longitudinal cohorts for real-time monitoring of exposure and biological effects in users of ENDS, HTPs

and other emerging products, to provide regulators with the relative risks of such products and the potential effects of evolving product characteristics.

- Conduct systematic monitoring of various types of biomarkers and indicators to better understand how differences in exposure with various product types translate into biological effects, preclinical symptoms and disease risk.
- Conduct research on communication strategies to inform populations about the exposures and effects of ENDS and HTPs; prevent their dual use with cigarettes; and prevent misperceptions of product risks.

5.9 Relevant policy recommendations

The following regulatory recommendations are proposed for consideration by policy-makers, researchers and the public health community, as appropriate.

- Take into account biomarker-based findings (from all countries) when making policy decisions on ENDS, HTPs and other new and emerging tobacco and nicotine products, relying on data obtained independently of the tobacco or ENDS industry and considering the limitations of biomarkers.
- Prioritize and support independent research, including building capacity for measuring biomarkers and for epidemiological studies to address the research gaps and priorities related to the public health impact of ENDS, HTPs and other new and emerging tobacco and nicotine products.
- In countries with the necessary capacity, monitor the recommended panel of biomarkers in users of ENDS, HTPs and other new and emerging tobacco and nicotine products.
- Clearly communicate to health-care professionals and the general public the current absence of evidence that use of HTPs reduces harm.
- Given the rapid pace at which new products are introduced and the time lag in scientific research on exposure and effect factors, Member States are strongly encouraged to consider requiring that the following information be provided by manufacturers before allowing marketing of any of these products in their country: (i) levels of emission of selected harmful chemicals and (ii) levels of the recommended panel of biomarkers in users.

References

1. Hecht SS, Murphy SE, Carmella SG, Li S, Jensen J, Le C et al. Similar uptake of lung carcinogens by smokers of regular light and ultra-light cigarettes. Cancer Epidemiol Biomarkers Prev. 2005;14:693–8. doi: 10.1158/1055-9965.EPI-04-0542.
2. Harris JE, Thun MJ, Mondul AM, Calle EE. Cigarette tar yields in relation to mortality from lung cancer in the cancer prevention study II prospective cohort 1982–8. BMJ. 2004;328(7431):72–9. doi:10.1136/bmj.37936.585382.44.
3. Burns DM, Dybing E, Gray N, Hecht S, Anderson C, Sanner T et al. Mandated lowering of toxicants in cigarette smoke: a description of the World Health Organization TobReg proposal. Tob Control. 2008;17(2):132–41. doi:10.1136/tc.2007.024158.
4. Benowitz NL, Bernert JT, Foulds J, Hecht SS, Jacob P, Jarvis M. J et al. Biochemical verification of tobacco use and abstinence: 2019 update. Nicotine Tob. Res. 2020 22 (7) 1086-1097. doi:10.1093/ntr/ntz132.
5. Mead AM, Geller AB, Teutsch SM, editors. Premium cigars: patterns of use, marketing, and health effects. Washington DC: National Academies Press; 2022. doi:10.17226/26421.
6. Murphy SE. Biochemistry of nicotine metabolism and its relevance to lung cancer. J Biol Chem. 2021;296:100722. doi:10.1016/j.jbc.2021.100722.
7. Hoffmann D, Hecht SS, Ornaf RM, Wynder EL. N′-Nitrosonornicotine in tobacco. Science. 1974;186:265–7. doi:10.1126/science.186.4160.265.
8. Hoffmann D, Hecht SS. Nicotine-derived N-nitrosamines and tobacco related cancer: current status and future directions. Cancer Res. 1985;45:935–44. PMID:3882226.
9. Hecht SS. Biochemistry biology and carcinogenicity of tobacco-specific N-nitrosamines. Chem Res Toxicol. 1998;11:559–603. doi:10.1021/tx980005y.
10. Smokeless tobacco and some tobacco-specific N-nitrosamines (IARC Monographs on the Evaluation of Carcinogenic Risks to Humans, vol. 89). Lyon: International Agency for Research on Cancer; 2007 (https://publications.iarc.fr/Book-And-Report-Series/Iarc-Monographs-On-The-Identification-Of-Carcinogenic-Hazards-To-Humans/Smokeless-Tobacco-And-Some-Tobacco-specific-Em-N-Em--Nitrosamines-2007).
11. Personalhabitsandindoorcombustions.In:IARCMonographsontheEvaluationofCarcinogenic Risks to Humans, vol. 100E. Lyon: International Agency for Research on Cancer; 2012:319–31 (https://publications.iarc.fr/Book-And-Report-Series/Iarc-Monographs-On-The-Identification-Of-Carcinogenic-Hazards-To-Humans/Personal-Habits-And-Indoor-Combustions-2012).
12. Balbo S, James-Yi S, Johnson CS, O'Sullivan G, Stepanov I, Wang M et al. (S)-N′-Nitrosonornicotine a constituent of smokeless tobacco is a powerful oral cavity carcinogen in rats. Carcinogenesis. 2013;34:2178–83. doi:10.1093/carcin/bgt162.
13. Balbo S, Johnson CS, Kovi RC, James-Yi SA, O'Sullivan MG, Wang M et al. Carcinogenicity and DNA adduct formation of 4-(methylnitrosamino)-1-(3-pyridyl)-1-butanone and enantiomers of its metabolite 4-(methylnitrosamino)-1-(3-pyridyl)-1-butanol in F-344 rats. Carcinogenesis. 2014;35(12):2798–806. doi:10.1093/carcin/bgu204.
14. Hecht SS, Hatsukami DK Smokeless tobacco and cigarette smoking: chemical mechanisms and cancer prevention. Nature Rev. Cancer 22 143-155, 2022.
15. YuanJM, ButlerLM, StepanovI, HechtSS. Urinarytobaccosmoke-constituentbiomarkersforassessing riskoflungcancer. CancerRes. 2014;74(2):401–11. doi:10.1158/0008-5472.CAN-13-3178.
16. Stepanov I, Sebero E, Wang R, Gao YT, Hecht SS, Yuan JM. Tobacco-specific N-nitrosamine exposures and cancer risk in the Shanghai Cohort Study: remarkable coherence with rat tumor sites. Int J Cancer. 2014;134(10):2278–83. doi:10.1002/ijc.28575.
17. Bustamante G, Ma B, Yakovlev G, Yershova K, Le C, Jensen J et al. Presence of the carcinogen N′-nitrosonornicotine in saliva of e-cigarette users. Chem Res Toxicol. 2018;31(8):731–8.

doi:10.1021/acs.chemrestox.8b00089.

18. Snook ME, Severson RF, Arrendale RF, Higman HC, Chortyk OT. Multi-alkyated polynuclear aromatic hydrocarbons of tobacco smoke: separation and identification. Beitr Tabakforsch. 1978;9:222–47. doi:10.2478/cttr-2013-0452.
19. Rodgman A, Perfetti T. The chemical components of tobacco and tobacco smoke. Boca Raton (FL): CRC Press; 2009 (https://www.routledge.com/The-Chemical-Components-of-Tobacco-and-Tobacco-Smoke/Rodgman-Perfetti/p/book/9781466515482).
20. Li Y, Hecht SS. Carcinogenic components of tobacco and tobacco smoke: A 2022 update. Food Chem Toxicol. 2022;165:113179. doi:10.1016/j.fct.2022.113179.
21. Scherer G, Pluym N, Scherer M. Intake and uptake of chemicals upon use of various tobacco/nicotine products: Can users be differentiated by single or combinations of biomarkers? Contrib Tob Nicotine Res. 2022;30:167–98. doi:10.2478/cttr-2021-0014.
22. Hecht SS. Tobacco smoke carcinogens and lung cancer. In: Penning TM, editor. Chemical carcinogenesis – current cancer research. Cham: Springer; 2011:53–74.
23. Some non-heterocyclic polycyclic aromatic hydrocarbons and some related exposures (IARC Monographs on the Evaluation of Carcinogenic Risks to Humans, vol. 92). Lyon: International Agency for Research on Cancer; 2010:35–818 (https://publications.iarc.fr/Book-And-Report-Series/Iarc-Monographs-On-The-Identification-Of-Carcinogenic-Hazards-To-Humans/Some-Non-heterocyclic-Polycyclic-Aromatic-Hydrocarbons-And-Some-Related-Exposures-2010).
24. Jain RB. Contributions of dietary demographic disease lifestyle and other factors in explaining variabilities in concentrations of selected monohydroxylated polycyclic aromatic hydrocarbons in urine: Data for US children adolescents and adults. Environ Pollut. 2020;266(1):115178. doi:10.1016/j.envpol.2020.115178.
25. Conney AH. Induction of microsomal enzymes by foreign chemicals and carcinogenesis by polycyclic aromatic hydrocarbons: GHA Clowes Memorial Lecture. Cancer Res. 1982;42:4875–917. PMID:6814745.
26. Zhong Y, Carmella SG, Hochalter JB, Balbo S, Hecht SS. Analysis of r-t-8,9 c-10-tetrahydroxy-7,8,9,10-tetrahydrobenzo[a]pyrene in human urine: a biomarker for directly assessing carcinogenic polycyclic aromatic hydrocarbon exposure plus metabolic activation. Chem Res Toxicol. 2011;24:73–80. doi:10.1021/tx100287n.
27. Agents classified by the IARC Monographs volumes 1–132. Lyon: International Agency for Research on Cancer; 2022 (https://monographs.iarc.who.int/agents-classified-by-the-iarc/, accessed 29 Novermber 2022).
28. Larsen ST, Nielsen GD. Effects of methacrolein on the respiratory tract in mice. Toxicol Lett. 2000;114(1–3):197–202. doi:10.1016/s0378-4274(99)00300-8.
29. Morgan DL, Price HC, O'Connor RW, Seely JC, Ward SM, Wilson RE et al. Upper respiratory tract toxicity of inhaled methylvinyl ketone in F344 rats and B6C3F1 mice. Toxicol Sci. 2000;58(1):182–94. doi:10.1093/toxsci/58.1.182.
30. Acrolein, crotonaldehyde and arecoline (IARC Monographs on the Identification of Carcinogenic Hazards to Humans, vol. 128). Lyon: International Agency for Research on Cancer; 2021 (https://publications.iarc.fr/Book-And-Report-Series/Iarc-Monographs-On-The-Identification-Of-Carcinogenic-Hazards-To-Humans/Acrolein-Crotonaldehyde-And-Arecoline-2021).
31. Haussmann HJ. Use of hazard indices for a theoretical evaluation of cigarette smoke composition. Chem Res Toxicol. 2012;25(4):794–810. doi:10.1021/tx200536w.
32. Burcham PC. Acrolein and human disease: Untangling the knotty exposure scenarios accompanying several diverse disorders. Chem Res Toxicol. 2017;30(1):145–61. doi:10.1021/acs.chemrestox.6b00310.
33. Mathias PI, B'Hymer C. Mercapturic acids: recent advances in their determination by liquid

chromatography/mass spectrometry and their use in toxicant metabolism studies and in occupational and environmental exposure studies. Biomarkers. 2016;21(4):293–315. doi:10.3109/1354750X.2016.1141988.

34. Chen M, Carmella SG, Li Y, Zhao Y, Hecht SS. Resolution and quantitation of mercapturic acids derived from crotonaldehyde, methacrolein and methyl vinyl ketone in the urine of smokers and nonsmokers. Chem Res Toxicol. 2020;33(2):669–77. doi:10.1021/acs.chemrestox.9b00491.
35. Scherer G, Urban M, Engl J, Hagedorn HW, Riedel K. Influence of smoking charcoal filter tipped cigarettes on various biomarkers of exposure. Inhal Toxicol. 2006;18(10):821–9. doi:10.1080/08958370600747945.
36. Scherer G, Urban M, Hagedorn HW, Feng S, Kinser RD, Sarkar M et al. Determination of two mercapturic acids related to crotonaldehyde in human urine: influence of smoking. Hum Exp Toxicol. 2007;26(1):37–47. doi:10.1177/0960327107073829.
37. Carmella SG, Chen M, Han S, Briggs A, Jensen J, Hatsukami DK et al. Effects of smoking cessation on eight urinary tobacco carcinogen and toxicant biomarkers. Chem Res Toxicol. 2009;22(4):734–41. doi:10.1021/tx800479s.
38. Alwis KU, Blount BC, Britt AS, Patel D, Ashley DL. Simultaneous analysis of 28 urinary VOC metabolites using ultra high performance liquid chromatography coupled with electrospray ionization tandem mass spectrometry (UPLC-ESI/MSMS). Anal Chim Acta. 2012;750:152–60. doi:10.1016/j.aca.2012.04.009.
39. Wei B, Alwis KU, Li Z, Wang L, Valentin-Blasini L, Sosnoff CS et al. Urinary concentrations of PAH and VOC metabolites in marijuana users. Environ Int. 2016;88:1–8. doi:10.1016/j.envint.2015.12.003.
40. Pluym N, Gilch G, Scherer G, Scherer M. Analysis of 18 urinary mercapturic acids by two high-throughput multiplex-LC-MS/MS methods. Anal Bioanal Chem. 2015;407(18):5463–76. doi:10.1007/s00216-015-8719-x.
41. Bagchi P, Geldner N, de Castro BR, De Jesus VR, Park SK, Blount BC. Crotonaldehyde exposure in US tobacco smokers and nonsmokers: NHANES 2005–2006 and 2011–2012. Environ Res. 2018;163:1–9. doi:10.1016/j.envres.2018.01.033.
42. Hatsukami DK, Luo X, Jensen JA, al'Absi M, Allen SS, Carmella SG et al. Effect of immediate vs gradual reduction in nicotine content of cigarettes on biomarkers of smoke exposure: a randomized cinical trial. JAMA. 2018;320(9):880–91. doi:10.1001/jama.2018.11473.
43. Frigerio G, Mercadante R, Polledri E, Missineo P, Campo L, Fustinoni S. An LC-MS/MS method to profile urinary mercapturic acids metabolites of electrophilic intermediates of occupational and environmental toxicants. J Chromatogr B Anal Technol Biomed Life Sci. 2019;1117:66–76. doi:10.1016/j.jchromb.2019.04.015.
44. Kotapati S, Esades A, Matter B, Le C, Tretyakova N. High throughput HPLC-ESI--MS/MS methodology for mercapturic acid metabolites of 1,3-butadiene: biomarkers of exposure and bioactivation. Chem Biol Interact. 2015;241:23–31. doi:10.1016/j.cbi.2015.02.009.
45. Ashley DL, Bonin MA, Cardinali FL, McCraw JM, Wooten JV. Measurement of volatile organic compounds in human blood. Environ Health Perspect. 1996;104(Suppl_5):871–7. doi:10.1289/ehp.96104s5871.
46. Ashley DL, Bonin MA, Hamar B, McGeehin MA. Removing the smoking confounder from blood volatile organic compounds measurements. Environ Res. 1995;71(1):39–45. doi:10.1006/enrs.1995.1065.
47. Luo X, Carmella SG, Chen M, Jensen JA, Wilkens LR, Le Marchand L et al. Urinary cyanoethyl mercapturic acid, a biomarker of the smoke toxicant acrylonitrile, clearly distinguishes smokers from non-smokers. Nicotine Tob Res. 2020;22:1744–7. doi:10.1093/ntr/ntaa080.
48. Cancelada L, Sleiman M, Tang X, Russell ML, Montesinos VN, Litter MI et al. Heated tobacco products: volatile emissions and their predicted impact on indoor air quality. Environ Sci

Technol. 2019;53(13):7866–76. doi:10.1021/acs.est.9b02544.

49 Farsalinos KE, Yannovits N, Sarri T, Voudris V, Poulas K, Leischow SJ. Carbonyl emissions from a novel heated tobacco product (IQOS): comparison with an e-cigarette and a tobacco cigarette. Addiction. 2018;113(11):2099–106. doi:10.1111/add.14365.

50. Pappas RS, Fresquez MR, Martone N, Watson CH. Toxic metal concentrations in mainstream smoke from cigarettes available in the USA. J Anal Toxicol. 2014;38(4):204–11. doi:10.1093/jat/bku013.

51. Marano KM, Naufal ZS, Kathman SJ, Bodnar JA, Borgerding MF, Garner CD et al. Cadmium exposure and tobacco consumption: biomarkers and risk assessment. Regul Toxicol Pharmacol. 2012;6 (2):243–52. doi:10.1016/j.yrtph.2012.07.008.

52. Hoffmann K, Becker K, Friedrich C, Helm D, Krause C, Seifert B. The German Environmental Survey 1990/1992 (GerES II): cadmium in blood urine and hair of adults and children. J Expo Anal Environ. Epidemiol. 2000;10:126–35. doi:10.1038/sj.jea.7500081.

53. Arsenic metals fibres and dusts. In: IARC Monographs on the Evaluation of Carcinogenic Risks to Humans, vol. 100C. Lyon: International Agency for Research on Cancer; 2012:121–45 (https://monographs.iarc.who.int/wp-content/uploads/2018/06/mono100C.pdf).

54. Toxicological profile for lead. Atlanta (GA): Agency for Toxic Substances and Disease Registry; 2020 (https://www.atsdr.cdc.gov/toxprofiles/tp13.pdf).

55. Belushkin M, Tafin Djoko D, Esposito M, Korneliou A, Jeannet C, Lazzerini M et al. Selected harmful and potentially harmful constituents levels in commercial e-cigarettes. Chem Res Toxicol. 2020;33(2):657–68. doi:10.1021/acs.chemrestox.9b00470.

56. Margham J, McAdam K, Forster M, Liu C, Wright C, Mariner D et al. Chemical composition of aerosol from an e-cigarette: a quantitative comparison with cigarette smoke. Chem Res Toxicol. 2016;29(10):1662–78. doi:10.1021/acs.chemrestox.6b00188.

57. Eshraghian EA, Al-Delaimy WK. A review of constituents identified in e-cigarette liquids and aerosols. Tob Prev Cessat. 2021;7:10. doi:10.18332/tpc/131111.

58. Gaur S, Agnihotri R. Health effects of trace metals in electronic cigarette aerosols – a systematic review. Biol Trace Elem Res. 2019;188(2):295–315. doi:10.1007/s12011-018-1423-x.

59. Paschal DC, Burt V, Caudill SP, Gunter EW, Pirkle JL, Sampson EJ et al. Exposure of the US population aged 6 years and older to cadmium: 1988–1994. Arch Environ Contam Toxicol. 2000;38:377–83. doi:10.1007/s002449910050.

60. Akiyama Y, Sherwood N. Systematic review of biomarker findings from clinical studies of electronic cigarettes and heated tobacco products. Toxicol Rep. 2021;8:282–94. doi:10.1016/j.toxrep.2021.01.014.

61. Oliveri D, Liang Q, Sarkar M. Real-world evidence of differences in biomarkers of exposure to select harmful and potentially harmful constituents and biomarkers of potential harm between adult e-vapor users and adult cigarette smokers. Nicotine Tob Res. 2020;22(7):1114–22. doi:10.1093/ntr/ntz185.

62. Hatsukami DK, Meier E, Lindgren BR, Anderson A, Reisinger SA, Norton KJ et al. A randomized clinical trial examining the effects of instructions for electronic cigarette use on smoking-related behaviors and biomarkers of exposure. Nicotine Tob. Res. 2020;22(9):1524–32. doi:10.1093/ntr/ntz233.

63. McRobbie H, Phillips A, Goniewicz ML, Smith KM, Knight-West O, Przulj D et al. Effects of switching to electronic cigarettes with and without concurrent smoking on exposure to nicotine carbon monoxide and acrolein. Cancer Prev Res. 2015;8(9):873–8. doi:10.1158/1940-6207.CAPR-15-0058.

64. Czoli CD, Fong GT, Goniewicz ML, Hammond D. Biomarkers of exposure among "dual users" of tobacco cigarettes and electronic cigarettes in Canada. Nicotine Tob Res. 2019;21(9):1259–66. doi:10.1093/ntr/nty174.

65. O'Connell G, Graff DW, D'Ruiz CD. Reductions in biomarkers of exposure (BoE) to harmful or potentially harmful constituents (HPHCs) following partial or complete substitution of cigarettes with electronic cigarettes in adult smokers. Toxicol Mech Meth. 2016;26(6):443–54. doi:10.1080/15376516.2016.1196282.
66. Cravo AS, Bush J, Sharma G, Savioz R, Martin C, Craige S et al. A randomised parallel group study to evaluate the safety profile of an electronic vapour product over 12 weeks. Regul Toxicol Pharmacol. 2016;81)Suppl–1):S1–14. doi:10.1016/j.yrtph.2016.10.003.
67. Morris P, McDermott S, Chapman F, Verron T, Cahours X, Stevenson M et al. Reductions in biomarkers of exposure to selected harmful and potentially harmful constituents following exclusive and partial switching from combustible cigarettes to myblu() electronic nicotine delivery systems (ENDS). Intern Emerg Med. 2022;17(2):397–410. doi:10.1007/s11739-021-02813-w.
68. Gale N, McEwan M, Eldridge AC, Fearon IM, Sherwood N, Bowen E et al. Changes in biomarkers of exposure on switching from a conventional cigarette to tobacco heating products: a randomized controlled study in healthy Japanese subjects. Nicotine Tob Res. 2019;21(9):1220–7. doi:10.1093/ntr/nty104.
69. Haziza C, de La Bourdonnaye G, Donelli A, Poux V, Skiada D, Weitkunat R et al. Reduction in exposure to selected harmful and potentially harmful constituents approaching those observed upon smoking abstinence in smokers switching to the menthol tobacco heating system 2.2 for 3 months (Part 1). Nicotine Tob Res. 2020;22(4):539–48. doi:10.1093/ntr/ntz013.
70. Round EK, Chen P, Taylor AK, Schmidt E. Biomarkers of tobacco exposure decrease after smokers switch to an e-cigarette or nicotine gum. Nicotine Tob. Res. 2019;21(9):1239–47. doi:10.1093/ntr/nty140.
71. Shahab L, Goniewicz ML, Blount BC, Brown J, McNeill A, Alwis KU et al. Nicotine, carcinogen and toxin exposure in long-term e-cigarette and nicotine replacement therapy users: a cross-sectional study. Ann Intern Med. 2017;166(6):390–400. doi:10.7326/M16-1107.
72. Ludicke F, Picavet P, Baker G, Haziza C, Poux V, Lama N et al. Effects of switching to the tobacco heating system 2.2 menthol smoking abstinence or continued cigarette smoking on biomarkers of exposure: a randomized controlled open-label multicenter study in sequential confinement and ambulatory settings (Part 1). Nicotine Tob Res. 2018;20(2):161–72. doi:10.1093/ntr/ntx028.
73. Goniewicz ML, Smith DM, Edwards KC, Blount BC, Caldwell KL, Feng J et al. Comparison of nicotine and toxicant exposure in users of electronic cigarettes and combustible cigarettes. JAMA Netw Open. 2018;1(8):e185937. doi:10.1001/jamanetworkopen.2018.5937.
74. Dai H, Benowitz NL, Achutan C, Farazi PA, Degarege A, Khan AS. Exposure to toxicants associated with use and transitions between cigarettes, e-cigarettes and no tobacco. JAMA Netw Open. 2022;5(2):e2147891. doi:10.1001/jamanetworkopen.2021.47891.
75. Hecht SS, Carmella SG, Chen M, Koch JFD, Miller AT, Murphy SE et al. Quantitation of urinary metabolites of a tobacco-specific lung carcinogen after smoking cessation. Cancer Res. 1999;59:590–96. PMID:9973205.
76. Goniewicz ML, Havel CM, Peng MW, Jacob P III, Dempsey D, Yu L et al. Elimination kinetics of the tobacco-specific biomarker and lung carcinogen 4-(methylnitrosamino)-1-(3-pyridyl)-1-butanol. Cancer Epidemiol Biomarkers Prev. 2009;18(12):3421–5. doi:10.1158/1055-9965.EPI-09-0874.
77. Hecht SS, Carmella SG, Murphy SE, Akerkar S, Brunnemann KD, Hoffmann D. A tobacco-specific lung carcinogen in the urine of men exposed to cigarette smoke. N Engl J Med. 1993;329:1543–6. doi:10.1056/NEJM199311183292105.
78. Kotandeniya D, Carmella SG, Pillsbury ME, Hecht SS. Combined analysis of N′-nitrosonornicotine and 4-(methylnitrosamino)-1-(3-pyridyl)-1-butanol in the urine of cigarette smokers

and e-cigarette users. J Chromatogr B Analyt Technol Biomed Life Sci. 2015;1007:121–6. doi:10.1016/j.jchromb.2015.10.012.

79. Scherer G, Scherer M, Mutze J, Hauke T, Pluym N. Assessment of the exposure to tobacco-specific nitrosamines and minor tobacco alkaloids in users of various tobacco/nicotine products. Chem Res Toxicol. 2022;35(4):684–93. doi:10.1021/acs.chemrestox.2c00020.
80. Hecht SS, Carmella SG, Kotandeniya D, Pillsbury ME, Chen M, Ransom BW et al. Evaluation of toxicant and carcinogen metabolites in the urine of e-cigarette users versus cigarette smokers. Nicotine Tob Res. 2015;17(6):704–9. doi:10.1093/ntr/ntu218.
81. Scherer G, Pluym N, Scherer M. Comparison of urinary mercapturic acid excretions in users of various tobacco/nicotine products. Drug Test Anal. 2022. doi:10.1002/dta.3372.
82. Chen M, Carmella SG, Lindgren BR, Luo X, Ikuemonisan J, Niesen B et al. Increased levels of the acrolein metabolite 3-hydroxypropyl mercapturic acid in the urine of e-cigarette users. Chem Res Toxicol. 2022. doi:10.1021/acs.chemrestox.2c00145.
83. Prokopowicz A, Sobczak A, Szula-Chraplewska M, Ochota P, Kosmider L. Exposure to cadmium and lead in cigarette smokers who switched to electronic cigarettes. Nicotine Tob Res. 2019;21(9):1198–205. doi:10.1093/ntr/nty161.
84. Prokopowicz A, Sobczak A, Szdzuj J, Grygoyc K, Kosmider L. Metal concentration assessment in the urine of cigarette smokers who switched to electronic cigarettes: a pilot study. Int J Environ Res Public Health. 2020;17(6):1877. doi:10.3390/ijerph17061877.
85. Hiler M, Weidner AS, Hull LC, Kurti AN, Mishina EV. Systemic biomarkers of exposure associated with ENDS use: A scoping review. Tob Control. 2021. doi:10.1136/tobaccocontrol-2021-056896.
86. Baig SA, Giovenco DP. Behavioral heterogeneity among cigarette and e-cigarette dual-users and associations with future tobacco use: findings from the Population Assessment of Tobacco and Health Study. Addict Behav. 2020;104:106263. doi:10.1016/j.addbeh.2019.106263.
87. Smith DM, Shahab L, Blount BC, Gawron M, Kosminder L, Sobczak A et al. Differences in exposure to nicotine, tobacco-specific nitrosamines, and volatile organic compounds among electronic cigarette users, tobacco smokers, and dual users from three countries. Toxics. 2020;8(4):88. doi:10.3390/toxics8040088.
88. Keith RJ, Fetterman JL, Orimoloye OA, Dardari Z, Lorkiewicz PK, Hamburg NM et al. Characterization of volatile organic compound metabolites in cigarette smokers, electronic nicotine device users, dual users, and nonusers of tobacco. Nicotine Tob Res. 2020;22(2):264–72. doi:10.1093/ntr/ntz021.
89. Piper ME, Baker TB, Benowitz NL, Kobinsky KH, Jorenby DE. Dual users compared to smokers: demographics, dependence and biomarkers. Nicotine Tob Res. 2019;21(9):1279–84. doi:10.1093/ntr/nty231.
90. Anic GM, Rostron BL, Hammad HT, van Bemmel DM, Del Valle-Pinero AY, Christensen CH et al. Changes in biomarkers of tobacco exposure among cigarette smokers transitioning to ENDS use: the Population Assessment of Tobacco and Health Study 2013–2015. Int J Environ Res Public Health. 2022;19(3):1462. doi:10.3390/ijerph19031462.
91. Scherer G. Suitability of biomarkers of biological effects (BOBEs) for assessing the likelihood of reducing the tobacco related disease risk by new and innovative tobacco products: a literature review. Regul Toxicol Pharmacol. 2018;94:203–33. doi:10.1016/j.yrtph.2018.02.002.
92. Chang CM, Cheng YC, Cho M, Mishina E, Del Valle-Pinero AY, van Bemmel D et al. Biomarkers of potential harm: summary of an FDA-sponsored public workshop. Nicotine Tob Res. 2019;21(1):3–13. doi:10.1093/ntr/ntx273.
93. Chang JT, Vivar JC, Tam J, Hammad HT, Christensen CH, van Bemmel DM et al. Biomarkers of potential harm among adult cigarette and smokeless tobacco users in the PATH Study wave 1 (2013–2014): a cross-sectional analysis. Cancer Epidemiol Biomarkers Prev. 2021;30(7):1320–7. doi:10.1158/1055-9965.EPI-20-1544.

94. Stratton K, Shetty P, Wallace R, Bondurant S. Clearing the smoke: assessing the science base for tobacco harm reduction. Washington DC: Institute of Medicine; 2001. doi:10.17226/10029.
95. Gregg EO, Fisher AL, Lowe F, McEwan M, Massey ED. An approach to the validation of biomarkers of harm for use in a tobacco context. Regul Toxicol Pharmacol. 2006;44(3):262–7. doi:10.1016/j.yrtph.2005.12.006.
96. Leemans CR, Snijders PJF, Brakenhoff RH. The molecular landscape of head and neck cancer. Nat Rev Cancer. 2018;18(5):269–82. doi:10.1038/nrc.2018.11.
97. Hecht SS. Tobacco smoke carcinogens and lung cancer. J Natl Cancer Inst. 1999;91:1194–210.
98. Basu AK. DNA damage, mutagenesis and cancer. Int J Mol Sci. 2018;19(4):970. doi:10.3390/ijms19040970.
99. Hwa Yun B, Guo J, Bellamri M, Turesky RJ. DNA adducts: formation biological effects and new biospecimens for mass spectrometric measurements in humans. Mass Spectrom Rev. 2020;39(1–2):55–82. doi:10.1002/mas.21570.
100. Geacintov NE, Broyde S. Repair-resistant DNA lesions. Chem Res Toxicol. 2017;30(8)1517–48. doi:10.1021/acs.chemrestox.7b00128.
101. Ma B, Stepanov I, Hecht SS. Recent studies on DNA adducts resulting from human exposure to tobacco smoke. Toxics. 2019;7(1):16. doi:10.3390/toxics7010016.
102. Tobacco smoke and involuntary smoking (IARC Monographs on the Evaluation of Carcinogenic Risks to Humans, vol. 83). Lyon: International Agency for Research on Cancer; 2004:35–102 (https://publications.iarc.fr/Book-And-Report-Series/Iarc-Monographs-On-The-Identification-Of-Carcinogenic-Hazards-To-Humans/Tobacco-Smoke-And-Involuntary-Smoking-2004).
103. Phillips DH, Venitt S. DNA and protein adducts in human tissues resulting from exposure to tobacco smoke. Int J Cancer. 2012;131(12):2733–53. doi:10.1002/ijc.27827.
104. Boysen G, Hecht SS. Analysis of DNA and protein adducts of benzo[a]pyrene in human tissues using structure-specific methods. Mutat Res. 2003;543:17–30. doi:10.1016/s1383-5742(02)00068-6.
105. Hecht SS. Oral cell DNA adducts as potential biomarkers for lung cancer susceptibility in cigarette smokers. Chem Res Toxicol. 2017;30(1):367–75. doi:10.1021/acs.chemrestox.6b00372.
106. Paiano V, Maertens L, Guidolin V, Yang J, Balbo S, Hecht SS. Quantitative liquid chromatography–nanoelectrospray ionization–high-resolution tandem mass spectrometry analysis of acrolein–DNA adducts and etheno-DNA adducts in oral cells from cigarette smokers and nonsmokers. Chem Res Toxicol. 2020;33(8):2197–207. doi:10.1021/acs.chemrestox.0c00223.
107. Frost-Pineda K, Liang Q, Liu J, Rimmer L, Jin Y, Feng S et al. Biomarkers of potential harm among adult smokers and nonsmokers in the total exposure study. Nicotine Tob Res. 2011;13(3):182–93. doi:10.1093/ntr/ntq235.
108. Shiels MS, Pfeiffer RM, Hildesheim A, Engels EA, Kemp TJ, Park JH et al. Circulating inflammation markers and prospective risk for lung cancer. J Natl Cancer Inst. 2013;105:1871–80. doi:10.1093/jnci/djt309.
109. Holvoet P, Jenny NS, Schreiner PJ, Tracy RP, Jacobs DR, Multi-Ethnic Study of Atherosclerosis. The relationship between oxidized LDL and other cardiovascular risk factors and subclinical CVD in different ethnic groups: the Multi-Ethnic Study of Atherosclerosis (MESA). Atherosclerosis. 2007;194(1):245–52. doi:10.1016/j.atherosclerosis.2006.08.002.
110. Libby P. Inflammation during the life cycle of the atherosclerotic plaque. Cardiovasc Res. 2021;117(13):2525–36. doi:10.1093/cvr/cvab303.
111. Mendall MA, Patel P, Asante M, Ballam L, Morris J, Strachan DP et al. Relation of serum cytokine concentrations to cardiovascular risk factors and coronary heart disease. Heart. 1997;78(3):273–7. doi:10.1136/hrt.78.3.273.
112. Tuut M, Hense HW. Smoking other risk factors and fibrinogen levels. Evidence of effect modification. Ann Epidemiol. 2001;11(4):232–8. doi:10.1016/s1047-2797(00)00226-x.

113. Luc G, Arveiler D, Evans A, Amouyel P, Ferrieres J, Bard JM et al. Circulating soluble adhesion molecules ICAM-1 and VCAM-1 and incident coronary heart disease: the PRIME Study. Atherosclerosis. 2003;170(1):169–76. doi:10.1016/s0021-9150(03)00280-6.
114. Tracy RP, Psaty BM, Macy E, Bovill EG, Cushman M, Cornell ES et al. Lifetime smoking exposure affects the association of C-reactive protein with cardiovascular disease risk factors and subclinical disease in healthy elderly subjects. Arterioscler Thromb Vasc Biol. 1997;17(10):2167–76. doi:10.1161/01.atv.17.10.2167.
115. Finkelstein R, Fraser RS, Ghezzo H, Cosio MG. Alveolar inflammation and its relation to emphysema in smokers. Am J Respir Crit Care Med. 1995;152(5:1):1666–72. doi:10.1164/ajrccm.152.5.7582312.
116. Sullivan AK, Simonian PL, Falta MT, Mitchell JD, Cosgrove GP, Brown K et al. Oligoclonal CD4+ T cells in the lungs of patients with severe emphysema. Am J Respir Crit Care Med. 2005;172(5):590–6. doi:10.1164/rccm.200410-1332OC.
117. Majo J, Ghezzo H, Cosio MG. Lymphocyte population and apoptosis in the lungs of smokers and their relation to emphysema. Eur Respir J. 2001;17(5):946–53. doi:10.1183/09031936.01.17509460.
118. Moy ML, Teylan M, Weston NA, Gagnon DR, Danilack VA, Garshick E. Daily step count is associated with plasma C-reactive protein and IL-6 in a US cohort with COPD. Chest. 2014;145(3):542–50. doi:10.1378/chest.13-1052.
119. Celli BR, Locantore N, Yates J, Tal-Singer R, Miller BE, Bakke P et al. Inflammatory biomarkers improve clinical prediction of mortality in chronic obstructive pulmonary disease. Am J Respir Crit Care Med. 2012;185(10):1065–72. doi:10.1164/rccm.201110-1792OC.
120. Dickens JA, Miller BE, Edwards LD, Silverman EK, Lomas DA, Tal-Singer R et al. COPD association and repeatability of blood biomarkers in the ECLIPSE cohort. Respir Res. 2011;12:146. doi:10.1186/1465-9921-12-146.
121. Agusti A, Edwards LD, Rennard SI, MacNee W, Tal-Singer R, Miller BE et al. Persistent systemic inflammation is associated with poor clinical outcomes in COPD: a novel phenotype. PLoS One. 2012;7(5):e37483. doi:10.1371/journal.pone.0037483.
122. Pinto-Plata V, Casanova C, Mullerova H, de Torres JP, Corado H, Varo N et al. Inflammatory and repair serum biomarker pattern: association to clinical outcomes in COPD. Respir Res. 2012;13:71. doi:10.1186/1465-9921-13-71.
123. Tsai JJ, Liao EC, Hsu JY, Lee WJ, Lai YK. The differences of eosinophil- and neutrophil-related inflammation in elderly allergic and non-allergic chronic obstructive pulmonary disease. J Asthma. 2010;47(9):1040–4. doi:10.1080/02770903.2010.491145.
124. Man SF, Xuekui Z, Vessey R, Walker T, Lee K, Park D et al. The effects of inhaled and oral corticosteroids on serum inflammatory biomarkers in COPD: an exploratory study. Ther Adv Respir Dis. 2009;3(2):73–80. doi:10.1177/1753465809336697.
125. Thomsen M, Ingebrigtsen TS, Marott JL, Dahl M, Lange P, Vestbo J et al. Inflammatory biomarkers and exacerbations in chronic obstructive pulmonary disease. JAMA. 2013;309(22):2353–61. doi:10.1001/jama.2013.5732.
126. van Durme YM, Verhamme KM, Aarnoudse AJ, Van Pottelberge GR, Hofman A, Witteman JC et al. C-reactive protein levels haplotypes and the risk of incident chronic obstructive pulmonary disease. Am J Respir Crit Care Med. 2009;179(5):375–82. doi:10.1164/rccm.200810-1540OC.
127. Thomsen M, Dahl M, Lange P, Vestbo J, Nordestgaard BG. Inflammatory biomarkers and comorbidities in chronic obstructive pulmonary disease. Am J Respir Crit Care Med. 2012;186(10):982–8. doi:10.1164/rccm.201206-1113OC.
128. Higashimoto Y, Iwata T, Okada M, Satoh H, Fukuda K, Tohda Y. Serum biomarkers as predictors of lung function decline in chronic obstructive pulmonary disease. Respir Med. 2009;103(8):1231–8. doi:10.1016/j.rmed.2009.01.021.

129. Mannino DM, Valvi D, Mullerova H, Tal-Singer R. Fibrinogen COPD and mortality in a nationally representative US cohort. COPD. 2012;9(4):359–66. doi:10.3109/15412555.2012.668249.
130. Cockayne DA, Cheng DT, Waschki B, Sridhar S, Ravindran P, Hilton H et al. Systemic biomarkers of neutrophilic inflammation tissue injury and repair in COPD patients with differing levels of disease severity. PLoS One. 2012;7(6):e38629. doi:10.1371/journal.pone.0038629.
131. Dahl M, Tybjaerg-Hansen A, Vestbo J, Lange P, Nordestgaard BG. Elevated plasma fibrinogen associated with reduced pulmonary function and increased risk of chronic obstructive pulmonary disease. Am J Respir Crit Care Med. 2001;164(6):1008–11. doi:10.1164/ajrccm.164.6.2010067.
132. Milne GL, Yin H, Hardy KD, Davies SS, Roberts LJ. Isoprostane generation and function. Chem Rev. 2011;111(10):5973–96. doi:10.1021/cr200160h.
133. How tobacco smoke causes disease: the biology and behavioral basis for smoking-attributable disease: a report of the Surgeon General. Atlanta (GA): Centers for Disease Control and Prevention, Office on Smoking and Health; 2010 (https://www.ncbi.nlm.nih.gov/books/NBK53017/).
134. Yuan JM, Carmella SG, Wang R, Tan YT, Adams-Haduch J, Gao YT et al. Relationship of the oxidative damage biomarker 8-epi-prostaglandin F2 alpha to risk of lung cancer development in the Shanghai Cohort Study. Carcinogenesis. 2018;39(7):948–54. doi:10.1093/carcin/bgy060.
135. Christensen CH, Chang JT, Rostron BL, Hammad HT, van Bemmel DM, Del Valle-Pinero AY et al. Biomarkers of inflammation and oxidative stress among adult former smoker current e-cigarette users –results from wave 1 PATH Study. Cancer Epidemiol Biomarkers Prev. 2021;30(10):1947–55. doi:10.1158/1055-9965.EPI-21-0140.
136. McElroy JP, Carmella SG, Heskin AK, Tang MK, Murphy SE, Reisinger SA et al. Effects of cessation of cigarette smoking on eicosanoid biomarkers of inflammation and oxidative damage. PLoS One. 2019;14(6):e0218386. doi:0.1371/journal.pone.0218386.
137. Guo J, Hecht SS. DNA damage in human oral cells induced by use of e-cigarettes. Drug Test Anal. 2022. doi:10.1002/dta.3375.
138. Cheng G, Guo J, Carmella SG, Lindgren B, Ikuemonisan J, Niesen B et al. Increased acrolein–DNA adducts in buccal brushings of e-cigarette users. Carcinogenesis. 2022;43(5):437–44. doi:10.1093/carcin/bgac026.
139. Guo J, Ikuemonisan J, Hatsukami DK, Hecht SS. Liquid chromatography–nanoelectrospray ionization–high-resolution tandem mass spectrometry analysis of apurinic/apyrimidinic sites in oral cell DNA of cigarette smokers e-cigarette users and nonsmokers. Chem Res Toxicol. 2021;34(12):2540–8. doi:10.1021/acs.chemrestox.1c00308.
140. Okafor CN, Okafor N, Kaliszewski C, Wang L. Association between electronic cigarette and combustible cigarette use with cardiometabolic risk biomarkers among US adults. Ann Epidemiol. 2022;71:44–50. doi:10.1016/j.annepidem.2022.02.002.
141. Benowitz NL, St Helen G, Nardone N, Addo N, Zhang JJ, Harvanko AM et al. Twenty-four-hour cardiovascular effects of electronic cigarettes compared with cigarette smoking in dual users. J Am Heart Assoc. 2020;9(23):e017317. doi:10.1161/JAHA.120.017317.
142. Perez MF, Mead EL, Atuegwu NC, Mortensen EM, Goniewicz M, Oncken C. Biomarkers of toxicant exposure and inflammation among women of reproductive age who use electronic or conventional cigarettes. J Womens Health (Larchmt). 2021;30(4):539–50. doi:10.1089/jwh.2019.8075.
143. Majid S, Keith RJ, Fetterman JL, Weisbrod RM, Nystoriak J, Wilson T et al. Lipid profiles in users of combustible and electronic cigarettes. Vasc Med. 2021;26(5):483–8. doi:10.1177/1358863X211009313.
144. Rao P, Liu J, Springer ML. JUUL and combusted cigarettes comparably impair endothelial function. Tob Regul Sci. 2020;6(1):30–7. doi:10.18001/TRS.6.1.4.

145. Chaumont M, van de Borne P, Bernard A, Van Muylem A, Deprez G, Ullmo J et al. Fourth generation e-cigarette vaping induces transient lung inflammation and gas exchange disturbances: results from two randomized clinical trials. Am J Physiol Lung Cell Mol Physiol. 2019;316(5):L705–19. doi:10.1152/ajplung.00492.2018.
146. Muthumalage T, Lamb T, Friedman MR, Rahman I. E-cigarette flavored pods induce inflammation epithelial barrier dysfunction and DNA damage in lung epithelial cells and monocytes. Sci Rep. 2019;9(1):19035.
147. Moheimani RS, Bhetraratana M, Yin F, Peters KM, Gornbein J, Araujo JA et al. Increased cardiac sympathetic activity and oxidative stress in habitual electronic cigarette users: implications for cardiovascular risk. JAMA Cardiol. 2017;2(3):278–84.
148. Metzen D, M'Pembele R, Zako S, Mourikis P, Helten C, Zikeli D et al. Platelet reactivity is higher in e-cigarette vaping as compared to traditional smoking. Int J Cardiol. 2021.343:146–8. doi:10.1016/j.ijcard.2021.09.005.
149. Mohammadi L, Han DD, Xu F, Huang A, Derakhshandeh R, Rao P et al. Chronic e-cigarette use impairs endothelial function on the physiological and cellular levels. Arterioscler Thromb Vasc Biol. 2022;42(11):1333–50.
150. Verma A, Anand K, Bhargava M, Kolluri A, Kumar M, Palve DH. Comparative evaluation of salivary biomarker levels in e-cigarette smokers and conventional smokers. J Pharm Bioallied Sci. 2021;13(Suppl_2):S1642–5.
151. Patskan G, Reininghaus W. Toxicological evaluation of an electrically heated cigarette. Part 1: Overview of technical concepts and summary of findings. J Appl Toxicol. 2003;23(5):323–8.
152. Tricker AR, Kanada S, Takada K, Martin Leroy C, Lindner D, Schorp MK et al. Reduced exposure evaluation of an electrically heated cigarette smoking system. Part 6: 6-day randomized clinical trial of a menthol cigarette in Japan. Regul Toxicol Pharmacol. 2012;64(2–Suppl):S64–73.
153. Martin Leroy C, Jarus-Dziedzic K, Ancerewicz J, Lindner D, Kulesza A, Magnette J. Reduced exposure evaluation of an electrically heated cigarette smoking system. Part 7: a one-month randomized ambulatory controlled clinical study in Poland. Regul Toxicol Pharmacol. 2012;64(2–Suppl):S74–84.
154. Lüdicke F, Ansari SM, Lama N, Blanc N, Bosilkovska M, Donelli A et al. Effects of switching to a heat-not-burn tobacco product on biologically relevant biomarkers to assess a candidate modified risk tobacco product: a randomized trial. Cancer Epidemiol Biomarkers Prev. 2019;28(11):1934–43.
155. Gale N, McEwan M, Camacho OM, Hardie G, Proctor CJ, Murphy J. Changes in biomarkers after 180 days of tobacco heating product use: a randomised trial. Intern Emerg Med. 2021;16(8):2201–12.
156. Gale N, McEwan M, Hardie G, Proctor CJ, Murphy J. Changes in biomarkers of exposure and biomarkers of potential harm after 360 days in smokers who either continue to smoke switch to a tobacco heating product or quit smoking. Intern Emerg Med. 2022;17(7):2017–30.
157. Sakaguchi C, Nagata Y, Kikuchi A, Takeshige Y, Minami N. Differences in levels of biomarkers of potential harm among users of a heat-not-burn tobacco product, cigarette smokers, and never-smokers in Japan: a post-marketing observational study. Nicotine Tob Res. 2021;23(7):1143–52.
158. Ogden MW, Marano KM, Jones BA, Morgan WT, Stiles MF. Switching from usual brand cigarettes to a tobacco-heating cigarette or snus: Part 3. Biomarkers of biological effect. Biomarkers. 2015;20(6–7):404–10.
159. Biondi Zoccai G, Carnevale R, Vitali M, Tritapepe L, Martinelli O, Macrina F et al. A randomized trial comparing the acute coronary systemic and environmental effects of electronic vaping cigarettes versus heat-not-burn cigarettes in smokers of combustible cigarettes undergoing invasive coronary assessment: rationale and design of the SUR-VAPES 3 trial. Minerva Cardio-

angiol. 2020;68(6):548–55.

160. Haziza C, de La Bourdonnaye G, Donelli A, Skiada D, Poux V, Weitkunat R et al. Favorable changes in biomarkers of potential harm to reduce the adverse health effects of smoking in smokers switching to the menthol tobacco heating system 2.2 for 3 months (Part 2). Nicotine Tob Res. 2020;22(4):549–59.
161. Singh KP, Lawyer G, Muthumalage T, Maremanda KP, Khan NA, McDonough SR et al. Systemic biomarkers in electronic cigarette users: implications for noninvasive assessment of vaping-associated pulmonary injuries. ERJ Open Res. 2019;5(4):00182–2019. doi:10.1183/23120541.00182-2019.
162. Sakamaki-Ching S, Williams M, Hua M, Li J, Bates SM, Robinson AN et al. Correlation between biomarkers of exposure effect and potential harm in the urine of electronic cigarette users. BMJ Open Respir Res. 2020;7(1). doi:10.1136/bmjresp-2019-000452.
163. Leidy NK, Rennard SI, Schmier J, Jones MK, Goldman M. The breathlessness cough and sputum scale: the development of empirically based guidelines for interpretation. Chest. 2003;124(6):2182–91. doi:10.1378/chest.124.6.2182. doi:10.1038/npjpcrm.2016.83.
164. Leidy NK, Schmier JK, Jones MK, Lloyd J, Rocchiccioli K. Evaluating symptoms in chronic obstructive pulmonary disease: validation of the Breathlessness Cough and Sputum Scale. Respir Med. 2003;97(Suppl_A):S59–70. PMID:12564612.
165. Miller MR, Hankinson J, Brusasco V, Burgos F, Casaburi R, Coates A et al. Standardisation of spirometry. Eur Respir J. 2005;26(2):319–38. doi:10.1183/09031936.05.00034805.
166. Miller MR, Crapo R, Hankinson J, Brusasco V, Burgos F, Casaburi R et al. General considerations for lung function testing. Eur Respir J. 2005;26(1):153–61. doi:10.1183/09031936.05.00034505.
167. Martinez-Morata I, Sanchez TR, Shimbo D, Navas-Acien A. Electronic cigarette use and blood pressure endpoints: a systematic review. Curr Hypertens Rep. 2020;23(1):2. doi:10.1007/s11906-020-01119-0.
168. Keith R, Bhatnagar A. Cardiorespiratory and immunologic effects of electronic cigarettes. Curr Addict Rep. 2021;8(2:336–46. doi:10.1007/s40429-021-00359-7.
169. Shields PG, Song MA, Freudenheim JL, Brasky TM, McElroy JP, Reisinger SA et al. Lipid laden macrophages and electronic cigarettes in healthy adults. EBioMedicine. 2020;60:102982. doi:10.1016/j.ebiom.2020.102982.
170. Dai H, Khan AS. A longitudinal study of exposure to tobacco-related toxicants and subsequent respiratory symptoms among US adults with varying e-cigarette use status. Nicotine Tob Res. 2020;22(Suppl–1):S61–9. doi:10.1093/ntr/ntaa180.
171. Xie W, Kathuria H, Galiatsatos P, Blaha MJ, Hamburg NM, Robertson RM et al. Association of electronic cigarette use with incident respiratory conditions among US adults from 2013 to 2018. JAMA Netw Open. 2020;3(11):e2020816. doi:10.1001/jamanetworkopen.2020.20816.
172. Goniewicz ML, Miller CR, Sutanto E, Li D. How effective are electronic cigarettes for reducing respiratory and cardiovascular risk in smokers? A systematic review. Harm Reduct J. 2020;17(1):91. doi:10.1186/s12954-020-00440-w.
173. Rezk-Hanna M, Gupta R, Nettle CO, Dobrin D, Cheng CW, Means A et al. Differential effects of electronic hookah vaping and traditional combustible hookah smoking on oxidation inflammation and arterial stiffness. Chest. 2022;161(1):208–18. doi:10.1016/j.chest.2021.07.027.
174. The health consequences of smoking – 50 years of progress: a report of the Surgeon General. Rockville (MD): United States Department of Health and Human Services; 2014. PMID:24455788.
175. Luoto R, Uutela A, Puska P. Occasional smoking increases total and cardiovascular mortality among men. Nicotine Tob Res. 2000;2(2):133–9. doi:10.1080/713688127.
176. Bjartveit K, Tverdal A. Health consequences of smoking 1–4 cigarettes per day. Tob Control. 2005;14(5):315–20. doi:10.1136/tc.2005.011932.

177. Pope CA 3rd, Burnett RT, Krewski D, Jerrett M, Shi Y, Calle EE et al. Cardiovascular mortality and exposure to airborne fine particulate matter and cigarette smoke: shape of the exposure–response relationship. Circulation. 2009;120(11):941–8. doi:10.1161/CIRCULATIONAHA.109.857888.
178. Lee A, Lee SY, Lee KS. The use of heated tobacco products is associated with asthma, allergic rhinitis and atopic dermatitis in Korean adolescents. Sci Rep. 2019;9(1):17699. doi:10.1038/s41598-019-54102-4.
179. Kamada T, Yamashita Y, Tomioka H. Acute eosinophilic pneumonia following heat-not-burn cigarette smoking. Respirol Case Rep. 2016;4(6):e00190. doi:10.1002/rcr2.190.
180. Aokage T, Tsukahara K, Fukuda Y, Tokioka F, Taniguchi A, Naito H et al. Heat-not-burn cigarettes induce fulminant acute eosinophilic pneumonia requiring extracorporeal membrane oxygenation. Respir Med Case Rep. 2019;26:87–90. doi:10.1016/j.rmcr.2018.12.002.
181. Liu Y, Sanoff HK, Cho H, Burd CE, Torrice C, Ibrahim JG et al. Expression of p16(INK4a) in peripheral blood T-cells is a biomarker of human aging. Aging Cell. 2009;8(4):439–48. doi:10.1111/j.1474-9726.2009.00489.x.
182. Lu AT, Quach A, Wilson JG, Reiner AP, Aviv A, Raj K et al. DNA methylation GrimAge strongly predicts lifespan and healthspan. Aging (Albany NY). 2019;11(2):303–27. doi:10.18632/aging.101684.
183. Richmond RC, Sillero-Rejon C, Khouja JN, Prince C, Board A, Sharp G et al. Investigating the DNA methylation profile of e-cigarette use. Clin Epigenetics. 2021;13(1):183. doi:10.1186/s13148-021-01174-7.
184. Song MA, Freudenheim JL, Brasky TM, Mathe EA, McElroy JP, Nickerson QA et al. Biomarkers of exposure and effect in the lungs of smokers, nonsmokers and electronic cigarette users. Cancer Epidemiol Biomarkers Prev. 2020;29(2):443–51. doi:10.1158/1055-9965.EPI-19-1245.
185. Philibert R, Hollenbeck N, Andersen E, Osborn T, Gerrard M, Gibbons FX et al. A quantitative epigenetic approach for the assessment of cigarette consumption. Front Psychol. 2015;6:656. doi:10.3389/fpsyg.2015.00656.
186. Andersen A, Reimer R, Dawes K, Becker A, Hutchens N, Miller S et al. DNA methylation differentiates smoking from vaping and non-combustible tobacco use. Epigenetics. 2022;17(2):178–90. doi:10.1080/15592294.2021.1890875.
187. Caliri AW, Caceres A, Tommasi S, Besaratinia A. Hypomethylation of LINE-1 repeat elements and global loss of DNA hydroxymethylation in vapers and smokers. Epigenetics. 2020;15(8):816–29. doi:10.1080/15592294.2020.1724401.
188. Ohmomo H, Harada S, Komaki S, Ono K, Sutoh Y, Otomo R et al. DNA methylation abnormalities and altered whole transcriptome profiles after switching from combustible tobacco smoking to heated tobacco products. Cancer Epidemiol Biomarkers Prev. 2022;31(1):269–79. doi:10.1158/1055-9965.EPI-21-0444.
189. Spira A, Beane J, Shah V, Liu G, Schembri F, Yang X et al. Effects of cigarette smoke on the human airway epithelial cell transcriptome. Proc Natl Acad Sci U S A. 2004;101(27):10143–8. doi:10.1073/pnas.0401422101.
190. Beane J, Sebastiani P, Liu G, Brody JS, Lenburg ME, Spira A. Reversible and permanent effects of tobacco smoke exposure on airway epithelial gene expression. Genome Biol. 2007;8(9):R201. doi:10.1186/gb-2007-8-9-r201.
191. Pavel AB, Campbell JD, Liu G, Elashoff D, Dubinett S, Smith K et al. Alterations in bronchial airway miRNA expression for lung cancer detection. Cancer Prev Res (Phila). 2017;10(11):651–9. doi:10.1158/1940-6207.CAPR-17-0098.
192. Martin EM, Clapp PW, Rebuli ME, Pawlak EA, Glista-Baker E, Benowitz NL et al. E-cigarette use results in suppression of immune and inflammatory-response genes in nasal epithelial cells similar to cigarette smoke. Am J Physiol Lung Cell Mol Physiol. 2016;311(1):L135–44.

doi:10.1152/ajplung.00170.2016.

193. Tommasi S, Caliri AW, Caceres A, Moreno DE, Li M, Chen Y et al. Deregulation of biologically significant genes and associated molecular pathways in the oral epithelium of electronic cigarette users. Int J Mol Sci. 2019;20(3):738. doi:10.3390/ijms20030738.
194. Hamad SH, Brinkman MC, Tsai YH, Mellouk N, Cross K, Jaspers I et al. Pilot study to detect genes involved in DNA damage and cancer in humans: potential biomarkers of exposure to e-cigarette aerosols. Genes (Basel). 2021;12(3):448. doi:10.3390/genes12030448.
195. Staudt MR, Salit J, Kaner RJ, Hollmann C, Crystal RG. Altered lung biology of healthy never smokers following acute inhalation of e-cigarettes. Respir Res. 2018;19(1):78. doi:10.1186/s12931-018-0778-z.
196. Gaeckle NT, Pragman AA, Pendleton KM, Baldomero AK, Criner GJ. The oral–lung axis: the impact of oral health on lung health. Respir Care. 2020;65(8):1211–20. doi:10.4187/respcare.07332.
198. Dickson RP, Huffnagle GB. The lung microbiome: new principles for respiratory bacteriology in health and disease. PLoS Pathog. 2015;11(7):e1004923. doi:10.1371/journal.ppat.1004923.
198. Mathieu E, Escribano-Vazquez U, Descamps D, Cherbuy C, Langella P, Riffault S et al. Paradigms of lung microbiota functions in health and disease particularly in asthma. Front Physiol. 2018;9:1168. doi:10.3389/fphys.2018.01168.
199. Hajishengallis G, Darveau RP, Curtis MA. The keystone-pathogen hypothesis. Nat Rev Microbiol. 2012;10(10):717–25. doi:10.1038/nrmicro2873.
200. Shi J, Yang Y, Xie H, Wang X, Wu J, Long J et al. Association of oral microbiota with lung cancer risk in a low-income population in the Southeastern USA. Cancer Causes Control. 2021;32(12):1423–32. doi:10.1007/s10552-021-01490-6.
201. Hosgood HD, Cai Q, Hua X, Long J, Shi J, Wan Y et al. Variation in oral microbiome is associated with future risk of lung cancer among never-smokers. Thorax. 2021;76(3):256–63. doi:10.1136/thoraxjnl-2020-215542.
202. Hayes RB, Ahn J, Fan X, Peters BA, Ma Y, Yang L et al. Association of oral microbiome with risk for incident head and neck squamous cell cancer. JAMA Oncol. 2018;4(3)358–65. doi:10.1001/jamaoncol.2017.4777.
203. Mammen MJ, Scannapieco FA, Sethi S. Oral–lung microbiome interactions in lung diseases. Periodontology. 2000;83(1):234–41. doi:10.1111/prd.12301.
204. Plachokova AS, Andreu-Sanchez S, Noz MP, Fu J, Riksen NP. Oral microbiome in relation to periodontitis severity and systemic inflammation. Int J Mol Sci. 2021;22(11):5876. doi:10.3390/ijms22115876.
205. Schulz S, Reichert S, Grollmitz J, Friebe L, Kohnert M, Hofmann B et al. The role of Saccharibacteria (TM7) in the subgingival microbiome as a predictor for secondary cardiovascular events. Int J Cardiol. 2021;331:255–61. doi:10.1177/0022034519831671.
206. Kato-Kogoe N, Sakaguchi S, Kamiya K, Omori M, Gu YH, Ito Y et al. Characterization of salivary microbiota in patients with atherosclerotic cardiovascular disease: a case–control study. J Atheroscler Thromb. 2022;29(3):403–21. doi:10.5551/jat.60608.
207. Goh CE, Trinh P, Colombo PC, Genkinger JM, Mathema B, Uhlemann AC et al. Association between nitrate-reducing oral bacteria and cardiometabolic outcomes: results from ORIGINS. J Am Heart Assoc. 2019;8(23):e013324. doi:10.1161/JAHA.119.013324.
208. Wu J, Peters BA, Dominianni C, Zhang Y, Pei Z, Yang L et al. Cigarette smoking and the oral microbiome in a large study of American adults. ISME J. 2016;10(10):2435–46. doi:10.1038/ismej.2016.37.
209. Droemann D, Goldmann T, Tiedje T, Zabel P, Dalhoff K, Schaaf B. Toll-like receptor 2 expression is decreased on alveolar macrophages in cigarette smokers and COPD patients. Respir Res. 2005;6:68. doi:10.1186/1465-9921-6-68.

210 Kulkarni R, Antala S, Wang A, Amaral FE, Rampersaud R, Larussa SJ et al. Cigarette smoke increases Staphylococcus aureus biofilm formation via oxidative stress. Infect Immun. 2012;80(11):3804–11. doi:10.1128/IAI.00689-12.
211. Ganesan SM, Joshi V, Fellows M, Dabdoub SM, Nagaraja HN, O'Donnell B et al. A tale of two risks: smoking diabetes and the subgingival microbiome. ISME J. 2017;11(9):2075–89. doi:10.1038/ismej.2017.73.
212. Mueller DC, Piller M, Niessner R, Scherer M, Scherer G. Untargeted metabolomic profiling in saliva of smokers and nonsmokers by a validated GC-TOF-MS method. J Proteome Res. 2014;13(3):1602–13. doi:10.1021/pr401099r.
213. Valles Y, Inman CK, Peters BA, Ali R, Wareth LA, Abdulle A et al. Types of tobacco consumption and the oral microbiome in the United Arab Emirates Healthy Future (UAEHFS) pilot study. Sci Rep. 2018;8(1):11327. doi:10.1038/s41598-018-29730-x.
214. Gopinath D, Wie CC, Banerjee M, Thangavelu L, Kumar RP, Nallaswamy D et al. Compositional profile of mucosal bacteriome of smokers and smokeless tobacco users. Clin Oral Investig. 2022;26(2):1647–56. doi:10.1007/s00784-021-04137-7.
215. Chopyk J, Bojanowski CM, Shin J, Moshensky A, Fuentes AL, Bonde SS et al. Compositional differences in the oral microbiome of e-cigarette users. Front Microbiol. 2021;12:599664. doi:10.3389/fmicb.2021.599664.
216. Pushalkar S, Paul B, Li Q, Yang J, Vasconcelos R, Makwana S et al. Electronic cigarette aerosol modulates the oral microbiome and increases risk of infection. iScience. 2020;23(3):100884. doi:10.1016/j.isci.2020.100884.
217. Ganesan SM, Dabdoub SM, Nagaraja HN, Scott ML, Pamulapati S, Berman ML et al. Adverse effects of electronic cigarettes on the disease-naive oral microbiome. Sci Adv. 2020;6(22):eaaz0108. doi:10.1126/sciadv.aaz0108.
218. Goolam Mahomed T, Peters RPH, Allam M, Ismail A, Mtshali S, Goolam Mahomed A et al. Lung microbiome of stable and exacerbated COPD patients in Tshwane South Africa. Sci Rep. 2021;11(1):19758. doi:10.1038/s41598-021-99127-w.
219. Schulz-Weidner N, Weigel M, Turujlija F, Komma K, Mengel JP, Schlenz MA et al. Microbiome analysis of carious lesions in pre-school children with early childhood caries and congenital heart disease. Microorganisms. 2021;9(9):1904. doi:10.3390/microorganisms9091904.
220. Rocas IN, Siqueira JF Jr. Culture-independent detection of Eikenella corrodens and Veillonella parvula in primary endodontic infections. J Endod. 2006;32(6):509–12. doi:10.1016/j.joen.2005.07.004.
221. Kato I, Vasquez AA, Moyerbrailean G, Land S, Sun J, Lin HS et al. Oral microbiome and history of smoking and colorectal cancer. J Epidemiol Res. 2016;2(2):92–101. doi:10.5430/jer.v2n2p92.
222. Tishchenko OV, Kryvenko LS, Gargina VV. Influence of smoking heating up tobacco products and e-cigarettes on the microbiota of dental plaque. Pol Merkur Lekarski. 2022;50(295):16–20. PMID:35278292.
223. Benowitz NL, Swan GE, Jacob P 3rd, Lessov-Schlaggar CN, Tyndale RF. CYP2A6 genotype and the metabolism and disposition kinetics of nicotine. Clin Pharmacol Ther. 2006;80(5):457–67. doi:10.1016/j.clpt.2006.08.011.
224. Benowitz NL, Hukkanen J, Jacob P 3rd. Nicotine chemistry metabolism kinetics and biomarkers. Handb Exp Pharmacol. 200;192:29–60. doi:10.1007/978-3-540-69248-5_2.
225. Lerman C, Schnoll RA, Hawk LW Jr, Cinciripini P, George TP, Wileyto EP et al. Use of the nicotine metabolite ratio as a genetically informed biomarker of response to nicotine patch or varenicline for smoking cessation: a randomised double-blind placebo-controlled trial. Lancet Respir Med. 2015;3(2):131–8. doi:10.1016/S2213-2600(14)70294-2.
226. Sosnoff CS, Caron K, Akins JR, Dortch K, Hunter RE, Pine BN et al. Serum concentrations of cotinine and trans-3´-hydroxycotinine in US adults: results from wave 1 (2013–2014) of the

Population Assessment of Tobacco and Health Study. Nicotine Tob Res. 2022;24(5):736–44. doi:10.1093/ntr/ntab240.

227. Yuan JM, Nelson HH, Carmella SG, Wang R, Kuriger-Laber J, Jin A et al. CYP2A6 genetic polymorphisms and biomarkers of tobacco smoke constituents in relation to risk of lung cancer in the Singapore Chinese Health Study. Carcinogenesis. 2017;38(4):411–8. doi:10.1093/carcin/bgx012.
228. Yuan JM, Gao YT, Wang R, Chen M, Carmella SG, Hecht SS. Urinary levels of volatile organic carcinogen and toxicant biomarkers in relation to lung cancer development in smokers. Carcinogenesis. 2012;33:804–9. doi:10.1093/carcin/bgs026.
229. Yuan JM, Gao YT, Murphy SE, Carmella SG, Wang R, Zhong Y et al. Urinary levels of cigarette smoke constituent metabolites are prospectively associated with lung cancer development in smokers. Cancer Res. 2011;71(21):6749–57. doi:10.1158/0008-5472.CAN-11-0209.
230. Yuan JM, Koh WP, Murphy SE, Fan Y, Wang R, Carmella SG et al. Urinary levels of tobacco-specific nitrosamine metabolites in relation to lung cancer development in two prospective cohorts of cigarette smokers. Cancer Res. 2009;69:2990–5. doi:10.1158/0008-5472.CAN-08-4330.
231. Church TR, Anderson KE, Caporaso NE, Geisser MS, Le CT, Zhang Y et al. A prospectively measured serum biomarker for a tobacco-specific carcinogen and lung cancer in smokers. Cancer Epidemiol Biomarkers Prev. 2009;18(1):260–6. doi:10.1158/1055-9965.EPI-08-0718.
232. Verplaetse TL, Peltier MR, Roberts W, Moore KE, Pittman BP, McKee SA. Associations between nicotine metabolite ratio and gender with transitions in cigarette smoking status and e-cigarette use: findings across waves 1 and 2 of the Population Assessment of Tobacco and Health (PATH) study. Nicotine Tob Res. 2020;22(8):1316–21. doi:10.1093/ntr/ntaa022.
233. Znyk M, Jurewicz J, Kaleta D. Exposure to heated tobacco products and adverse health effects, a systematic review. Int J Environ Res Public Health. 2021;18(12):6651. doi:10.3390/ijerph18126651.
234. Murphy SE, Park SS, Thompson EF, Wilkens LR, Patel Y, Stram DO et al. Nicotine N-glucuronidation relative to N-oxidation and C-oxidation and UGT2B10 genotype in five ethnic/racial groups. Carcinogenesis. 2014;35(11):2526–33. doi:10.1093/carcin/bgu191.
235. Carmella SG, Ming X, Olvera N, Brookmeyer C, Yoder A, Hecht SS. High throughput liquid and gas chromatography–tandem mass spectrometry assays for tobacco-specific nitrosamine and polycyclic aromatic hydrocarbon metabolites associated with lung cancer in smokers. Chem Res Toxicol. 2013;26(8):1209–17. doi:10.1021/tx400121n.
236. Chen M, Carmella SG, Sipe C, Jensen J, Luo X, Le CT et al. Longitudinal stability in cigarette smokers of urinary biomarkers of exposure to the toxicants acrylonitrile and acrolein. PLoS One. 2019;14(1):e0210104. doi.:10.1371/journal.pone.0210104.
237. Yan W, Byrd GD, Ogden MW. Quantitation of isoprostane isomers in human urine from smokers and nonsmokers by LC-MS/MS. J Lipid Res. 2007;48(7):1607–17. doi:10.1194/jlr.M700097-JLR200.
238. Chen HJ, Lin WP. Quantitative analysis of multiple exocyclic DNA adducts in human salivary DNA by stable isotope dilution nanoflow liquid chromatography–nanospray ionization tandem mass spectrometry. Anal Chem. 2011;83(22):8543–51. doi:10.1021/ac201874d.
239. Hartmann-Boyce J, Butler AR, Theodoulou A, Onakpoya IJ, Hajek P, Bullen C et al. Biomarkers of potential harm in people switching from smoking tobacco to exclusive e-cigarette use dual use or abstinence: secondary analysis of Cochrane systematic review of trials of e-cigarettes for smoking cessation. Addiction. 2023;118(3):539–45. doi:10.1111/add.16063.
240. De Jesus VR, Bhandari D, Zhang L, Reese C, Capella K, Tevis D et al. Urinary biomarkers of exposure to volatile organic compounds from the Population Assessment of Tobacco and Health Study wave 1 (2013–2014). Int J Environ Res Public Health. 2020;17(15):5408. doi:10.3390/ijerph17155408.

241. Majeed B, Linder D, Eissenberg T, Tarasenko Y, Smith D, Ashley D. Cluster analysis of urinary tobacco biomarkers among US adults: Population Assessment of Tobacco and Health (PATH) biomarker study (2013–2014). Prev Med. 2020;140:106218. doi:10.1016/j.ypmed.2020.106218.
242. Goniewicz ML, Smith DM, Edwards KC, Blount BC, Caldwell KL, Feng J et al. Comparison of nicotine and toxicant exposure in users of electronic cigarettes and combustible cigarettes. JAMA Netw Open. 2018;1(8):e185937. doi:10.1001/jamanetworkopen.2018.5937
243. Wei B, Goniewicz ML, O'Connor RJ, Travers MJ, Hyland AJ. Urinary metabolite levels of flame retardants in electronic cigarette users: a study using the data from NHANES 2013–2014. Int J Environ Res Public Health. 2018;15(2):201. doi:10.3390/ijerph15020201.
244. Tehrani MW, Newmeyer MN, Rule AM, Prasse C. Characterizing the chemical landscape in commercial e-cigarette liquids and aerosols by liquid chromatography–high-resolution mass spectrometry. Chem Res Toxicol. 2021;34(10):2216–26. doi:10.1021/acs.chemrestox.1c00253.
245. Smith DM, Christensen C, van Bemmel D, Borek N, Ambrose B, Erives G et al. Exposure to nicotine and toxicants among dual users of tobacco cigarettes and e-cigarettes: Population Assessment of Tobacco and Health (PATH) study 2013–2014. Nicotine Tob Res. 2021;23(5):790–7. doi:10.1093/ntr/ntaa252.
246. Luo X, Carmella SG, Chen M, Jensen JA, Wilkens LR, Le Marchand L et al. Urinary cyanoethyl mercapturic acid a biomarker of the smoke toxicant acrylonitrile clearly distinguishes smokers from nonsmokers. Nicotine Tob Res. 2020;22(10):1744–7. doi:10.1093/ntr/ntaa080.
247. Schick SF, Blount BC, Jacob PR, Saliba NA, Bernert JT, El Hellani A et al. Biomarkers of exposure to new and emerging tobacco delivery products. Am J Physiol Lung Cell Mol Physiol. 2017;313(3):L425–52. doi: 10.1152/ajplung.00343.2016.
248. Pisinger C, Rasmussen SKB. The health effects of real-world dual use of electronic and conventional cigarettes versus the health effects of exclusive smoking of conventional cigarettes: a systematic review. Int J Environ Res Public Health. 2022;19(20):13687. doi:10.3390/ijerph192013687.
249. Rudasingwa G, Kim Y, Lee C, Lee J, Kim S, Kim S. Comparison of nicotine dependence and biomarker levels among traditional cigarette, heat-not-burn cigarette, and liquid e-cigarette users: results from the Think Study. Int J Environ Res Public Health. 2021;18(9):4777. doi:10.3390/ijerph18094777.

6. Internet, influencer and social media marketing of tobacco and non-therapeutic nicotine products and associated regulatory considerations[3]

Becky Freeman, Prevention Research Collaboration, School of Public Health, Faculty of Medicine and Health, University of Sydney, Sydney, Australia

Pamela Ling, MD, Department of Medicine and Center for Tobacco Control Research and Education, University of California, San Francisco (CA), USA

Stella Aguinaga Bialous, DrPH, School of Nursing and Center for Tobacco Control Research and Education, University of California, San Francisco (CA), USA

Contents

Abstract

Bans on tobacco advertising, promotion and sponsorship (TAPS) are a cornerstone of comprehensive tobacco control laws. Global progress in implementing TAPS bans has been facilitated by adoption of the WHO Framework Convention on Tobacco Control (WHO FCTC). Enforcement of bans on TAPS is, however, over-reliant on self-regulation by producers of entertainment and digital content and online platforms. TAPS laws must maintain pace with the changing media landscape, which includes monitoring and reporting TAPS that cross international borders, primarily through online digital media platforms. TAPS laws must also keep pace with rapid changes in newer non-therapeutic nicotine devices, as well

3 This paper draws on and includes sections from recently published works by the authors *(1,2)*. The paper also draws on reports and findings from the WHO Expert Advisory Committee on Cross-border Tobacco Advertising and Promotion *(3)*, chaired by Becky Freeman, and the report and findings of the WHO FCTC Article 13 Working Group on Tobacco Advertising, Promotion and Sponsorship *(4)*.

as new tobacco products. These include electronic nicotine delivery systems (ENDS), electronic non-nicotine delivery systems (ENNDS), personal vaporizers, heated tobacco products, nicotine salt, other nicotine products resembling nicotine replacement therapy, and various vitamin and cannabis products with the same delivery devices or marketing channels as tobacco products. Many of these products are not regulated, as the manufacturers exploit loopholes in the definition of nicotine and/or tobacco products or are in a regulatory grey area where authority is unclear. Policies are required that anticipate changes in tobacco, nicotine and related products and also in marketing and evolving online and digital media.

Keywords: tobacco advertising, social media, online digital media, marketing, regulation

6.1 Background

In most parts of the world, there have long been bans on direct tobacco advertising in traditional mass media – broadcast television, radio and print – and channels such as billboards. Laws on tobacco advertising, promotion and sponsorship (TAPS) must, however, continue to progress to address the seemingly endless ways in which the tobacco industry attempts to promote its products, maintain current customers, lure back those that quit smoking, and entice new users *(1)*. When only some forms of TAPS are regulated, the industry redirects its promotional efforts and budget to promotions that are exempt from regulation *(5)*. A TAPS ban that is heralded as comprehensive and progressive can quickly be outdated if it is not updated to cover innovations in both promotional opportunities and product offerings. The rapid change to a predominantly digital media environment, including the explosive rise and dominance of online social media, has also enabled the tobacco industry to exploit and develop new forms of promotion *(6)*. The combination of significant circumvention of TAPS laws, new forms of media and new products *(2)* is not intractable: although challenging, it is possible to adopt novel policy approaches to further limit TAPS and exposure.

The WHO FCTC definition of tobacco advertising and promotion is "any form of commercial communication, recommendation or action with the aim, effect or likely effect of promoting a tobacco product or tobacco use either directly or indirectly", and tobacco sponsorship is defined as "any form of contribution to any event, activity or individual with the aim, effect or likely effect of promoting a tobacco product or tobacco use either directly or indirectly" *(7)*. Parties are also encouraged to consider including ENDS and ENNDS in any TAPS regulatory approaches. The definitions are intentionally broad to ensure that they encompass the myriad ways in which the tobacco industry promotes its products. Parties to the WHO FCTC are required to ensure that their TAPS laws are comprehensive and, barring any constitutional impediments, ban all forms of TAPS. This paper focuses on how manufacturers of tobacco and other nicotine delivery products that are non-therapeutic are using online digital media sharing platforms, particularly social media, to market their products, although we acknowledge that TAPS extends beyond online environments. Additionally, the eighth report of the WHO study group on tobacco product regulation *(8)* included "Global marketing and promotion of novel and emerging nicotine and tobacco products and their impacts". This paper does not duplicate but complements and builds on that work.

The WHO FCTC guidelines for implementation of Article 13 state that the depiction of tobacco in entertainment media, such as films, online videos and computer games, is a form of TAPS *(9)*. It is the commercial nature of these forms of entertainment media that defines the tobacco depictions they contain as TAPS, regardless of any tobacco industry involvement in the creation or funding of the content. Much of this entertainment media content is accessed through social media and streaming platforms on personal Internet-enabled devices, such as smartphones. This type of content can also be created, uploaded or broadcast in one country and then viewed and shared in another. Cross-border digital media consumption provides more channels through which the tobacco industry can circumvent TAPS bans.

Some forms of online pro-tobacco messaging might be considered "legitimate expression" if there is no associated commercial link to the message. For example, a user who posts an image of themselves using a tobacco product on social media would not be considered to be making a "commercial communication" if they did not receive any financial or other benefit (such as free products) for posting the image. Most content on social media platforms is not commercial in nature, but social media platforms rely on commercial content in order to generate revenue. Commercial, paid content is then placed in social media feeds, often targeting specific types of users according to their demographics and interests. Commercial entities also post so-called "unpaid" content onto their social media platform accounts, which are often referred to as "organic" posts *(10)*. The account owner does not pay for "organic" posts to appear in user social media feeds but, instead, crafts posts that are likely to appeal to users. Such "organic" posts readily meet the definition of a commercial communication, as they are posted as part of strategic marketing plans on behalf of manufacturers. Illustrations of these concepts are provided below.

6.2 Impact of online and social media marketing on tobacco and ENDS use

Both exposure to and interactions with social media tobacco content have a significant impact on the patterns of ENDS and tobacco use by adolescents. Due to the amount of time adolescents spend engaging with online content, social media may be a critical place in which to intervene, possibly with anti-tobacco or tobacco prevention messages *(11)*.

A systematic review and meta-analysis of the association between exposure to tobacco content on social media and lifetime tobacco use, use of tobacco in the past 30 days and susceptibility to use of tobacco by never users found that participants who were exposed to tobacco content on social media had higher odds of reporting lifetime tobacco use than those who were not exposed (odds raio [OR], 2.18; 95% CI, 1.54 ; 3.08; $I^2 = 94\%$), past 30-day tobacco use (OR, 2.19; 95% CI, 1.79 ; 2.67; $I^2 = 84\%$), and susceptibility to use of tobacco by never users (OR, 2.08; 95% CI, 1.65 ; 2.63; $I^2 = 73\%$). Subgroup analyses showed similar associations for tobacco promotion, active engagement, passive engagement, lifetime exposure to tobacco content, exposure to tobacco content on more than two platforms, and exposure of adolescents and young adults to tobacco content *(12)*.

6.2.1 ENDS use by young adults

Tobacco product advertising has long been established as a cause of young people starting to use tobacco products *(13)*. Much of the current advertising of ENDS and other nicotine products is based on approaches and themes similar to those used in the past to promote conventional tobacco products *(14)*.

Exposure of young adults to marketing of both tobacco and ENDS and engagement with pro-tobacco and ENDS information increases their likelihood of using ENDS products *(15)*. The increased likelihood of use remains even after adjustment for baseline e-cigarette use and the feedback loop from e-cigarette use to exposure to information and engagement. In contrast, engagement in anti-tobacco and anti-ENDS information reduced their probability of e-cigarette use. These findings not only stress the importance of regulating promotional and marketing information about nicotine and tobacco products on social media but also suggest that social media could be used as a cost-efficient platform for disseminating anti-tobacco, anti-ENDS campaign messages to prevent young adults from using ENDS products *(15)*.

6.2.2 Illustrative examples of TAPS in online and digital media

Several online libraries of examples of digital media TAPS include both tobacco and nicotine products. The site of Stanford Research into the Impact of Tobacco Advertising has an extensive collection (60 000 examples) of all types of TAPS that cover all forms of tobacco and nicotine products, including not only traditional cigarettes but also ENDS, nicotine pouches, waterpipes, smokeless tobacco and cigars *(16)*.

Direct, paid tobacco advertisements on online media are the easiest form of online TAPS to recognize, monitor and enforce. It may, however, be difficult to distinguish between direct, paid tobacco promotion and content with no commercial connection. For example, an investigation in 2018 found that tobacco companies were engaging popular social media influencers to promote tobacco products through their highly viewed social media profiles *(17)*. The influencers' posts did not disclose that they were advertising tobacco or that they had received tobacco industry incentives to post tobacco depictions and branding on their profile feeds and pages. Influencers are also heavily involved in the promotion of electronic cigarettes and other novel nicotine devices *(18)*.

In addition to TAPS and tobacco depictions in entertainment and online media, tobacco industry corporate communication campaigns are a well-documented source of pro-tobacco messaging. These promotions sidestep TAPS laws and TAPS definitions. Examples include corporate social responsibility messaging *(19)*, industry-funded "Foundation" campaigns *(20)*, industry funding of science and research, political and lobbying activities and promotions, including paid editorials (advertorials) in news media *(21)*, and unpaid posts on social media from both company branded accounts and employees *(22)*. These corporate communications, which often fall under "legitimate expression" exemptions *(7)*, nonetheless have the same media platforms and serve the same purpose as direct advertising. Regulators should consider intervening when companies use the "legitimate expression" exemption inappropriately, such as when their public relations communications that appear in mass media channels (e.g. newspaper advertorials) consist of thinly veiled product promotion.

1. Digital media-sharing platforms

- *Direct product promotion through paid advertisements.* Such direct promotion is often signalled by inclusion of the words "paid sponsorship", "paid partnership" or "#ad".

- *Influencer promotions.* The tobacco industry and those working to further its interests incentivize or sponsor individuals to post content online featuring products or brands. Social media influencers, who have thousands or even millions of followers, are compensated by the brands and are coached by influencer marketing companies about when to post for maximum exposure and how to avoid posting content that looks like a staged advertisement. Strategies may also include organizing parties and contests with brand sponsorships and encouraging participants to post on their own social media accounts. Influencers may be instructed to amplify their promotional social media posts via hashtags, both related and unrelated to the brand but with enormous viewership (e.g. #love, #art, #fashion).

  Example: TakeAPart has conducted in-depth reporting on tobacco influencer marketing, including the images posted and the messages associated with different brands and products *(17)*.

- *Commercial promotions of posts by consumers of their own tobacco use.* Consumers who use tobacco products may share content that depicts tobacco use and may also comment directly on content that advocates tobacco consumption or recommends particular brands or products. Depending on the context, this may constitute legitimate expression, such as unpaid personal communication. Other parties working in the interests of the tobacco industry can then choose to increase the reach of this content by paying digital media communication platforms to broadcast it to other audiences, turning these personal, legitimate expression posts into commercial promotions.

- *Event promotion.* Participants or teams in an event are sponsored by tobacco companies and social media, and audiovisual sharing platforms broadcast the event and/or images from the event. In the case of major sporting events such as motor racing, the reach can be global, as these events are widely broadcast, including in traditional media.

  Example: see reference *22.*

- *Corporate and campaign promotions.* Tobacco companies, or those working to further their interests, promote a corporate or campaign brand rather than a tobacco product brand and operate social media accounts that promote the corporate or campaign brand. Corporate promotion campaigns and actions often portray tobacco companies as innovative performers and socially responsible actors and advance novel tobacco products as "less harmful alternatives" to traditional cigarettes, often despite lack of independent scientific evidence to support such claims.

  Examples includes the Philip Morris "Unsmoke" campaign *(24).*

- Propaganda crusades by Philip Morris International and Altria: "Smoke-free Future" and "Moving Beyond Smoke" campaigns exposing the hypocrisy of the claim: "A tobacco company that actually cares about health" *(25).*

- *Tobacco use depictions embedded in commercial content in which those depictions are not legitimate expression.* While the bulk of the content on social media is not commercial in nature, commercial content draws a large amount of user traffic (for example, music videos, short films, web series) or is linked to a content creator that generates revenue from user traffic and users purchasing the products featured or reviewed. Music videos, for example, are widely viewed and shared,

and popular content on audiovisual sharing sites are also a major global source of exposure to tobacco depictions.
Example: Cranwell et al. *(26)*.

- *Product integration.* Tobacco companies, or those working to further their interests, work with producers, production companies and screenwriters to build storylines involving their products and integrate them seamlessly into their productions.

- *Sponsored news or "infotainment" content.* The tobacco industry, or those working to further its interests, offers facility visits to news or current affairs journalists or editors, pitch story ideas, or sponsor news stories on related or unrelated topics.
Example: Meade *(21)*.

- *Device advertising promotion and sponsorship.* Advertising or promoting a device or devices for consumption of tobacco products may directly or indirectly advertise or promote tobacco products themselves.

2. Tobacco companies and those working to further their interests operate social media accounts and websites with content that is broadcast across borders. These sites are frequently used not only for legitimate expression but also to promote the corporate brands of a company, to promote specific products or disseminate brand messaging under the guise of providing information to consumers, or as an exercise in so-called corporate social responsibility. Social networking sites and corporate websites are used by the tobacco industry to reinvent itself as a modern, socially responsible, sustainable industry and to dissociate itself from the harm caused by its products. Multiple transnational tobacco companies are using paid full page "public relations" announcements to resume brand promotion in prestigious newspapers and magazines that have long banned tobacco advertising from their pages.
Examples: Foundation for a Smoke Free World *(27)* and Freeman et al. *(28)*

3. *Films, television and streaming content* are significant sources of tobacco depictions. Content that is appealing to young people, such as reality television programming, has been found to contain high amounts of tobacco depictions.

Example: Barker et al. *(29)*

The online database of films https://smokefreemedia.ucsf.edu/sfm-media gives the incidence of tobacco use in cinema.

4. *Streaming television programmes.* With viewership of traditional television decreasing and online streaming and paid subscription increasing, streamed content is a growing source of tobacco promotion. Globally, young people (aged 18–34 years) are much more likely to be users of the Internet and smartphones than those aged 35 and older in both high- and lower-income countries. Tobacco depictions in popular streamed content are more prevalent than in traditional broadcast or cable programming. Many countries have long banned tobacco advertising via "mass media", typically defined as broadcast distribution (television and radio). Today, social media is a potent new mass media distribution channel, which is skewed notably to youthful audiences.

   Examples: Reference *30* provides a detailed report and analysis showing that a global streaming giant's programmes depicted more smoking imagery than broadcast shows. See also Barker et al. *(31)*.

5. *Video and computer games.* Both packaged and online video games are popular among young people, and very few controls are in place to protect or prevent users from exposure to tobacco depictions embedded within games or in-game or in-app purchases. Age restrictions may not take tobacco use into account and are easily avoided by younger players.

   Example of games featuring tobacco use are described in reference *32*.

6. *Smartphone applications.* Some smartphone applications, or "apps" as they are popularly known, show images of cigarette brands or images that resemble existing brands. Pro-smoking apps include approaches such as a cartoon game and an opportunity to simulate a high-quality smoking experience, free apps or apps that facilitate the sale of tobacco products, as well as novel and emerging tobacco products, including devices designed for consuming such products.

   Example: BinDhim et al. *(33)*.

6.2.3 Social media platform tobacco advertising policies

Popular social media platforms, including Facebook *(34)*, Instagram *(35)* and Twitter *(36)*, have adopted policies that prohibit paid tobacco advertising. These policies do not, however, apply to political and corporate messaging ads sponsored by the tobacco industry and do not restrict tobacco companies from using hashtags to attract social media post attention *(37)*, nor do they prevent tobacco companies from operating unpaid "organic" accounts on these platforms, which serve as popular conduits for brand advertisements. PMI, for example, operates a Facebook page that has more than one million followers *(38)*. Google also has an advertising policy on dangerous products or services and prohibits tobacco or any products containing tobacco; products that form a component of a tobacco product, as well as products and services that directly facilitate or promote tobacco consumption; and products designed to simulate tobacco smoking *(39)*. Google searches for tobacco retailers, however, provide localized results and direct links to sales outlets.

A study of e-cigarette and e-cigarette use-related posts on TikTok, a social media platform that has a large adolescent user base, in 2020 showed that the majority of posts positively framed e-cigarettes and use of these products *(40)*. In 2022, TikTok updated its community guidelines and claimed that it banned content that offers the purchase, sale, trade or solicitation of drugs or other controlled substances, alcohol or tobacco products (including e-cigarettes), smokeless or conventional tobacco products, synthetic nicotine products, e-cigarettes, and other ENDS. The new policy further specifies that

> content depicting the use of tobacco products by adults, or mentioning controlled substances, is not eligible for recommendation. Please remember that content which suggests, depicts, imitates, or promotes the possession or consumption of alcoholic beverages, tobacco, or drugs by a minor is prohibited. Content that offers instruction targeting minors on how to buy, sell, or trade alcohol, tobacco, or controlled substances is prohibited per our Community Guidelines as well. *(41)*

An evaluation of social media policies related to tobacco product promotion and sales on 11 sites in the USA that are popular with young people was conducted in May 2021 *(42)*. Nine of the 11 sites prohibited "paid advertising" for tobacco products; however, only three of them prohibited "sponsored content" that promotes tobacco. Six platforms restricted content that "sells tobacco products", and three claimed to "prohibit underage access" to content that promotes or sells tobacco products. Although most platform policies prohibited paid tobacco advertising, few addressed less direct strategies, such as sponsored or influencer content, and few had age-gating to prevent access by young people.

There is no evidence that these voluntary policies lead to reduced exposure to TAPS. This rapidly evolving media environment, coupled with lax regulation of social media communication platforms, including the over-reliance on platform self-regulation *(43)*, complicates extension of comprehensive TAPS bans to truly include online media. A mandate that all social media platforms ban all forms of tobacco, e-cigarettes and novel product advertising, both paid and organic, and prohibit the use of influencers, is crucial. Implementation of these policies should specify mechanisms for reporting noncompliance, provide for periodic audits, and require platforms to report on how they are ensuring that the law is being enforced on their sites. Currently, it is largely tobacco control stakeholders that are monitoring the amount and type of TAPS on social media platforms *(44)*. More of that burden should be shifted to the social media companies themselves. Social media companies could largely automate identification of tobacco promotion via sophisticated artificial intelligence systems for overseeing content.

For example, in an analysis of 4526 unique Instagram users who had created 19 951 IQOS-related posts, nearly half of the users (42.1%) were business accounts authorized by Instagram, of which 59.0% belonged to personal goods and general merchandise stores and 18.1% to creators and celebrities. Most active accounts in the network were directly associated with IQOS (e.g. containing "IQOS" in the user name) or related to the tobacco business as self-identified in the account biography description. These results show clearly that current self-regulation by social media platforms is far from enough *(45)*.

6.2.4 Global status of tobacco advertising laws

Article 13 of the WHO FCTC recognizes the crucial role of TAPS bans in effective tobacco control and includes banning cross-border TAPS as part of a comprehensive approach *(9)*. Parties to the WHO FCTC recognize the continuing difficulty of monitoring and enforcing cross-border TAPS bans and are preparing an addendum to the Article 13 guidelines to reflect the dramatic changes in the media landscape since the guidelines were adopted in 2008 *(3)*. Parties have also called for a mechanism for more effective global cooperation in managing cross-border TAPS *(3)*. Countries can more easily ban online TAPS that originate in their own countries, but, without international co-operation, it is more difficult to ban those that originate from another country and then "leak" across digital borders. The European Union, for example, requires all its Member States to ban cross-border tobacco advertising and sponsorship and actively monitors and enforces those provisions *(46)*.

6.3 Discussion

About half (91/180, 50.6%) of countries that report to the WHO FCTC on TAPS regulations stated that their TAPS ban included the domestic Internet *(47)*. The cross-border nature of online TAPS presents an additional challenge to regulators. Its nature is cross-border whenever content created, uploaded or broadcast in one country may be consumed or shared in another, thereby crossing geographical borders. The service providers may also be located in different countries from the country in which the service is provided. Content may also cross "digital" borders, as access is not always effectively limited to one geographical location. Cross-border digital media consumption provides new and emerging channels through which the tobacco industry and those acting to further its interests can circumvent controls on TAPS.

Entertainment media may cross borders through Internet-enabled devices (computers, smartphones, tablets, smart televisions) that:

- facilitate online streaming of films, television series or shows, video games, music videos, sporting, news, music, dance and other entertainment events;
- enable access to electronic versions of international and domestic newspapers and magazines;
- facilitate access to social media posts, including commercial and user-generated content and website pages;
- provide opportunities for engagement between consumers and commercial entities through social media; and
- may contain tobacco depictions or deliver embedded advertising content.

While the tobacco industry may not sponsor depictions of tobacco in films in countries with comprehensive TAPS bans, the policies rarely extend to unsponsored depictions in entertainment media. To escape such limitations, tobacco companies could provide free product samples to the prop masters of production companies in the hope that they will appear in the hands of actors. Since 2012, India has required that films depicting smoking should be accompanied by a 10-s Government-issued anti-smoking advertisement and that a static health warning at the bottom of the screen be visible for the duration of the tobacco depiction *(48)*. Any tobacco product brand names that appear on screen must be blurred out. Other countries, including China and Thailand, regulate the smoking and tobacco content permissible in television and films. Although the association between smoking depictions in films and the increased risk of smoking uptake by young people has been replicated in several studies *(49)*, there has been no research or evaluation on the impact of policy interventions to reduce tobacco depiction *(48)*.

Global media content producers and streaming services such as Disney *(50)* and Netflix *(51)* have made public commitments to reduce the frequency of tobacco depictions in new content, particularly that aimed at younger audiences. The move came only after it was revealed that the number of tobacco depictions on Netflix shows popular with young people had increased over time *(30)*. An online database of smoking depictions in media maintained by the University of California San Francisco (USA) documents the continued promotion of tobacco use in both Disney and Netflix content, among all other major media companies *(52)*.

6.3.1 Cross-border advertising

Countries that are committed to ending the promotion of tobacco products must not only strengthen their domestic TAPS bans but work with and support other nations in reducing cross-border TAPS. This will require more effective global cooperation and a commitment by all countries to update TAPS regulations regularly in response to new media and communications platforms and consumption patterns and also the evolving industry tactics that merge political interference, advertising and product development. In order to meet the highest global standards, improving and updating TAPS laws must be continuous, coupled with leadership for regulatory innovations, such as a complete end to the retail sale, including online, of tobacco products *(53)*.

Legal experts have proposed possible ways in which WHO FCTC Parties could act together to reduce cross-border TAPS *(54)*. They include:

- establishing mechanisms through which Parties can report to other Parties instances of tobacco advertising that originate on the other Parties' territory (either directly through specified contact persons or through a central public health body);
- agreeing to take appropriate action upon receiving reports from other Parties of cross-border tobacco advertising originating in their territory and to inform the reporting Party about the action taken;
- agreeing to provide assistance to other Parties in the investigation, preparation and prosecution of offences or possible offences, such as by facilitating access to relevant evidence and witnesses;
- agreeing to enforce, against individuals or organizations residing or with assets in their territory, judgements under FCTC-implementing laws made in the territory of another Party or to give reasons for refusing to do so;
- establishing mechanisms through which Parties report to one another on their experiences in respect of cross-border advertising;

- establish mechanisms through which Parties can discuss the effectiveness of cooperative measures adopted to meet jurisdictional and enforcement challenges and enter into new arrangements, as required; and
- establish mechanisms for sharing experiences and expertise in respect of relevant developments in technology.

6.3.2 Regulation of online marketing of other harmful products

Other commercial determinants of health, such as alcohol, food and non-alcoholic beverages, and gambling face challenges similar to those posed by tobacco in effective regulation of online marketing. As for tobacco, social media platforms have their own ill-defined, poorly enforced polices on the promotion of gambling and alcohol *(55,56)*. Food and non-alcoholic beverages are subject to even weaker restrictions, and social media sites are flooded with promotions of unhealthy processed food and energy and soft drinks *(57)*. Government action on the many forms of unhealthy advertising varies widely.

6.3.3 Challenges to regulation of tobacco advertising

One limitation in assessing the global state of TAPS bans is the limited body of work on implementation and enforcement of TAPS laws and regulations *(58)*. While countries that are Parties to the WHO FCTC Convention report on the scope of their TAPS laws, the reports do not include details of enforcement activities, and exemptions and the limits of policy reach are not well described.

Other challenges to effective regulation of TAPS in entertainment and online media are the following *(4)*.

- The popularity of content-sharing platforms, including social media, allows users to create and share content. People can view and share digital media freely, easily and quickly. This situation has blurred the lines between consumer and brand owner and poses a challenge to controlling cross-border TAPS.
- The changed media landscape and types of TAPS mean that regulations might have to be updated and made "future-proof" against emerging TAPS.
- Countries ban only cross-border TAPS that originate in their own countries and not those that are broadcast into the country from outside.
- It is difficult to distinguish between paid and unpaid depictions of tobacco use and brands.
- It is difficult to identify the origin, both country and creator or owner, of TAPS content, particularly online.

- Systematic documentation and capture of both tobacco industry promotional activities and tobacco depictions in entertainment media is difficult.
- Countries that have not ratified the WHO FCTC might be sources of cross-border TAPS.
- Young adults and adolescents are a highly desirable target population for this type of TAPS; however, only limited research, resources and policy action have been directed to protect this age group from exposure.

6.3.4 New products and associated challenges to online marketing

Enforcement of regulations and advertising control policies is a global challenge. For example, although the sale and import of all ENDS products are banned in Thailand *(59)*, they are sold illegally online *(60)*. The same is true in Brazil, where the marketing, advertising and importation of ENDS are not allowed, but they are sold illegally at e-cigarette shops, tobacco shops, on the Internet and through delivery apps *(61)*. In South Africa, ENDS are licensed to be sold only by prescription, but they are widely advertised as smoking cessation products and sold without prescription *(62)*.

In the Republic of Korea, heated tobacco product devices were considered to be electronics rather than tobacco products *(63)*; therefore, they have been advertised with lifestyle appeal, including a social media campaign by British American Tobacco in 2019 featuring hip-hop musicians popular among young people *(64,65)*. The music video was not subject to age restrictions because it contained images only of the heating devices and not the tobacco pods; it accrued more than a million views *(64)*.

New tobacco, nicotine and other aerosolized products are continuing to enter markets, with aggressive promotion both in high-income countries, where the prevalence of cigarette smoking is decreasing and where consumers can afford expensive new products, and in lower- and middle-income countries, circumventing bans on tobacco advertising *(2)*.

6.4 Conclusions

A wide range of media outlets, including social media, expose users to TAPS. The global media landscape has changed substantially since adoption of the "Guidelines for implementation of Article 13" in 2008 and continues to change and develop. Entertainment media content is increasingly available at regional and global level, including through the Internet, which can result in cross-border exposure to TAPS. The consequence of this shift in technology is that current approaches to controlling TAPS are insufficient.

The new, rapidly evolving media environment coupled with lax regulation of social media communication platforms, including over-reliance on self-regulation, means that extending comprehensive tobacco bans to effectively include cross-border TAPS, while challenging, is a high priority. In addition to strong domestic regulatory action, international action, both regionally and globally, including through cooperation between countries, will be required to reduce TAPS *(58)*. As media platforms are rapidly evolving in ways that may create regulatory loopholes that allow resumption of tobacco brand promotion, regulatory limits must be forward-looking to anticipate likely technological evolution, and regulators should be empowered to act swiftly in response to changing circumstances.

The current self-regulation led by social media platforms is not sufficient. More refined, well-enforced regulatory action, especially to limit marketing by official accounts, online retailers and celebrities, is necessary to restrict the proliferation of promotional content for tobacco products.

6.5 Research gaps and priorities

Published research tends to focus on why laws on TAPS are necessary *(66)*; the impact of exposure to TAPS *(67)*; how TAPS laws affect exposure to TAPS *(68,69)* and the potential impact on smoking attitudes, beliefs and behaviour *(70)*; and how the tobacco industry acts in the face of newly adopted TAPS laws *(71)* and subverts existing TAPS laws *(72)*. There are few data on how TAPS bans are implemented and then monitored and enforced after implementation. Evaluation and assessment of how countries can most effectively collaborate to control cross-border TAPS are also necessary. While the types and modes of online TAPS are increasingly being monitored systematically, data from monitoring the tobacco industry should be collected continuously, especially in light of new tobacco and nicotine product development and the continuously evolving media landscape. Assessment of the comprehensiveness of TAPS bans, particularly in terms of capturing online forms of TAPS and monitoring how loopholes in regulations are exploited are essential.

6.6 Policy recommendations

Regulators should develop a comprehensive strategy to reduce the amount of advertising, promotion and sponsorship of tobacco and nicotine products on social media platforms and online digital entertainment media. Such a strategy could reduce the exposure of adolescents and young adults to tobacco content and, ultimately, tobacco use.

- Ensure that TAPS laws are comprehensive, cover online digital media platforms, including social media, and are sufficiently flexible to encompass new media and platforms.

- The cross-border nature of online digital TAPS requires international cooperation for effective monitoring and enforcement.
- Require the tobacco industry to disclose all TAPS activities, including any activities on online digital media platforms, to government authorities in order to strengthen monitoring and enforcement.
- Include novel and emerging nicotine and tobacco products in comprehensive laws to ban tobacco and non-therapeutic nicotine products advertising promotion and sponsorship.
- Conduct ongoing surveillance of the evolution of both online digital media platforms and novel and emerging nicotine and tobacco products to ensure that TAPS laws remain comprehensive, including prohibition of advertising themes such as lifestyle, fashion, creativity, identity, pleasure and socializing.

References

1. Freeman B, Watts C, Astuti PAS. Global tobacco advertising, promotion and sponsorship regulation: What's old, what's new and where to next? Tob Control. 2022;31:216–21. doi:10.1136/tobaccocontrol-2021-056551.
2. Ling PM, Kim M, Egbe CO, Patanavanich R, Pinho M, Hendlin Y. Moving targets: how the rapidly changing tobacco and nicotine landscape creates advertising and promotion policy challenges. Tob Control. 2022;31(2):222–8. doi:10.1136/tobaccocontrol-2021-056552.
3. Tobacco advertising, promotion and sponsorship: depiction of tobacco in entertainment media. Report by the Convention Secretariat (FCTC/COP/7/38). Conference of the Parties to the WHO Framework Convention on Tobacco Control, seventh session. Geneva: World Health Organization; 2016 (FCTC_COP_7_38_EN.pdf).
4. Conference of the Parties to the WHO Framework Convention on Tobacco Control. Decision FCTC/COP7(5). Tobacco advertising, promotion and sponsorship: depiction of tobacco in entertainment media. Geneva: World Health Organization; 2016 (https://fctc.who.int/docs/librariesprovider12/meeting-reports/fctc_cop7_5_en.pdf?sfvrsn=f2653e3c_16&download=true).
5. Blecher E. The impact of tobacco advertising bans on consumption in developing countries. J Health Econ. 2008;27(4):930–42. doi:10.1016/j.jhealeco.2008.02.010.
6. Freeman B. New media and tobacco control. Tob Control. 2012;21(2):139–44. doi:10.1136/tobaccocontrol-2011-050193.
7. WHO Framework Convention on Tobacco Control. Introduction. Article 1. Use of terms. Geneva: World Health Organization; 2003 (https://apps.who.int/iris/rest/bitstreams/50793/retrieve).
8. WHO Study Group on Tobacco Product Regulation. Report on the scientific basis of tobacco product regulation: eigth report of a WHO study group. Geneva: World Health Organization; 2021 (https://www.who.int/publications/i/item/9789240022720).
9. WHO Framework Convention on Tobacco Control Secretariat. Guidelines for implementation of Article 13 of the WHO Framework Convention on Tobacco Control (Tobacco advertising, promotion and sponsorship) 2008 (https://www.who.int/fctc/guidelines/article_13.pdf?ua=1).
10. Zelefsky V. The differences between paid and organic content on social media. Jersey City (NJ): Forbes; 2022 (https://www.forbes.com/sites/forbescommunicationscouncil/2022/05/06/

the-differences-between-paid-and-organic-content-on-social-media/?sh=3ddfe0661526).
11. Cavazos-Rehg P, Li X, Kasson E, Kaiser N, Borodovsky JT, Grucza R et al. Exploring how social media exposure and interactions are associated with ENDS and tobacco use in adolescents from the PATH study. Nicotine Tob Res. 2020;23(3):487–94. doi:10.1093/ntr/ntaa113.
12. Donaldson SI, Dormanesh A, Perez C, Majmundar A, Allem JP. Association between exposure to tobacco content on social media and tobacco use: a systematic review and meta-analysis. JAMA Pediatr. 2022;176(9):878–85. doi:10.1001/jamapediatrics.2022.2223.
13. The role of the media in promoting and reducing tobacco use (Tobacco Control Monograph No. 19. NIH Pub. No. 07-6242). Bethesda (MD): National Cancer Institute; 2008 (https://cancer-control.cancer.gov/brp/tcrb/monographs/monograph-19).
14. E-cigarette use among youth and young adults. A report of the Surgeon General. Atlanta (GA): Centers for Disease Control and Prevention, National Center for Chronic Disease Prevention and Health Promotion, Office on Smoking and Health; 2016 (https://www.cdc.gov/tobacco/sgr/e-cigarettes/index.htm).
15. Yang Q, Clendennen SL, Loukas A. How does social media exposure and engagement influence college students' use of ENDS Pproducts? A cross-lagged longitudinal study. Health Commun. 2021:1–10. doi:10.1080/10410236.2021.1930671.
16. Ad collections. Palo Alto (CA): Stanford Research into the Impact of Tobacco Advertising; 2023 (https://tobacco.stanford.edu/).
17. Where there's smoke. Washington DC: Campaign for Tobacco-Free Kids; undated (https://www.takeapart.org/wheretheressmoke/).
18. Klein EG, Czaplicki L, Berman M, Emery S, Schillo B. Visual attention to the use of #ad versus #sponsored on e-cigarette influencer posts on social media: a randomized experiment. J Health Commun. 2020;25(12):925–30. doi:10.1080/10810730.2020.1849464.
19. Greenland S, Lužar K, Low D. Tobacco CSR, sustainability reporting, and the marketing paradox. In: Crowther D, Seifi S, editors. The Palgrave Handbook of Corporate Social Responsibility. London: Palgrave Macmillan; 2020:1–27 (https://link.springer.com/referenceworkentry/10.1007/978-3-030-22438-7_67-1).
20. Legg T, Peeters S, Chamberlain P, Gilmore AB. The Philip Morris-funded Foundation for a Smoke-Free World: tax return sheds light on funding activities. Lancet. 2019;393(10190):2487–8. doi:10.1016/S0140-6736(19)31347-9.
21. Meade A. Philip Morris-sponsored articles in the Australian could breach tobacco advertising laws. The Guardian, 18 November 2020 (https://www.theguardian.com/media/2020/nov/19/philip-morris-sponsored-articles-in-the-australian-could-breach-tobacco-advertising-laws).
22. Watts C, Hefler M, Freeman B. "We have a rich heritage and, we believe, a bright future": how transnational tobacco companies are using Twitter to oppose policy and shape their public identity. Tob Control. 2019;28(2):227–32. doi:10.1136/tobaccocontrol-2017-054188.
23. Driving addiction. Tobacco sponsorship in Formula One, 2021. New York (NY): STOP, Bloomberg Philanthropies; 2021 (https://exposetobacco.org/wp-content/uploads/TobaccoSponsorshipFormula-One-2021.pdf).
24. Davies M, Stockton B, Chapman M, Cave T. The "unsmoke" screen: the truth behind PMI's cigarette-free future. London: Bureau of Investigative Journalism; 2020 (https://www.thebureauinvestigates.com/stories/2020-02-24/the-unsmoke-screen-the-truth-behind-pmis-cigarette-free-future).
25. Jackler RK. Propagands crusades by Philip Morris International & Altria: "Smoke-free future" and "Moving beyond smoke" campaigns. Exposing the hypocrisy of the claim: "A tobacco company that actually cares about health". Palo Alto (CA): Stanford University School of Medicine; 2022 (https://tobacco-img.stanford.edu/wp-content/uploads/2022/03/02103210/PMI-SFF-White-Paper-3-2-2022F-.pdf).

26. Cranwell J, Opazo-Breton M, Britton J. Adult and adolescent exposure to tobacco and alcohol content in contemporary YouTube music videos in Great Britain: a population estimate. J Epidemiol Community Health. 2016;70:488–92. doi:10.1136/jech-2015-206402.
27. Foundation for a Smoke-free World. Bath: University of Bath, Tobacco Tactics; 2022 (https://tobaccotactics.org/wiki/foundation-for-a-smoke-free-world).
28. Freeman B, Hefler M, Hunt D. Philip Morris International's use of Facebook to undermine Australian tobacco control laws. Public Health Res Pract. 2019;29(3):e2931924. doi:10.17061/phrp2931924.
29. Barker AB, Opazo Breton M, Cranwell J, Britton J, Murray RL. Population exposure to smoking and tobacco branding in the UK reality show 'Love Island'. Tob Control. 2018;27:709–11. doi:10.1136/tobaccocontrol-2017-054125.
30. While you were streaming: smoking on demand. A surge in tobacco imagery is putting youth at risk. Washington DC: Truth Initiative; 2019 (https://truthinitiative.org/research-resources/tobacco-pop-culture/while-you-were-streaming-smoking-demand).
31. Barker AB, Smith J, Hunter A, Britton J, Murray RL. Quantifying tobacco and alcohol imagery in Netflix and Amazon Prime instant video original programming accessed from the United Kingdom of Great Britain and Northern Ireland: a content analysis. BMJ Open. 2019;9:e025807. doi:10.1136/bmjopen-2018-025807.
32. Some video games glamorize smoking so much that cigarettes can help players win. Washington DC: Truth Initiative; 2018 (https://truthinitiative.org/research-resources/tobacco-pop-culture/some-video-games-glamorize-smoking-so-much-cigarettes-can).
33. BinDhim NF, Freeman B, Trevena L. Pro-smoking apps for smartphones: the latest vehicle for the tobacco industry? Tob Control. 2014;23:e4. doi:10.1136/tobaccocontrol-2012-050598.
34. Advertising policies. Prohibited content. Menlo Park (CA): Facebook; 2021 (https://www.facebook.com/policies/ads/#).
35. Ads on Instagram. Menlo Park (CA): Instagram; 2021 (https://help.instagram.com/1415228085373580).
36. Twitter ads policies. San Francisco (CA): Twitter; 2021 (https://business.twitter.com/en/help/ads-policies.html).
37. O'Brien EK, Hoffman L, Navarro MA, Ganz O. Social media use by leading US e-cigarette, cigarette, smokeless tobacco, cigar and hookah brands. Tob Control. 2020;29(e1):e87–97. doi:10.1136/tobaccocontrol-2019-055406.
38. Freeman B, Hefler M, Hunt D. Philip Morris International's use of Facebook to undermine Australian tobacco control laws. Public Health Res Pract. 2019;29:e2931924 (https://www.phrp.com.au/issues/september-2019-volume-29-issue-3/philip-morris-internationals-use-facebook-undermine-australian-tobacco-control-laws/).
39. Advertising policies help. List of ad policies: dangerous products or services. Mountain View (CA): Google; 2021(https://support.google.com/adwordspolicy/answer/6014299?hl=en).
40. Sun T, Lim CCW, Chung J, Cheng B, Davidson L, Tisdale C et al. Vaping on TikTok: a systematic thematic analysis. Tob Control. 2023;32(2):2514. doi:10.1136/tobaccocontrol-2021-056619.
41. TikTok. Community guidelines. Illegal activities and regulated goods. Beijing: ByteDance; 2023 (https://www.tiktok.com/community-guidelines?lang=en#32).
42. Kong G, Laestadius L, Vassey J, Majmundar A, Stroup AM, Meissner HI et al. Tobacco promotion restriction policies on social media. Tob Control. 2022 Nov 3;tobaccocontrol-2022-057348. doi:10.1136/tc-2022-057348.
43. Gosh D. Are we entering a new era of social media regulation? Harvard Business Review, 14 January 2021 (https://hbr.org/2021/01/are-we-entering-a-new-era-of-social-media-regulation).
44. About STOP. It's time to shine the light on the tobacco industry. New York (NY): Bloomberg Philanthropies; 2021 (https://exposetobacco.org/about/).

45. Gu J, Abroms LC, Broniatowski DA, Evans WD. An investigation of influential users in the promotion and marketing of heated tobacco products on Instagram: a social network analysis. Int J Environ Res Public Health. 2022;19(3):1686. doi:10.3390/ijerph19031686.
46. Ban on cross-border tobacco advertising and sponsorship. Brussels: European Commission; 2021 (https://ec.europa.eu/health/tobacco/advertising_en).
47. WHO Framework Convention on Tobacco Control Secretariat. Article 13. Demand reduction measures. C2722 – Ban covering the domestic internet. Report charts. Geneva: World Health Organization; 2020 (https://untobaccocontrol.org/impldb/indicator-report/?wpdtvar=3.2.7.2.b).
48. Smoke-free movies: from evidence to action, third edition. Geneva: World Health Organization; 2015 (https://apps.who.int/iris/rest/bitstreams/850394/retrieve).
49. Leonardi-Bee J, Nderi M, Britton J. Smoking in movies and smoking initiation in adolescents: systematic review and meta-analysis. Addiction. 2016;111(10):1750–63. doi:10.1111/add.13418.
50. Barnes B. There's no smoking in Disney films. What about when it owns Fox? New York Times, 25 April 2018 (https://www.nytimes.com/2018/04/25/business/media/smoking-movies-disney-fox.html).
51. Romo V. Netflix promises to quit smoking on (most) original programming. National Public Radio, 4 July 2019 (https://www.npr.org/2019/07/04/738719658/netflix-promises-to-quit-smoking-on-most-original-programming).
52. Onscreen tobacco database. San Francisco (CA): Smokefree Media; 2021 (https://smokefreemedia.ucsf.edu/sfm-media).
53. Smith EA, Malone RE. An argument for phasing out sales of cigarettes. Tob Control. 2020;29(6):703–8. doi: doi:10.1136/tobaccocontrol-2019-055079.
54. Kenyon AT, Liberman J. Controlling cross-border tobacco advertising, promotion and sponsorship – implementing the FCTC. Melbourne: Centre for Media and Communications Law; 2006 (https://www.researchgate.net/publication/228192102_Controlling_Cross-Border_Tobacco_Advertising_Promotion_and_Sponsorship_-_Implementing_the_FCTC).
55. Carah N, Brodmerkel S. Alcohol marketing in the era of digital media platforms. J Stud Alcohol Drugs. 2021;82(1):18–27. PMID:33573719.
56. Torrance J, John B, Greville J, O'Hanrahan M, Davies N, Roderique-Davies G. Emergent gambling advertising; a rapid review of marketing content, delivery and structural features. BMC Public Health. 2021;21(1):718. doi:10.1186/s12889-021-10805-w.
57. McCarthy CM, de Vries R, Mackenbach JD. The influence of unhealthy food and beverage marketing through social media and advergaming on diet-related outcomes in children – A systematic review. Obesity Rev. 2022;23(6):e13441. doi:10.1111/obr.13441.
58. Kennedy RD, Grant A, Spires M, Cohen JE. Point-of-sale tobacco advertising and display bans: policy evaluation study in five Russian cities. JMIR Public Health Surveillance. 2017;3(3):e52. doi:10.2196/publichealth.6069.
59 Patanavanich R, Glantz S. Successful countering of tobacco industry efforts to overturn Thailand's ends ban. Tob Control. 2021;30(e1):e10–9. doi:10.1136/tobaccocontrol-2020-056058.
60. Phetphum C, Prajongjeep A, Thawatchaijareonying K, Wongwuttiyan T, Wongjamnong M, Yossuwan S et al. Personal and perceptual factors associated with the use of electronic cigarettes among university students in northern Thailand. Tob Induc Dis. 2021;19:31. doi:10.18332/tid/133640.
61. de Pinho MCM, Riva MPR, de Souza Cury L, Andreis M. A promoção de novos produtos de tabaco nas redes sociais à luz da pandemia [Promotion of new tobacco products on social media in the pandemic]. Rev Bras Cancerol. 2020;66. doi:10.32635/2176-9745.RBC.2020v66nTemaAtual.1108.
62. Agaku I, Egbe CO, Ayo-Yusuf O. Associations between electronic cigarette use and quitting

behaviours among South African adult smokers. Tob Control. 2022;31(3):464–72. doi:10.1136/tobaccocontrol-2020-056102.

63. Kong J, Chu S, Park K, Lee S. The tobacco industry and electronic cigarette manufacturers enjoy a loophole in the legal definition of tobacco in South Korean law. Toba Control. 2021;30(4):471–2.
64. Yi J, Kim J, Lee S. British American Tobacco's 'Glo Sens' promotion with K-pop. Tobacco Control. 2021;30(5):594-6. doi:10.1136/tobaccocontrol-2020-055662.
65. Comprehensive ban on cross-border tobacco advertising, promotions and sponsorship in ASEAN region. Bangkok: Southeast Asia Tobacco Control Alliance; undated (https://seatca.org/dmdocuments/Cross-border_final.pdf).
66. Chido-Amajuoyi OG, Mantey DS, Clendennen SL, Pérez A. Association of tobacco advertising, promotion and sponsorship (TAPS) exposure and cigarette use among Nigerian adolescents: implications for current practices, products and policies. BMJ Glob Health. 2017;2(3). doi:10.1136/bmjgh-2017-000357.
67. Tan AS, Hanby EP, Sanders-Jackson A, Lee S, Viswanath K, Potter J. Inequities in tobacco advertising exposure among young adult sexual, racial and ethnic minorities: examining intersectionality of sexual orientation with race and ethnicity. Tob Control. 2021;30(1):84–93. doi:10.1136/tobaccocontrol-2019-055313.
68. Kahnert S, Demjén T, Tountas Y, Trofor AC, Przewoźniak K, Zatoński WA et al. Extent and correlates of self-reported exposure to tobacco advertising, promotion and sponsorship in smokers: Findings from the EUREST-PLUS ITC Europe surveys. Tob Induc Dis. 2018;16. doi:10.18332/tid/94828.
69. Li L, Borland R, Yong HH, Sirirassamee B, Hamann S, Omar M et al. Impact of point-of-sale tobacco display bans in Thailand: findings from the international tobacco control (ITC) Southeast Asia survey. Int J Environ Res Public Health. 2015;12(8):9508–22. doi:10.3390/ijerph120809508.
70. Nicksic NE, Bono RS, Rudy AK, Cobb CO, Barnes AJ. Smoking status and racial/ethnic disparities in youth exposure to tobacco advertising. J Ethn Subst Abuse. 2020;21(3):959–74. doi:10.1080/15332640.2020.1815113.
71. da Silva ALO, Grilo G, Branco PAC, Fernandes AMMS, Albertassi PGD, Moreira JC. Tobacco industry strategies to prevent a ban on the display of tobacco products and changes to health warning labels on the packaging in Brazil. Tob Prev Cessat. 2020;6:66. doi:10.18332/tpc/128321.
72. Astuti PAS, Assunta M, Freeman B. Raising generation 'A': a case study of millennial tobacco company marketing in Indonesia. Tob Control. 2018;27(e1):e41–9. doi:10.1136/tobaccocontrol-2017-054131.

7. Overall recommendations

The WHO Study Group on Tobacco Product Regulation publishes a series of reports to provide a scientific basis for tobacco product regulation. They are a WHO technical product (formerly known as a global public health good); these, as noted above, are goods (or resources) developed by WHO that are of benefit globally or to many countries in many regions *(1)*. The TobReg reports, in line with Articles 9 and 10 of the WHO Framework Convention on Tobacco Control (FCTC) *(2)*, provide evidence-based approaches to regulation of the contents, emissions and design features of nicotine and tobacco products. The previous report *(3)*, on the deliberations of the tenth meeting of the Study Group, provides recommendations substantiated by sound science on novel and emerging nicotine and tobacco products. The recommendations of the Study Group are relevant in various contexts according to the national regulatory environment, the prevalence of use of tobacco and non-therapeutic nicotine products and other relevant factors that have implications for regulation of these products, such as policy goals and capacity for regulation, including bans.

The deliberations, outcomes and recommendations of the ninth meeting of TobReg reported here addressed new ways in which non-therapeutic nicotine, particularly in nicotine products, is promoted and delivered to people of different ages, including children and adolescents. The WHO FCTC *(2)*, which is the first international public health treaty, negotiated under the auspices of WHO to combat the tobacco epidemic, has saved many lives over the past 20 years. Countries are increasingly introducing strong tobacco control policy measures, including supply-and-demand measures, in line with the FCTC, to protect their citizens and reduce the prevalence of tobacco product use. This has led to decreasing sales of cigarettes and other tobacco products globally. Therefore, the tobacco industry is leveraging technology and using innovative means to boost its profits by introducing new ways of marketing and promoting nicotine and tobacco products *(3,4)*, including strategies and tactics to make tobacco and non-therapeutic nicotine products attractive, especially to children and adolescents *(5,6)* to sustain use of the products. Given its mandate, TobReg, assisted by subject matter experts, synthesizes comprehensive evidence from the published literature to develop evidence-based recommendations on product regulation. These are made available to countries, through the WHO Director-General, to assist countries in addressing regulatory challenges in tobacco control, which remains a global priority.

Regulators are reminded that tobacco kills more than 8 million people a year *(7,8)*. More than 7 million of those deaths are attributed directly to tobacco use and about 1.3 million to exposure of non-smokers to second-hand smoke *(9,10)*. Tobacco also eventually kills up to half of its users and remains a global health emergency *(7)*. A further complication for tobacco control, posing challenges to regulators, is use of synthetic nicotine, which is not necessarily be covered under some tobacco control laws. Online marketing of nicotine and tobacco products and the introduction and promotion, including to children and adolescents, of nicotine pouches adds further complications. Thus, the recommendations of this report, should not be considered in isolation but seen in the context of wider tobacco control to complement the recommendations of the Study Group in other reports on tobacco product regulation *(11–18)*, which address cigarettes, smokeless tobacco, waterpipe tobacco, design features, flavours, as well as novel and emerging products.

The tobacco control community is well aware of the deliberate efforts by the tobacco industry to undermine tobacco control and slow implementation of the WHO FCTC. The industry aggressively markets and promotes novel nicotine and tobacco products, which pose a serious threat to tobacco control, and uses covert strategies to advertise and promote its products online. The Study Group, having examined the requests by Member States for technical assistance on topics of interest to regulators, as considered in this report, made several recommendations. TobReg nevertheless reiterates its conclusion in its previous report *(18)* that regulators should maintain a focus on wider tobacco control and should not allow themselves to be distracted by the tactics of tobacco and related industries, nor the aggressive promotional strategies used by those industries to sustain the use of tobacco and non-therapeutic nicotine products.

This report highlights the importance of:

- continued focus on tobacco control to decrease the prevalence of tobacco use;
- comprehensive tobacco control laws that apply to all tobacco products and all forms of tobacco and non-therapeutic nicotine products, without exception;
- international cooperation to address cross-border marketing of nicotine and tobacco products;
- comprehensive laws to regulate tobacco advertising, promotion and sponsorship laws, including new ways of promoting tobacco and non-therapeutic nicotine products;
- strengthen monitoring and enforcement of regulations on nicotine and tobacco products, including the activities of the tobacco and related industries;
- close regulatory gaps that could be exploited by the tobacco and related industries; and
- implement the recommendations of the Study Group.

Sections 2–6 of the report provide scientific information, evidence on online marketing and policy recommendations and guidance to bridge regulatory gaps in tobacco control. The report also identifies areas for further work and research, with a focus on the regulatory needs of countries, while accounting for regional differences, thus providing a strategy for continued, targeted technical support to all countries, especially WHO Member States. The main recommendations of the Study Group are outlined below.

7.1 Main recommendations

The main recommendations to policy-makers and all other interested parties are the following:

- noting the aggressive promotion of both tobacco and nicotine products globally, the Study Group urges Member States to ensure continuing focus on evidence-based measures to reduce tobacco use, as outlined in the WHO Framework Convention on Tobacco Control, and not to be distracted by the tobacco industry or other vested interests;
- to ensure that regulations on tobacco products are extended and applied to all forms of nicotine and tobacco products and not restricted to conventional cigarettes;
- to require manufacturers to disclose information on these products regarding:
  - emission levels of selected harmful chemicals and
  - levels of biomarkers in the panel used in pre-marketing evaluation;
- to ensure that laws on tobacco advertising, promotion and sponsorship are comprehensive and in line with the WHO Framework Convention on Tobacco Control as a minimum and that they encompass online digital media platforms, including social media and any other forms of direct or indirect marketing;
- to strengthen monitoring and enforcement and cooperate internationally to address cross-border practices of the tobacco and related industries, including online digital tobacco advertising, promotion and sponsorship;
- to require the tobacco and related industries to disclose to government authorities all advertising, promotion and sponsorship activities, including those on online digital media platforms;
- to address the content and emissions of tobacco products and support product evaluation, monitoring and disclosure, in keeping with Articles 9 and 10 of the WHO Framework Convention on Tobacco Control, when formulating, adopting or updating tobacco product regulations;

- to ban the addition of menthol and other ingredients that facilitate inhalation in non-therapeutic nicotine products and all tobacco products, including synthetic coolants with a chemical structure or physiological and sensory effects similar to those of menthol;
- to amend national tobacco control laws to fill any regulatory gap for synthetic nicotine products, to ensure that the full range of synthetic nicotine products fall within their scope, including pharmacologically similar analogues that are currently marketed and any products that may emerge in the future;
- to require uniform labelling rules for manufacturers of products containing synthetic nicotine or mixtures of nicotine from various sources, either natural or synthetic, so that the contents of different nicotine forms or analogues are declared separately;
- to establish or extend surveillance of products and their users, including demographics, use of other tobacco and related products, brand, type and flavour used in nicotine pouches to acquire knowledge and assess the prevalence of use and user profiles;
- to regulate nicotine pouches to prevent all forms of marketing and take all other action necessary to minimize: young people's access to them, their appeal to young people and initiation of use by young people;
- to regulate non-therapeutic nicotine products in the same manner as products of similar appearance, content and use;
- to ensure that nicotine pouches are not classified as pharmaceutical products unless they are proven to be nicotine replacement therapies by following stringent pharmaceutical pathways for licensing as nicotine replacement therapies, as prescribed by the appropriate national regulatory authority;
- to use industry-independent data on biomarkers and country experiences in making policy decisions on electronic nicotine delivery systems, heated tobacco products and other novel and emerging nicotine and tobacco products; and
- to implement the recommendations of the Study Group on specific challenges posed in regulating non-therapeutic nicotine and all forms of tobacco products.

Countries are urged to implement the above recommendations, as enough information is available about the topics under consideration for countries to act to protect the health of their populations, especially the younger generation. While the report acknowledges that still more is to be learnt about some topics, including synthetic nicotine and biomarkers for assessing ENDS, ENNDS and HTPs, continued independent research is necessary to build further information on these topics. The data required include the prevalence of use of nicotine pouches, the characteristics of those products, the use of synthetic nicotine in nicotine products and their availability, the science of synthetic nicotine, and promotional strategies of the tobacco and related industries. Given that 1.3 billion people use tobacco globally, the tobacco control community should continue to accelerate use of evidence-based policies and recommendations, such as those in the WHO Framework Convention on Tobacco Control, measures outline in WHO MPOWER and the relevant reports of the Conference of the Parties to the WHO FCTC. Countries should thus implement proven policy measures and, in addition, consider implementing the recommendations in this report. Specific recommendations on each of the topics considered are available in sections 2.9, 2.10, 3.4, 4.11, 5.7, 5.8, 5.9 and 6.7.

7.2 Significance for public health policies

The Study Group's report provides guidance for understanding research and evidence on the scientific basis of the regulation of nicotine and tobacco products, including cigarettes, smokeless tobacco and waterpipe tobacco. The Eighth report of the Study Group *(18)* addressed novel and emerging nicotine and tobacco products, in particular electronic nicotine delivery systems, electronic non-nicotine delivery systems, and heated tobacco products. This ninth report highlights the effects of additives that facilitate inhalation; the public health implications of social and digital marketing; the challenges associated with the marketing of nicotine pouches and synthetic nicotine and the regulatory implications of marketing of these products; and current evidence on biomarkers of exposure, effect and susceptibility for assessing electronic nicotine delivery systems, electronic non-nicotine delivery systems and heated tobacco products. The report also considers the potential impact of introduction of these products on tobacco control, identifies research gaps and makes recommendations. The recommendations directly address some of the unique regulatory challenges faced by Member States by direct and indirect product market advertising and by penetration of several global markets of products such as nicotine pouches and products with synthetic nicotine. In addition, the report updates Member States' knowledge and provides guidance for formulation of effective strategies for regulating nicotine and tobacco products.

The Study Group, though its unique composition of regulatory, technical and scientific experts, navigates and distils complex data and research and synthesizes them into policy recommendations to inform policy development at country, regional and global levels. This report, by scientists with expertise in various disciplines relevant to the regulation of tobacco products, addresses the challenges faced by governments for effective regulation of tobacco and non-therapeutic nicotine products. Regulators, governments and other interested parties can rely on the science and evidence presented to formulate policies to strengthen tobacco control and close regulatory loopholes, as appropriate. The identification of gaps in policy and research on the topics considered, including on nicotine pouches, synthetic nicotine, online and digital marketing of nicotine and tobacco products, indicates areas in which there is insufficient information. In formulating their research agendas, countries could focus on areas pertinent to their policy goals, objectives and country context and regulatory environment. This is a critical role of the Study Group, especially for governments with inadequate resources or capacity to navigate technical information on tobacco product regulation.

The recommendations of the Study Group promote international coordination of regulatory work and adoption of best practices in regulating tobacco and non-therapeutic nicotine products, strengthen capacity for product regulation all six WHO regions, provide a ready resource to Member States based on sound science and support implementation of the WHO FCTC by its Parties. Given the aggressive promotion of nicotine and tobacco products globally, the Study Group urges Member States to ensure continued focus on evidence-based measures to reduce tobacco use as outlined in the WHO FCTC without distraction from the tobacco and related industries.

Tobacco product regulation complements other provisions of the WHO FCTC for reducing the demand for tobacco. The recommendations of the Study Group, if effectively implemented, would contribute to reducing the prevalence of tobacco use and improving health.

7.3 Implications for the Organization's programmes

The report fulfils the mandate of the WHO Study Group on Tobacco Product Regulation to provide the Director-General with scientifically sound, evidence-based recommendations for Member States about tobacco product regulation,[4] which is a highly technical area of tobacco control in which Member States face complex regulatory challenges. The outcomes of the Study Group's deliberations and main recommendations will improve Member States' understanding of conventional and newer products and the promotional strategies used by manufacturers.

The report's contribution to knowledge on regulating tobacco and non-therapeutic nicotine products will play a critical role in informing the work of the Secretariat, especially in providing technical support to Member States. It will also contribute to updating regulators, through the Global Tobacco Regulators Forum,

4 In November 2003, the Director-General formalized the status of the former Scientific Advisory Committee on Tobacco Product Regulation from a scientific advisory committee to a study group.

and Parties to the WHO FCTC through WHO's reports to the tenth session of the Conference of the Parties in November 2023, including on technical matters related to Articles 9 and 10 of the WHO FCTC. The report will include the key messages and recommendations of this ninth report of the Study Group. All these actions will contribute to meeting target 3.a of the Sustainable Development Goals (Strengthen implementation of the World Health Organization Framework Convention on Tobacco Control in all countries), as appropriate, and the triple billion targets of WHO's Thirteenth Global Programme of Work.

The report, a WHO technical product (a WHO global public health good *(1)*), is available to all countries to help drive impact nationally and globally to reducing tobacco use and improve overall public health.

References

1. Thirteenth General Programme of Work 2019–2023. Geneva: World Health Organization; 2019 (https://apps.who.int/iris/bitstream/handle/10665/324775/WHO-PRP-18.1-eng.pdf, accessed 29 December 2020).
2. WHO Framework Convention on Tobacco Control. Geneva: World Health Organization; 2003 (http://www.who.int/fctc/en/, accessed 10 January 2021).
3. Tobacco industry tactics: advertising, promotion and sponsorship. Geneva: World Health Organization; 2019 (https://applications.emro.who.int/docs/FS-TFI-202-2019-EN.pdf?ua=1).
4. Driving addiction: F1, Netflix and cigarette company advertising. STOP. A global tobacco industry watchdog. Paris: International Union Against Tuberculosis and Lung Disease; https://exposetobacco.org/wp-content/uploads/F1-Netflix-Driving-Addiction.pdf
5. Modern Addiction: Myths and Facts about how the Tobacco Industry Hooks Young Users. Brief, August 2021. STOP. A Global Tobacco Industry Watchdog. https://exposetobacco.org/wp-content/uploads/Modern-Addiction-Mythbuster.pdf
6. Flavours (including Menthol) in Tobacco Products. Brief, May 2022. STOP. A Global Tobacco Industry Watchdog. https://exposetobacco.org/wp-content/uploads/Flavors-Including-Menthol-In-Tobacco-Products-FINAL.pdf
7. WHO report on the global tobacco epidemic, 2019. Geneva: World Health Organization; 2019 (http://www.who.int/tobacco/global_report/en, accessed 19 December 2020).
8. GBD 2019 Risk Factors Collaborators. Global burden of 87 risk factors in 204 countries and territories, 1990–2019: a systematic analysis for the Global Burden of Disease Study 2019. Lancet. 2020;396:1223–49.
9. Tobacco. Fact sheet. Geneva: World Health Organization; 2020 (https://www.who.int/news-room/fact-sheets/detail/tobacco, accessed 23 December 2020).
10. Findings from the Global Burden of Disease Study 2017: GBD Compare. Seattle (WA): Institute for Health Metrics and Evaluation; 2018 (http://vizhub.healthdata.org/gbd-compare, accessed 19 December 2020).
11. The scientific basis of tobacco product regulation. Report of a WHO study group (WHO Technical Report Series, No. 945). Geneva: World Health Organization; 2007 (https://www.who.int/tobacco/global_interaction/tobreg/who_tsr.pdf, accessed 10 January 2021).
12. The scientific basis of tobacco product regulation. Second report of a WHO study group (WHO Technical Report Series, No. 951). Geneva: World Health Organization; 2008 (https://apps.who.int/iris/bitstream/handle/10665/43997/TRS951_eng.pdf?sequence=1, accessed 10 January 2021).

13. WHO Study Group on Tobacco Product Regulation. Report on the scientific basis of tobacco product regulation: Third report of a WHO study group (WHO Technical Report Series, No. 955). Geneva: World Health Organization; 2009 (https://apps.who.int/iris/bitstream/handle/10665/44213/9789241209557_eng.pdf?sequence=1, accessed 10 January 2021).
14. WHO Study Group on Tobacco Product Regulation. Report on the scientific basis of tobacco product regulation: Fourth report of a WHO study group (WHO Technical Report Series, No. 967). Geneva: World Health Organization; 2012 (https://apps.who.int/iris/bitstream/handle/10665/44800/9789241209670_eng.pdf?sequence=1, accessed 10 January 2021).
15. WHO Study Group on Tobacco Product Regulation. Report on the scientific basis of tobacco product regulation: Fifth report of a WHO study group (WHO Technical Report Series, No. 989). Geneva: World Health Organization; 2015 (https://apps.who.int/iris/bitstream/handle/10665/161512/9789241209892.pdf?sequence=1, accessed 10 January 2021).
16. WHO Study Group on Tobacco Product Regulation. Report on the scientific basis of tobacco product regulation: Sixth report of a WHO study group (WHO Technical Report Series, No. 1001). Geneva: World Health Organization; 2017 (https://apps.who.int/iris/bitstream/handle/10665/260245/9789241210010-eng.pdf?sequence=1, accessed 10 January 2021).
17. WHO Study Group on Tobacco Product Regulation. Report on the scientific basis of tobacco product regulation: Seventh report of a WHO study group (WHO Technical Report Series, No. 1015). Geneva: World Health Organization; 2019 (https://apps.who.int/iris/bitstream/handle/10665/329445/9789241210249-eng.pdf, accessed 10 January 2021).
18. WHO study group on tobacco product regulation. Report on the scientific basis of tobacco product regulation: eighth report of a WHO study group. Geneva: World Health Organization; 2021 (WHO Technical Report Series, No. 1029; https://www.who.int/publications/i/item/9789240022720).